Human physiological work capacity

THE INTERNATIONAL BIOLOGICAL PROGRAMME

The International Biological Programme was established by the International Council of Scientific Unions in 1964 as a counterpart of the International Geophysical Year. The subject of IBP was defined as 'The Biological Basis of Productivity and Human Welfare', and the reason for its establishment was recognition that the rapidly increasing human population called for a better understanding of the environment as a basis for the rational management of natural resources. This could be achieved only on the basis of scientific knowledge, which in many fields of biology and in many parts of the world was felt to be inadequate. At the same time it was recognized that human activities were creating rapid and comprehensive changes in the environment. Thus, in terms of human welfare, the reason for IBP lay in its promotion of basic knowledge relevant to the needs of man.

IBP provided the first occasion on which biologists throughout the world were challenged to work together for a common cause. It involved an integrated and concerted examination of a wide range of problems. The programme was co-ordinated through a series of seven sections representing the major subject areas of research. Four of these sections were concerned with the study of biological productivity on land, in freshwater, and in the seas, toegether with the processes of photosynthesis and nitrogen fixation. Three sections were concerned with adaptability of human populations, conservation of ecosystems and the use of biological resources.

After a decade of work, the programme terminated in June 1974 and this series of volumes brings together, in the form of syntheses, the results of national and international activities.

INTERNATIONAL BIOLOGICAL PROGRAMME 15

Human physiological work capacity

R. J. Shephard

Professor of Applied Physiology
Department of Preventive Medicine and Biostatistics
University of Toronto

CAMBRIDGE UNIVERSITY PRESS

CAMBRIDGE
LONDON · NEW YORK · MELBOURNE

CAMBRIDGE UNIVERSITY PRESS
Cambridge, New York, Melbourne, Madrid, Cape Town, Singapore, São Paulo, Delhi

Cambridge University Press
The Edinburgh Building, Cambridge CB2 8RU, UK

Published in the United States of America by Cambridge University Press, New York

www.cambridge.org
Information on this title: www.cambridge.org/9780521112642

First published 1978
This digitally printed version 2009

A catalogue record for this publication is available from the British Library

Library of Congress Cataloguing in Publication data

Shephard, Roy J

Human physiological work capacity.

(International biological programme; 15)

Includes bibliographic references and index.

1. Work – Physiological aspects. 2. Human biology.
I. Title. II. Series.
QP301.S49 612′.042 77-80847

ISBN 978-0-521-21781-1 hardback
ISBN 978-0-521-11264-2 paperback

Contents

Table des matières

Preface

The human adaptability project of the International Biological Programme (HA-IBP) is but one of seven sections of the programme, which has as its overall theme 'the biological basis of productivity and human welfare'. The peculiar emphasis of HA-IBP is upon *Homo sapiens*. Under this project a daring inter-disciplinary attempt has been made to explore interactions between man and his natural environment on a world-wide scale. Publication of the resulting data has proceeded in both 'horizontal' and 'vertical' directions; thus, 'horizontal' reports have covered varied aspects of adaptability in a single type of habitat such as the circumpolar territories, while 'vertical' reports have compared and contrasted smaller bodies of data collected on peoples living in a wide variety of environments – urban and rural, primitive and civilized, with exposure to extremes of heat, cold and high altitude.

The present volume is a 'vertical' analysis of data on human physiological work capacity. Among the substantial body of research workers attracted to HA-IBP were an industrious cohort of exercise physiologists – men and women with a particular concern for the measurement of physical activity and the determination of physiological work capacity. Such an expertise has obvious relevance to both the general theme of human ecology and the specific study of the biological basis of productivity. Some of the participants were investigators with a long-standing interest in human anthropology, but quite a proportion were city-dwellers who had not previously ventured far outside the urban environment. What attracted the latter to HA-IBP? Undoubtedly many factors were at work, not least the superb organization of those directing the programme. However, I suspect that a number of the exercise physiologists were also intrigued by the potential contrast between the current inactivity of 'civilized' man, 'Homo sedentarius' as he has been called, and the more active state of a primitive man totally dependent on his natural habitat. Certainly, this is a theme that will recur in the present volume. Is present day inactivity a recent and unfortunate consequence of diminishing evolutionary pressures? Are primitive people really more active than their city-dwelling counterparts, and if so, does their higher activity level have survival value? In general, answers to these and related questions are sought among experiments conducted within the framework of the HA-IBP project, but where appropriate, reference has been made to other information available in the world literature.

After a brief review of working capacity in the context of evolution, attention is directed to the problem of sampling. Traditionally, this topic has received little attention from exercise physiologists. It has too often been assumed that data obtained on ten medical students can be applied without question to the remainder of humanity. Correct sampling is of particular importance when it is intended to describe the characteristics of an entire population, to make comparisons with other entire populations, and to relate findings to the ecological stability of a given environment. Practical examples are thus given of biasses that can develop when relatively large segments of a population are examined.

A second issue affecting the validity of results is the comparability of methodology from one laboratory to another. The complex techniques used to measure human activity are particularly vulnerable to interlaboratory errors. Traditional exercise physiology has often side-stepped this problem by reporting results as the difference between two sets of data measured on the same piece of equipment. However, the question of standardization could not be avoided in the HA-IBP project, where many different laboratories were to collect information in relative independence. Steps in the development of a standard exercise-test protocol are thus reviewed, and a retrospective assessment is made of the procedures that were adopted.

Factors of climate, season, and geography are next examined in an attempt to determine the severity of local environments and their possible influence upon levels of physical activity and working capacity. Particular attention is directed to the performance of man in extremes of heat and cold. Circadian rhythms of work capacity are explored, and the possible effects of varying day-length are considered. In many communities, the socio-economic environment has more impact upon both potential and realized working capacity than does the physical environment. The succeeding chapter is thus devoted to the interaction of human performance with such variables as the quantity and quality of available nutrients, family size and habits of child care, the prevalence of disease, and patterns of habitual activity.

Central to the whole HA-IBP theme is the degree to which individual differences in physical working capacity can be inherited. This topic is reviewed on the basis of both formal genetic studies such as twin comparisons and more empirical contrasts, such as those between constitutionally different subjects who have lived in the same environment, or subjects of similar initial constitution who have lived for many generations in contrasting environments. Information is also presented on the physical and physiological characteristics of superb athletes, since such individuals are likely to indicate the full potential range of human variation that can be realized through a combination of inheritance and selection.

Final chapters consider the growth and ageing of working capacity. Environment and heredity each have a strong influence upon the maturation of working capacity in the growing child. It is also intriguing to

Содержание

Contenido

Foreword

Physical activity and the capacity for work are such fundamental determinants of human survival that it may come as a surprise that their exact measurement on a population scale has only been achieved quite recently. The relevance of such assessments for ecological analysis in general, and of productivity in particular – the two major aims of the IBP – is pointed out by Professor Shephard in his preface to the present volume. Within the HA Section of the International Biological Programme the acquisition of data on physiological work capacity was from the start seen as an objective of the highest scientific and practical desirability. Difficult as it is to carry out physiological studies on large population samples (and even more difficult in the field) many international HA teams achieved this aim with an impressive degree of success, as this book makes clear. To Professor Lange Anderson of Norway, who acted as theme coordinator, goes great credit for his role in securing the interest of physiologists in many countries in this aspect of the HA programme.

A strong and long lasting impetus to this endeavour was given during the planning stages of the HA Section when decisions had to be made on test procedures which would find wide acceptance both for laboratory and field studies. The comparison and the validation of different techniques were successfully accomplished by a team from some 7 countries who worked in Toronto in 1967 (see *Human Biology. A Guide to Field Methods*. IBP Handbook No. 9, ed. by J. S. Weiner and J. A. Lourie. Blackwells: Oxford). The success of this cooperative venture was due equally to the enthusiasm of the physiologists participating and to the outstandingly efficient organisation and leadership of Professor Shephard. During the operational phase of IBP Roy Shephard himself made many notable contributions in his studies of the Eskimo in the field, and of various Canadian groups in the urban environment. With the ending of IBP it was a matter for deep gratitude that Professor Shephard, an internationally recognised authority in this branch of physiology, undertook to survey and synthesise the knowledge available from inside and outside the IBP. He has now provided what must surely remain for a long time, not only a source-book, but an authoritative text on every important aspect of the subject of human physiological work capacity.

The scope and depth of his survey are very clearly indicated by Professor Shephard in his preface and introduction. It is gratifying that he has found it possible to assemble and examine data on a substantial array of communities from many different habitats and climates, and from a variety of social and economic backgrounds, and also to illuminate the influence of such facts as age, sex, physical constitution, nutritional status and fitness.

Foreword

Within the series of HA contributions to the IBP Synthesis Volumes being published by CUP, the present work neatly complements that by P. B. Eveleth and J. M. Tanner on *Worldwide Variation in Human Growth* (1976). These two studies provide a conspectus on a world-wide scale, of two major attributes of the human species, one morphological and the other physiological.

Distinct from these two are the volumes devoted to the comprehensive biological analysis – physiological, medical, demographic, genetic and developmental – of communities living in two of the most difficult biomes of the world. These are presented in *The Human Biology of Circumpolar Populations* ed. by Professor F. A. Milan, and *The Biology of High-Altitude Peoples* ed. by Professor P. T. Baker.

The volume edited by Professor G. A. Harrison, *Population Structure and Human Variation*, complements these four volumes in an interesting way. It comprises a series of case studies, each of which provides a vivid illustration of special aspects of population biological structure which enter into the 'vertical' and 'horizontal' surveys of the other volumes. These case studies focus on particular aspects of structure and functional constitution and on the forces, or agencies, which determine the physiological fitness in the short term, and reproductive fitness in the long term.

These volumes in the CUP series are not intended to, and cannot possibly, synthesise all the results from the HA Section. When in 1974 the IBP was brought to a close, over 50 countries had mounted some 250 projects. Several thousands of scientific papers have been published along with some 30 monographs on particular national projects. Particulars of all the contributing countries, their projects, the team personnel and their publications and reports are to be found in *Human Adaptability: A History and Compendium of Research within the IBP* by K. J. Collins and J. S. Weiner (Taylor and Francis, London, 1977).

J. S. WEINER
Convenor, HA Section, IBP

consider the problem of comparing data between groups of children having the same age, but differing in body size. Lastly, it is debated whether an increase of physical activity upon the part of sedentary city-dwellers can enhance their development as adolescents or slow their subsequent rate of ageing.

Already, the nature of our theme has forced the use of descriptions that are all too apt to arouse passions. At many subsequent points in the text, it will be necessary to categorize people as black or white, nations as less or more developed, and life-styles as primitive or civilized. However, I hasten to reassure the reader that such descriptions are made on a purely factual basis, and involve no value judgements as to the merits of colour, industrial development or contemporary civilization. It is dangerous, indeed, to attempt a judgement of alternatives on the basis of passing acquaintance with a community, although I must confess that at brief inspection the simple life of the hunter/gatherer has many attractions over the tumultuous greed of a busy metropolis.

As in other scientific monographs, occasional problems arise from inconsistencies in existing data. Sometimes, it is necessary to point the need for further research rather than attempt to draw premature conclusions. However, I have had several important advantages over writers who have explored physical working capacity from other points of view. Firstly, almost all of the HA-IBP research was completed within the space of five years, giving a clear starting point to my story, and a sharp ending. Furthermore, where material was not yet published, HA-IBP investigators were most generous in making available to me their personal files of data. Finally, I enjoyed throughout my task the warm and generous encouragement of Professor J. S. Weiner, Convenor of the HA-IBP project. I am most grateful to him for introducing me to this area of knowledge. My personal experiences, both with fellow investigators and with the people of the Canadian Arctic have been richly rewarding, while the synthesis of this varied information has been a fascinating assignment. Hopefully, the results will also have practical importance, as civilized man wrestles with the problems of urban inactivity and survival in an environment where natural resources are ever shrinking.

Toronto R.J.S.

1 Introduction

The concepts of physical fitness and working capacity

Physical fitness and working capacity are both elusive concepts. After several days of vigorous discussion, an expert committee of the World Health Organization (Andersen *et al.*, 1971) was able merely to state the interrelationship of the two variables: 'physical fitness is the ability to perform muscular work satisfactorily'. The present author has viewed fitness as an exercise in human ecology – the matching of the individual to his environment, physical, social and psychological (Shephard, 1969*a*, 1974*a*). As such, it stands central to the human adaptability theme of the International Biological Programme (IBP). In many primitive societies, physical fitness and an associated high level of working capacity have had survival value. Until the advent of western civilization, the strength and endurance of the individual provided the principal energy resource for the cultivation of crops and/or the capture of game; furthermore, unless the energy yields of field and chase matched the demands of growth, reproduction, body heating and the continuing quest for food, a community was doomed to early extinction.

Physical fitness and evolution

One might thus anticipate that an unacculturated primitive society would be marked by an evolutionary pressure favouring the survival, mating and successful rearing of children by the more resourceful males – those members of the tribe with a capacity for long and arduous physical work, able to perform the elemental tasks of agriculture, hunting and fishing in a particularly skillful and efficient manner. Anecdotal reports from the Canadian Arctic have indeed described the operation of such considerations in the choice of marriage partners by traditional Eskimos (Metayer, 1976). Nevertheless, the evolutionary importance of physical fitness has varied markedly with the severity of the climate, the nature of the terrain, the richness of the local fauna and flora, the ravages of disease, the technical expertise of the tribe, and the cultural expectations prevalent in its society. The harsh circumpolar environments apparently have taxed adaptive potential to permit the survival of quite small communities, whereas in more richly endowed geographic regions substantial settlements have lived quite comfortably despite only occasional leisurely use of their muscles.

Problems of cultural change

At the outset of the human adaptability project (Weiner, 1964), it was recognized that many primitive societies were facing 'imminent cultural disintegration and in some instances loss of physical identity in the face of advancing civilization'. The hope was that their status could be documented before this threat became a reality. Unfortunately, many communities have undergone great change over the decade marking the preparation and realization of the International Biological Programme. Often, governments have encouraged or even required concentration of villagers in much larger settlements than those dictated by centuries of experience of a given habitat. Immediate motives have been good – the provision of modern housing, education, primary health care and an opportunity for settled employment. However, the effects on those members of the tribe seeking to conserve a traditional life-style have been disastrous, forced migrations destroying the delicate balance between the energy costs of hunting and its rewards. Too frequently, children have been uprooted from the protective warmth of extended families to spend the critical years of physical, social, and emotional development in the sterile atmosphere of a residential school in a major urban centre. Here, they have acquired many of the habits and attitudes, with most of the vices of the 'civilized' person; if eventually they have returned to their birthplace, it has been as an alienated and embittered stranger lacking both the attitudes and skills necessary to a traditional life-style. Even in the absence of direct governmental intervention, hunting preserves have been progressively destroyed by the construction of highways, railways, oil pipelines and dams, afforestation, deforestation, and the use of agricultural chemicals (Bakács, 1972). The physical fitness and working capacity seen by IBP investigators thus in many instances has been that of a community undergoing rapid deterioration not only in its physical condition but also in its nutrition, health, morale, sobriety, and racial purity.

Physical fitness and industrial society

Highly developed industrialized societies present almost the end-stage in the deterioration of physical condition. Daily work is highly mechanized and often requires little more energy expenditure than the resting metabolism. Physical fatigue (Simonson, 1971) can arise from sustained 'static' effort (for example, supporting a heavy weight in an awkward posture, or controlling a heavy lever); it may also develop in industries that require a monotonously patterned high rate of rhythmic work. But more usually, intolerance of daily work has a sociological or a psychological basis. Can one thus conclude that physical fitness and a good working capacity no longer have survival value? In terms of the energy balance of the community, this seems true. However, there is also increasing evidence that failures of adaptation to the urban milieu – particularly the 'epidemic' of ischaemic heart disease (Morris, 1951; S. M. Fox, 1974) and the vast social problems of poor mental health and drug abuse (Bakács, 1972) – are associated with a lack of habitual activity and consequent deteriorations of physical fitness and working capacity. IBP investigators have thus been interested to compare the fitness levels of urban and rural communities, both in heavily industrialized countries such as Canada, Czechoslovakia and Japan, and also in regions where the migration to the cities has been quite recent (South and Central America). From the evolutionary standpoint, it is unfortunate that many of the medical problems associated with city life do not become obvious until middle age, when child-rearing is almost complete. There is thus little possibility that natural selection will help man to adapt his constitution to the urban milieu.

Specific genetic studies

A strong genetic theme runs through all of the human adaptability proposals. Nevertheless, early recommendations (Baker & Weiner, 1966) stressed the need to examine populations of specific genetic interest – in-breeding and cross-mating groups, mono- and dizygotic twins, old and recent migrants, international-class athletes and their immediate relatives. It was hoped that such research would clarify (i) the extent of the variability developed by a given genetic group when it was exposed to differing environments, and (ii) the similarity of adaptations that appeared when differing genetic groups were exposed to the same environment for short or long periods of time. Explorations of the first question include studies of the coastal and highland natives of New Guinea (Cotes *et al.*, 1973*a*), the lowland and highland natives of Ethiopia (Andersen, 1971) and the Eskimos of Alaska, Canada and Greenland (Simpson & McAlpine, 1976). Investigations of the second question are exemplified by comparisons of Kurds and Yemenites exposed to the heat of the Negev (C. T. M. Davies *et al.*, 1972), and comparisons among the Eskimos, Lapps, Ainu and white immigrants who have encountered the cold of the Arctic (Laughlin, 1966; Shephard, 1975*a*). Athletes are of particular interest as illustrating the limits of human adaptability in specific directions, attained by a combination of initial selection and subsequent training. The past decade has seen detailed investigation of many types of athlete, but there have as yet been relatively few examinations of close blood relatives.

Use of standard methodology

We shall return to the question of methodology in Chapter 3. However, it can be said here that one of the exciting prospects of the IBP was the possibility of making comparisons between data obtained in many parts of the world, confident that all participating laboratories had used a common methodology. In this respect, the IBP study followed the example of the International Geophysical Year, although it was recognized that physiological measurements were more difficult to standardize than physical observations. Proposals for the measurement of working capacity were examined by a working party, meeting in Toronto in 1967, and the resultant recommendations appeared in a subsequent monograph (Weiner & Lourie, 1969).

Aerobic and anaerobic power

Sustained physical activity implies the steady aerobic replenishment of muscle phosphagen reserves (adenosine triphosphate and creatine phosphate). Despite recent controversy (Kaijser, 1970; Keul, 1973; page 36), there is good evidence that this process normally is limited by the rate of transport of oxygen from the atmosphere to the working tissues (Shephard *et al.*, 1968*a*; Shephard, 1969*a*, 1974*a*). The most important determinant of endurance fitness is thus the *maximum oxygen intake* or *aerobic power.* This may be measured directly during exhausting exercise (Shephard *et al.*, 1968*a*), or it can be estimated indirectly from measurements of pulse rate and oxygen consumption or work load during submaximum effort (Shephard *et al.*, 1968*b*). Various authors have shown such data provides a useful guide to the likelihood of fatigue during sustained and heavy physical work (Christensen, 1953; I. Åstrand, 1960, 1967; Bonjer, 1968). Depending on environment, posture, the size of the muscles involved and the peak loadings that are anticipated, steady energy expenditures should not exceed 30–40% of aerobic power over an eight-hour working day. Further, if heavy work is unpaced, most employees choose to operate at about 40% of their aerobic power (Hughes & Goldman, 1970).

If very powerful efforts are required for periods of a few seconds, the maximum rate of working is determined by the ability to split phosphagen and translate the released energy into mechanical work, the so-called *anaerobic power.* Margaria (1966, 1967) has developed a field technique for estimating anaerobic power, but perhaps because most primitive communities favour deliberate rather than hurried patterns of activity, his method has not been used very widely.

Over periods of activity lasting from a few seconds to one minute, effort depends on tolerance of oxygen debt, the *anaerobic capacity*. There is no easy field method of measuring this quantity, and it has not been examined by the majority of IBP investigators.

Pulmonary function tests

Measurements of *forced expiratory volume* (FEV) and *vital capacity* were included in that section of the IBP methodology concerned with working capacity (Weiner & Lourie, 1969). Although correlations can sometimes be established between lung volumes and athletic performance (Shephard, 1969*a*, 1972*a*; Shephard, Godin & Campbell, 1973*a*) most authors have held that the lungs play a minor role in determining either oxygen transport or physical fitness, except in special circumstances such as at high altitudes or when swimming underwater (Shephard, 1971*a*). Nevertheless, the symptom of breathlessness (Schmidt & Comroe, 1944; Howell & Campbell, 1966; Shephard, 1974*b*) can induce a voluntary limitation of physical activity, and since unpleasant breathlessness of dyspnoea usually develops when the tidal volume is 50% or more of vital capacity, it becomes of interest to consider racial differences of lung volumes and other indices of pulmonary function. In some primitive communities, diseases such as tuberculosis have been rife until recently; thus simple measurements of pulmonary function also supplement radiographic observation in deciding the health of the subjects tested, particularly if direct communication is restricted by linguistic problems.

Anthropometry

Anthropometric data are of particular importance when comparing the working capacity of populations that differ in body size. In addition to the traditional indices of height, weight, and skinfold thicknesses, it is useful to add dimensions that will characterize the relative lengths of the trunk and the limbs, the shape of the chest, the body type (masculinity) and the body build (ectomorphic, mesomorphic, or endomorphic). There may also be specific assessments of body composition (for example, by soft-tissue radiography or hydrostatic weighing), and photography for somatotyping (Weiner & Lourie, 1969). Unfortunately, few IBP investigators exploited the potential for cooperation between anthropometrists and physiologists. The anthropometrists tended to concentrate on the growth and development of quite young children, whereas physiologists found adolescents and young adults more convenient subjects to examine. Often, the detailed anthropometric data needed for interpretation of physiological differences are lacking. Furthermore, there has been no general agreement on appropriate methods of data standardization. Should adult maximum oxygen intake values, for example, be expressed as ml/min per kilogram of body weight, per centimetre of standing height, per centimetre of sitting height, or per litre of leg volume? In growing children, is the best basis of standardization *height*, *height*^{2}, *height*^{3} (Eriksson, 1972; P. O. Åstrand & Rodahl, 1970), or some other exponent (Rode & Shephard, 1973*a*)?

Muscle force

Measurements of isometric muscle force are made quite readily by such devices as cable tensiometers and strain gauge dynamometers (Clarke, 1966). Unfortunately, the ancillary equipment needed for full immobilization of subjects during testing is both bulky and heavy. Perhaps for this reason, the important dimension of muscle strength (Weiner & Lourie, 1969) has been largely neglected by IBP investigators. As with aerobic power, the optimum basis of standardization also remains obscure. Should data be expressed as kilogram force, kilogram force per unit height or weight, torque, or torque per unit height or weight?

Tests of physical performance

From the viewpoint of human adaptability, the performance of the individual may seem of more practical importance than the underlying physiological attributes. Partly for this reason and partly to accomodate investigators with little equipment, the IBP proposals (Weiner & Lourie, 1969) included one form of the Harvard step test (Brouha, 1943) and a performance-test battery patterned after recommendations of the American Association for Health, Physical Education and Recreation (1965). Suggested measures were pull-ups, sit-ups, a flexed-arm hang, a shuttle run, a standing broad jump, a 50-yard dash, a softball throw, and a 600-yard run-walk.

Fairly extensive data have been obtained with the Harvard step test, particularly in athletes. The score for this test reflects a combination of cardio-respiratory endurance fitness and motivation. Perhaps on this account, the results have sometimes shown a closer correlation with athletic performance than that established for direct measurements of maximum oxygen intake (Ishiko, 1967).

The interpretation of performance-test scores has been discussed by G. R. Cumming & Keynes (1967) and Drake *et al.* (1968). In theory, the test battery covers many aspects of working capacity, and equations can be developed to predict maximum oxygen intake and other measures of fitness from the performance scores (Falls, Ismail & Macleod, 1966). However, when applied to general populations of children or adults, a large part of the variance of the data is attributable to age, simple anthropometric characteristics (height, weight and skinfold thickness) and recent familiarity with gymnasium procedures. Populations where gymnasia are unknown have a substantial handicap, and even within civilized groups much depends upon previous motor learning. US and Canadian children generally have performed more poorly than their European counterparts, but have out-distanced them in the peculiarly North American skill of throwing the softball (Knuttgen, 1961). A further indication of the unsatisfactory nature of performance tests is the large gain in the scores of US children between 1958 and 1965 (American Association for Health, Physical Education and Recreation, 1965); physiological measures of fitness have remained static or even have deteriorated over this time, and one must suspect that the gains of performance reflect practice of the required skills.

Habitual activity

The assessment of habitual activity patterns was an ambitious but also an important objective of the IBP programme (Weiner & Lourie, 1969). Many of the observed differences in physical fitness, in working capacity, and in the prevalence of cardio-vascular disease both between and within populations could reflect differences of habitual activity rather than more fundamental differences of constitution or environmental challenge. Further, a clear statement of the energy needs of a community seems essential to assess prospects for its nutrition, energy balance and survival (Godin & Shephard, 1973; Shephard & Godin, 1976).

The methods proposed for the IBP followed conventional wisdom (Durnin & Passmore, 1967; Godin & Shephard, 1973). They included the use of self-administered or retrospective questionnaires; diaries kept by the subject or an observer; formal time and motion studies; measurements of heart rate by a modified wrist-watch (Glasgov *et al.*, 1970; Masironi, 1971, 1974), tape-recorder (Holter, 1961; Andersen, 1966; Shephard 1967*b*), or electrochemical integrator (Wolff, 1966; Baker, Humphrey & Wolff, 1967); measurements of oxygen consumption by Douglas bag, Kofranyi–Michaelis respirometer (Kofranyi & Michaelis, 1949) or integrating motor pneumotachograph (IMP: Wolff, 1958); and the assessment of food intake (Durnin & Passmore, 1967).

Such procedures are relatively simple when applied to urban, industrialized communities who perform consistent types of work at fixed locations. However, there are many problems when observations are extended to primitive communities that have little idea of either a 24-hour day (Lobban, 1976) or a working week (Chapter 2). Such people are guided not by the Gregorian calendar but by the cycles of nature; their activity patterns respond to and exploit fully such signals as the running of the fish and the season of the young geese (Godin & Shephard, 1973; Shephard & Godin, 1976).

International cooperation

The IBP measurements of physical fitness and working capacity are fascinating in their range and depth. But perhaps the most exciting aspect of the IBP project has been the spirit of international cooperation underlying the accumulation of this wealth of data. More than fifty laboratories from twenty-six countries shared in the IBP studies of physical fitness and working capacity. Principal investigators have shown extreme generosity in making available not only published reports and monographs, but also the basic results vital to a detailed comparison of populations. To all who have thus shared in the task of IBP, the present author expresses his sincere appreciation.

Despite the vast and voluntary cooperative effort by the world community of applied physiologists, there are inevitably significant gaps in the IBP studies. Where appropriate, information from other sources has thus been included to provide a comprehensive synthesis of current knowledge on physical fitness, working capacity, and its importance to human adaptability.

2 Sampling and population studies

Physiologists sometimes devote considerable effort to the elaboration of precise methodology (Chapter 3). However, too little care is directed to the influence of population sampling upon the generality of the results thus obtained (Rose & Blackburn, 1968; Shephard & Andersen, 1971).

Sampling and the human adaptability project

In early discussion of the IBP project (Weiner, 1964), it was recognized there would be sampling problems when studying the populations of industrialized nations; thus in the context of physical fitness and working capacity, investigators were asked to record age, sex, physique, fitness, occupation, health status, and degree of urbanization. On the other hand, with respect to primitive communities, it was assumed that 'the relatively small size of these populations [would] render them more manageable'. Nevertheless, observers were asked to make a careful identification of the group studied in terms of linguistics, religious affiliation, nationality, ethnology, socio-cultural background (hunting, food-gathering, nomadic, pastoral, horticultural, simple agricultural, advanced agricultural, rural, urban, or industrial) and socio-economic status (in terms of accepted occupational and social indices). Investigators were also requested to state whether they thought the subjects tested were representative of the population under study; specific comment was to be made on sampling methods and any inhomogeneity within samples. Unfortunately, these instructions have been largely ignored, and in many instances information about sample composition is extremely limited.

Sample size

At the outset of the IBP project, it was assumed that application of normal-distribution statistics to relatively large samples would resolve many problems (Weiner, 1964; Weiner & Lourie, 1969). Thus, it was suggested that measurements be made on 390 individuals in order to specify the mean stature of a male population. With regard to measurements of aerobic power, it was assumed that one useful application of the results would be to substantiate a difference in levels of physical activity between populations (Weiner & Lourie, 1969). Since the likely effect of increased physical activity was to augment aerobic power by 20% and the coefficient of variation of age and sex specific maximum oxygen intake data within an industrialized society was about 16% (Shephard, 1969*a*), a minimum sample size of sixteen was established

($t_{15} = 20\% \cdot \surd(16)/16\% = 5.0$). Nevertheless, an element of 'overkill' was recommended; even where time and facilities were limited, investigators should study fifty male subjects between the ages of twenty and thirty years, with an increase of N to 100 per decade of adult life for the simpler performance-type tests.

Informed consent

Civilized populations generally have become suspicious of the complicated test procedures demanded by the exercise physiology laboratory. If due care is taken to preserve the principles of informed consent, without actual or implied coercion of potential subjects (Shephard, 1967*c*), the proportion of volunteers in Canadian cities seems about 35%, both in children randomly selected from school class lists (Shephard *et al.*, 1968*c*) and in adults randomly selected from the telephone directory (Bailey *et al.*, 1974*a*). Where higher success rates have been attained, there has often been some subtle coercion involving an authority figure such as a school teacher or a military commanding officer. However, more cooperation is found in small towns and rural areas, particularly if the investigating laboratory is well respected and has a history of community involvement.

Until recently, problems have been less in traditional societies. Unfortunately, the majority of primitive groups are rapidly becoming antagonistic to all forms of interaction with modern man. The IBP programme may even have accelerated this trend in some populations, as groups of thirty or more eager investigators descended upon small settlements. Many Canadian Indian and Eskimo groups now complain they are 'tired' of being studied. But five or six years ago, this was not so. The majority of IBP scientists were welcomed with extreme courtesy, friendship and cooperation.

The cooperation of athletes is again a separate question. The percentage of volunteers may range from less than 2% (when a letter of invitation is received from an unknown investigator) to near 100% (when the competitor is subjected to the modest coercion of test approbation by his coach, coupled with evaluation by a scientist who is himself an outstanding athelete). Particularly disappointing response rates have been encountered when testing has been staged in association with the Olympic and Pan American games; at such times, competitors are understandably reluctant to undertake any tests that may modify their subsequent performance.

Sampling bias in civilized populations

Here, the likely bias of volunteers depends somewhat on the aura of the test laboratory. Formal psychological evaluation of UK servicemen volunteering for physiological procedures at a military laboratory disclosed an above average number of individuals with neurotic and hysterical personality tendencies (Shephard & Kemp, unpublished report). A similar trend was noted when volunteers for an exercise programme were requested from a panel maintained by the Ontario Heart Foundation of middle-aged adults (Massie & Shephard, 1971). Neither of these groups of volunteers had a particularly high level of physical fitness. On the other hand, when a group of physical education-oriented investigators attempted to 'resynthesize' a random sample of the Canadian population, using volunteers from occupations listed by the Dominion Bureau of Statistics (Metivier & Orban, 1971), the working capacities thus established were so high that a bias towards unusually fit individuals must be suspected. Likewise, the initial random telephone recruitment in Saskatoon (Bailey *et al.*, 1974*a*) left a residual sample with a surprisingly large percentage of subjects who claimed to be physically active (14% of males and 9% of females reported four or five sessions of deliberate activity per week, such as walking a mile or more at a fast pace, jogging, tennis, squash, or swimming four or more lengths; a further 26% of men and 27% of women reported pursuing similar activities two or three times per week). Such levels of habitual activity far exceed the averages reported to the 1971 Canadian national census. Other programmes with a medical orientation have tended to recruit the health-conscious individual. Thus, volunteers for a pre-retirement test and exercise programme initiated at the University of Toronto (Sidney, 1973) contained almost no smokers. Among pre-pubertal volunteers, we found a small excess reported by their parents as having an 'above average' interest in sports and games (Shephard *et al.*, 1968*c*).

While it will be tempting to repeat IBP studies after the lapse of some years, the effects of selective sample attenuation must then be considered. Dehn & Bruce (1972) have suggested this can cause a faster rate of ageing in longitudinal than in cross-sectional studies, since the unhealthy are progressively eliminated from cross-sectional samples. However, over a longer period, the same type of difficulty can beset longitudinal studies, with progressive elimination of the sick, the poorly motivated, and the occupationally mobile. Personal experience (Kavanagh & Shephard, unpublished data) has shown selective retention of ectomorphs in a jogging programme, and of endomorphs in a hypnosis programme.

Sampling bias in primitive populations

It is not always easy to establish the percentage of a primitive population that agree to testing. Population lists such as the Eskimo 'disc list' can sometimes be obtained from government agencies, but if the population is nomadic such lists may fail to keep pace with migrations from one settlement to another, and even from one extended family to another. Precise identification of individuals is insecure when the language is verbal, uses an unfamiliar alphabet, and is poorly understood by the investigator; English transliterations of Christian names and surnames commonly differ between interpreters, and even between interviews.

Definition of geographic boundaries is equally difficult for the nomad. To quote one example, many of the Canadian IBP studies looked solely at Igloolik, a small island settlement near the tip of the Melville peninsula. However, some investigators included in their averaged statistics data on the smaller and sociologically different mainland community of Hall Beach, about a hundred kilometres distant.

Primitive people rarely refuse to participate in a test; nevertheless, reluctance is shown quite effectively by non-appearance at the laboratory, or even a sudden migration to another settlement. Factors having a positive impact on cooperation include the empathy of the investigator, a willingness to learn the native language and to share in its culture, the aura of a government-sponsored or approved health examination, and small personal incentives such as presentation of 'Polaroid' photographs to participants. Negative factors include the need to undress subjects in makeshift examining rooms (a particular problem with teenage girls educated in residential schools) and unexpected discomforts (such as a badly designed skin caliper used by one Canadian investigator). A little over 70% of the Igloolik villagers (224 subjects) agreed to undertake physiological tests. In many other IBP studies, the sample has been much smaller (Table 1).

One possible source of bias in primitive groups is season. Because of difficulties of transportation (and, in some cases, out of consideration for observer comfort), measurements have often been obtained during the long summer vacation. However, several IBP groups did take the precaution to compare summer and winter results (Chapter 4), and at least in the populations examined differences were not very large. A short sojourn can give an investigator a misleading example of a partially nomadic community, since the least acculturated and most active villagers are frequently absent on extended hunting trips. In any settlement, the first contact is usually with the most acculturated members of the community, if for no other reason than that they have clocks and understand the nature of appointments. More traditional subjects may only be contacted through a patient absorption of the native language and customs. This may require extended residence in a remote area and a willingness to face the privations of the nomad.

The coupling of physiological testing with the annual medical or dental examination increases the number of potential subjects, but at the same

Table 1. *Sample sizes for some of the populations where working capacity has been studied by IBP investigators*

Population	Population size	Sample size and sex	Author
Chile			
Aymara Indians	900	40M	Donoso *et al.* (1974*a*)
Ethiopia			
Amhara lowlanders	?	36M	Andersen (1971)
Amhara highlanders	?	49M	Andersen (1971)
India			
Ladakh	?	45M	Bharadway *et al.* (1973)
Tamil	?	47M	Bharadway *et al.* (1973)
Israel			
Kurds	800	19M; 12F	C. T. M. Davies *et al.* (1972)
Yemenites		20M; 12F	C. T. M. Davies *et al.* (1972)
Jamaica	?	19M; 12F	Miller *et al.* (1972)
Japan			
Ama	?	18F	Hirota *et al.* (1969)
New Guinea			
Coastal Kauls	?	27M; 26F	Cotes *et al.* (1973*a*)
Highland Lufas	?	26M; 26F	Cotes *et al.* (1973*a*)
Migrant Lufas	?	27M	Cotes *et al.* (1973*a*)
Nigeria			
Yaruba	?	63M; 8F	C. T. M. Davies & Van Haaren (1973)
South Africa			
Venda, Pedi			
Rural	?	200M	Wyndham (1973)
Urban	?	200M	Wyndham (1973)
Trinidad			
Caucasians	?	6M	Edwards *et al.* (1972)
East Indians	?	12M	Edwards *et al.* (1972)
Negroes	?	12M	Edwards *et al.* (1972)
Zaïre			
Hoto	?	27M	Ghesquière (1971)
Twa	?	23M	Ghesquière (1971)

Details for the circumpolar groups are given on page 17.
M = male; F = female.

time it may weight the sample with those having mild to moderate chronic medical disorders. On the other hand, a rigorous assessment of working capacity independent of the main medical examination may attract the fitter members of the community, while discouraging the attendance of women and older girls.

Even assuming the problems of sample selection are fully resolved within a single community, it must be stressed that generalizations from a given settlement to other superficially similar populations is rarely possible. Data from Igloolik, for example, cannot be applied to the entire population of Canadian Eskimos. Even in villages where genetic characteristics seem comparable, there are differences in the extent of contacts with civilized man. Differing missionary groups have sought to impose their characteristic cultures. In some settlements, many of the community have preserved their traditional nomadic life-style. In other areas, acculturation has proceeded apace, and the historic social fabric has been disrupted by alcoholism and associated problems of urban migration (Shephard & Itoh, 1976). A combination of natural game resources and imported food is still sufficient to avert malnutrition in many smaller villages, but in larger towns this is not always the case – particularly if government welfare payments are spent unwisely. Lastly, the prevalence of chronic disease varies quite widely from one region to another. For all of these reasons, it is often preferable to think of groups in terms of their village – Wainwright Eskimos rather than Alaskan Eskimos, and Igloolik Eskimos rather than Canadian Eskimos.

Practical observations on sampling bias

The types of bias that can develop when testing a primitive population are well illustrated by Canadian experience (Shephard, 1975*a*). Our physiology team examined 130 males and 94 females from the Igloolik settlement. Partly because of difficulties in crating and uncrating bulky apparatus, it was decided to exclude the Hall Beach population (page 12) from the physiological survey. For very practical reasons, we also made no attempt to carry out exercise tests on subjects under the age of nine years.

The sample showed significant age and sex bias, the percentages of the settlement volunteering for testing being as follows:

Age (yr)	Men (%)	Women (%)
< 20	77	65
20–39	67	56
> 40	41	48

There was a trend for villagers with chronic disease not to volunteer for physiological tests, as can be seen by comparing nursing station records between those who were and those who were not tested (Table 2). Ignoring such obvious contraindications to exercise as a congenital

Table 2. *A comparison of nursing station records between subjects volunteering for physiological tests and those not tested*

	Tested		Not tested	
	(*N*)	(%)	(*N*)	(%)
Normal health	176	78.6	49	44.1
Primary tuberculosis	9	4.0	12	10.8
Hilar calcification	11	4.9	22	19.8
Secondary or advanced tuberculosis	17	7.6	7	6.3
Fibrosis	9	4.0	13	11.7
Emphysema	2	0.9	2	1.8
Crippled	—	—	5	4.5
Recent 'coronary' attack	—	—	1	0.9
Total	224		111	

Data for Igloolik Eskimos from Shephard (1975*a*), by permission of the editor, *IBP Synthesis Volume: Circumpolar Peoples*, and Cambridge University Press.

dislocation of the hip, previous paralytic anterior poliomyelitis and a recent 'coronary' occlusion, there yet remained a significantly higher proportion ($\chi^2 = 33.6$, $P < 0.001$) of villagers with pulmonary disease among those not seen (56/105, 53.3%) than among those seen (48/176, 27.3%); this probably accounts for much of the age bias in our sample, since disease was more common among older members of the population. Many of the younger Eskimos who remained untested had only minor manifestations of tuberculosis (a Ghon focus, or hilar calcification), suggesting that, as in the civilized population, knowledge of a medical abnormality engendered a more cautious approach to life.

The tendency of villagers with chest disease not to volunteer for testing was confirmed when our sampling was checked against the case histories reported by the medical team (Dr J. Hildes and Dr O. Schaefer). On each of several criteria (hospital admissions for tuberculosis, history of other chest diseases, chronic cough, clubbing, and overall assessment of health), those tested by the physiology team were rated as healthier than those not tested (Table 3). Only for the more subjective measure of adventitious sounds was there no difference between the two samples.

The impact of sampling biasses on the mean values for physiological data can be illustrated by the measurements of forced expiratory volume and vital capacity (Table 4). The medical team had figures for forty-nine of the 223 subjects that we tested, and had sought out a further seventy-nine villagers that we had not seen. Mean results for subjects common to the two samples were remarkably similar (Table 4) and coefficients of

Table 3. *Comparison of medical examination results for patients seen by the physiology team with those for patients not seen by the team*

Severity of symptom or sign[a]	Hospital admission for TB		Other chest disease		Chronic cough		Clubbing		Adventitious sounds		Overall impression of health	
	Seen	Not seen	Seen	Not seen	Seen	Not seen	Seen	Not seen	Seen	Not seen	Seen	Not seen
0	74	66	—	17	34	15	150	218	1	3	6	10
1	14	24	18	19	20	13	0	10	137	198	9	142
2	4	21	1	4	17	31	1	3	26	46	4	12
3	2	2	6	17	—	—	0	3	7	24	0	1
χ^2	12.96		13.24		11.85		9.06				19.74	
P	$0.01 > P > 0.001$		$0.01 > P > 0.001$		$0.01 > P > 0.001$		$0.05 > P > 0.02$		n.s.[b]		$P < 0.001$	

Data for Igloolik Eskimos from Shephard (1975*a*), by permission of the editor, *Circumpolar Peoples*, and Cambridge University Press. Medical data kindly provided by Drs Schaefer and Hildes.

[a] The scales used by Drs Schaefer and Hildes have been compressed to give a reasonable sample size in each category.

[b] n.s. = not statistically significant.

Table 4. *The effects of sampling bias on mean values for one-second forced expiratory volume and vital capacity (l BTPS, mean±SD of data; number of subjects in parentheses)*

Age	Sex	Seen by both teams: Physiology team	Seen by both teams: Medical team	Seen by physiology team alone	Seen by medical team alone
		Forced vital capacity			
9–19	M	4.91 (2)	4.99 (2)	4.08 (44)	4.23 (2)[a]
9–19	F	—	—	3.59 (32)	—[b]
20–59	M	4.87±0.73 (41)	4.94±0.62 (41)	4.92 (28)	4.41±0.88 (41)
20–59	F	2.82±0.20 (5)	2.80±0.20 (5)	3.59 (44)	2.66±0.89 (24)
		One-second forced expiratory volume			
9–19	M	4.06 (2)	4.15 (2)	3.38 (44)	3.56 (2)[a]
9–19	F	—	—	3.02 (32)	—[b]
20–59	M	3.69±0.70 (41)	3.64±0.66 (41)	3.86 (28)	3.14±1.04 (41)
20–59	F	1.99±0.26 (5)	1.98±0.21 (5)	2.83 (44)	1.88±0.78 (24)

Data for Igloolik Eskimos from Shephard (1975*a*), by permission of the editor, *Circumpolar Peoples*, and Cambridge University Press. Medical data kindly provided by Drs Schaefer and Hildes.

BTPS = Body temperature and pressure, saturated with water pressure; SD = standard deviation.

[a] Average age higher than the 9- to 19-year-old children seen by physiology team alone.

[b] Data now missing.

correlation between the two sets of data were also satisfactory (for vital capacity, $r = 0.92$, and for one-second forced expiratory volume, $r = 0.93$). The additional men studied by the physiology team were apparently quite similar to the jointly studied population, but the medical investigators naturally sought out for their respiratory tests patients with respiratory symptoms leading to an equal restriction of both vital capacity and one-second forced expiratory volume. Unfortunately, few readings on women were common to the two series, but again those seen by the medical team alone had poorer results than those seen by the physiology team alone.

Bias in other IBP studies of working capacity

Bias in most other IBP studies of working capacity is unknown, since few investigators have defined their potential sample. C. T. M. Davies *et al.* (1972) made some observations on about 75% of the Kurdish and Yemenite Jews, but unfortunately were able to carry out exercise tests on only a tenth of this group. The method of sample reduction was unspecified. The present author has also estimated the representation of several circumpolar groups (Shephard, 1974*c*). Ikai *et al.* (1971) tested twenty-one Ainu males from a population of 275; the main occupations were fishing and farming, and the likely bias was towards healthy but sedentary subjects. Rennie *et al.* (1970) saw some 40% of the Alaskan Eskimos living at Wainwright, Alaska, although his sample (57M, 30F) was deficient in adult women. He later examined some 10% of the Point Hope and Point Barrow Eskimos (121M, 110F); this last group were highly acculturated, and by no means a genetic isolate. Lammert (1972) examined twelve men and three women; they cannot have been more than 20% of the Eskimos living in the Upernavik settlement. The IBP studies of the Lapps (Wilson, in preparation) are currently incomplete, but Andersen *et al.* (1962) have previously reported data for forty-nine men and twenty-one girls from a group of some 1000 Lapps resident in the Kautokeino region; his sample included some 10% of the older children and male adults, but no girls over the age of twenty-one. These examples are sufficient to indicate the shortcomings of most physiological studies; considerable caution is necessary before accepting such data as representative of a given region, and even more care is needed when comparing two samples that may have been biassed in differing directions.

Possible methods of generalizing from biassed data

Few physiologists seem to have considered the twin problems of preserving the free, informed consent of the subject and yet making valid generalizations from what are almost inevitably biassed data.

One important issue in primitive communities and among older subjects in general is the desirability of excluding diseased subjects from the population. Even among civilized communities, a significant proportion of a group is affected by chronic disease, often previously unsuspected. Statistics from the US, for example (US Department of Health, Education and Welfare, 1964–8), showed that in 1967–8, 23.2% of children under the age of seventeen had one or more chronic conditions ('diseases, illnesses, or impairments'). Exclusion of diseased individuals has some influence upon mean results, and a larger impact in reducing the variance of data.

The human adaptability handbook (Weiner & Lourie, 1969) preserved an open mind as to whether diseased subjects should be included in population statistics. Certainly, it is a difficult decision in a community where the majority of the inhabitants are affected by such systemic conditions as hookworm or malarial anaemia (page 94). Only healthy subjects can reveal the potential working capacity conferred by genetic constitution; on the other hand, this potential is so commonly restricted by disease that calculations based on healthy subjects may give a misleading impression of the working capacity of a settlement and its ability to adapt to the physical demands of the environment.

In the Canadian Eskimo community of Igloolik, 67.2% of the population over the age of nine had normal health (Table 2). Subjects who had a history of pulmonary disease were 0.7% shorter and 8.5% lighter than those who were healthy (Chapter 5). Muscle strengths were also reduced (hand-grip 2.6%, knee extension 5.8%), and there was a substantial deficit of maximum oxygen intake, averaging 12.1%. When calculating the possible work output of the villagers, it is important that healthy and diseased subjects be included in appropriate proportions. If the maximum oxygen intake of the healthy Eskimo is represented as 100%, that of the diseased Eskimo (32.8% of the community) becomes 87.9%, and the average maximum oxygen intake of all villagers is 96.0% of the value for those who are healthy. Many investigators have tested without regard to disease. It is likely that their samples have been biassed towards the healthy, but depending on the proportion of diseased subjects, reported values for a community such as Igloolik could lie anywhere on the continuum 87.9 to 100% of the maximum oxygen intake for a healthy population.

Similar considerations apply to socio-cultural factors. It is unfortunately excessively simplistic to classify present-day communities as nomads, hunters, or cultivators. The majority of the Igloolik Eskimos, for example, like to describe themselves as hunters, but when a careful inventory of sleds, dog-teams and snow-mobiles is made, it becomes obvious that not more than one man in five is a frequent hunter. Since hunting has a major influence upon working capacity (Rode & Shephard, 1971), it is important to base calculations of the average aerobic power of villagers on appropriate proportions of several distinct occupational classes; we estimated that a valid calculation for this community should include 6% very frequent hunters, 11% less frequent hunters, 64% transitional Eskimos, and 17% acculturated, wage-earning settlement workers.

Bias in activity measurements

The human adaptability handbook recommendation (Weiner & Lourie, 1969) was that activity measurements 'should cover a period of one week, including the weekend, and wherever possible should be repeated at least once, e.g. in summer and winter'. It was admitted that 'in some cases, such as agriculture, seasonal factors are so evident that an assessment of one day in the year can scarcely be adequate'. However, the enormous variety of occupations followed by the primitive hunter was given less than adequate recognition. Many forms of game are sought at different seasons (Fig. 1). Each type of hunt requires a wide range of characteristic activities, with highly specific rates of energy expenditure. The problem is compounded by rapid changes in hunting techniques with acculturation. Fur-

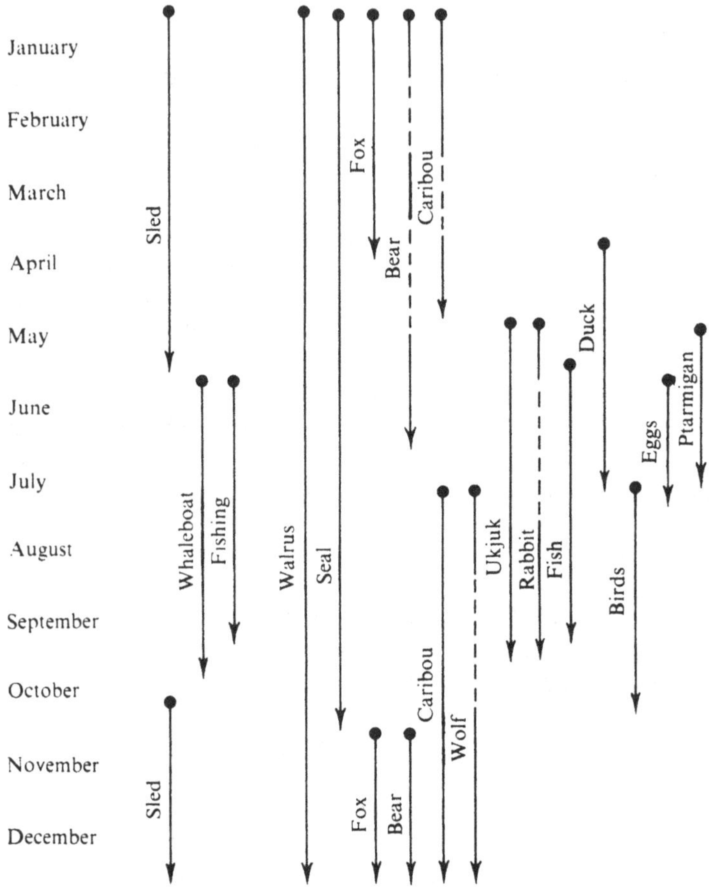

Fig. 1. Types of hunt undertaken at different seasons – a picture of life in the Canadian Eskimo community of Igloolik. The dotted sections indicate times when some hunting took place, but conditions for this type of hunt were less favourable than at the times shown by continuous lines. (From Godin & Shephard, 1973, courtesy of William Heinemann Medical Books Ltd.)

thermore, arduous journeys of several weeks duration may be needed in order to collect even a few items of data on a small party of hunters (Godin & Shephard, 1973; Shephard & Godin, 1976). Even the estimation of energy expenditures for the housewife is a major undertaking. The women of Igloolik spend much of their time in relatively standard wooden bungalows, but their activities (preparation of skins and repair of clothing) are substantially influenced by the hunting patterns of their mates. At some seasons, improvised domestic equipment is used at summer fishing camps and winter igloos. In both men and women, the relative proportions of settlement and camp life are markedly affected by acculturation. To obtain reasonably full and correctly biassed information on the energy expenditures of a single settlement would be a mammoth scientific undertaking. The best that we have been able to accomplish in Igloolik has been a description of summer and winter activities in those we regard as typical settlement workers, together with the measurement of energy expenditures incurred by a few traditional villagers as they engaged in twelve different forms of hunting (Godin & Shephard, 1973; Shephard & Godin, 1976). A few other investigators have made some measurements of energy expenditure in the communities they have studied, but the information reported is at best sketchy.

3 Methods for the measurement of physical fitness, working capacity and activity patterns

One of the avowed objectives of the IBP was to develop a common, standardized methodology that could be used by biological scientists in every corner of the world. The need for standardization was particularly evident with respect to the measurement of working capacity. Almost every laboratory had its own design of step or bicycle ergometer and its own characteristic exercise protocol. While many investigators claimed accuracies of 0.01–0.02% for their chemical methods of gas analysis, widely differing results were reported (Cotes & Woolmer, 1962) when laboratories were confronted by a single cylinder mixture (oxygen concentrations were estimated at 15.70–16.21%, and CO_2 concentrations at 6.01–6.28%). Perhaps mainly because of systematic errors of gas analysis (Cormack & Heath, 1974), IBP teams noted that maximum oxygen intakes differed by 10% (Bonjer, 1966) or even 25% (G. R. Cumming, 1970) when the same subjects were tested in different laboratories.

Approaches to the standardization of methodology

Recognizing that the choice of test procedure for the measurement of working capacity could prove a controversial issue, the IBP convened a bench-level working party in Toronto in the summer of 1967. The objectives were to evaluate a representative group of possible test procedures, to recommend the optimum protocol for inclusion in the human adaptability handbook, and to make a pilot trial of the chosen methodology on a randomly selected population of schoolchildren.

The results were quite widely publicized (Shephard *et al.* 1968*a*, *b*; Weiner & Lourie, 1969; Shephard, 1970, 1971*b*). However, the impact of this vital IBP endeavour was blunted by vested interests. Some laboratories were unwilling to abandon other methodologies by which they had already accumulated substantial quantities of data. Although the IBP was the sole organization to convene an experimentally oriented working party, the last few years of the 1960s saw other national, regional and international groups attempting to review or to legislate standard procedures, all operating with differing emphases. Whereas the IBP methodology was envisaged primarily as a basis for comparative physiological studies of primitive communities, industrial societies, and athletes, proposals from an expert committee of the World Health Organization (Andersen *et al.*, 1971) were aimed largely at the patient with cardio-vascular disease; the International Committee on the Standardization of Physical Fitness Tests (1969) in turn recommended procedures that the physical educator could apply to the athlete in the absence of medical supervision; the Internationales Seminar für Ergometrie (Mellerowicz, 1966) examined work output rather than physiological responses, and the International Labour Organization (Cotes, 1966) suggested a simple measure of fitness for the relatively light physical loads encountered in modern industrial work. The 1970s saw further proposals from US medical groups (Erb, 1970; Kattus, 1972) and exercise physiologists (Faulkner & Stoedefalke, 1975), unfortunately made with scant reference to preceding deliberations, while inevitably many textbooks of exercise physiology have yet to catch up with this ferment of activity.

Space hardly permits detailed review of methodology for all elements of working capacity. However, in view of the unique contribution of the IBP to the standardization of maximum oxygen intake measurements, and the centrality of such data to subsequent chapters, it seems appropriate to recapitulate briefly the findings of the working group, and to make a retrospective analysis of the validity of their conclusions in the light of discussions held by other international groups.

IBP recommendations for the measurement of working capacity

Technique

The accuracy and the validity of test measurements are influenced by many details of technique such as proper timing of gas sample collection (Phillips & Ross, 1967), correct calibration of the bicycle ergometer (G. R. Cumming & Alexander, 1968) and avoidance of systematic errors in gas analysis (Cotes & Woolmer, 1962; Cormack & Heath, 1974). The IBP thus suggested that where possible, procedures should be learnt by a preliminary visit to a reference laboratory.

Type of ergometer

While it was recognized that some laboratory studies of athletes might involve the use of arm ergometers (Mellerowicz, 1962) and various forms of free and tethered swimming (Magel, 1971; McArdle, Glaser & Magel, 1971; Holmér, 1972; Rennie *et al.*, 1973), the choice of ergometer for field studies plainly lay between the step test and the bicycle (the less correct but more frequently used term for the cycle ergometer), with only occasional possibilities for use of a treadmill. The relative merits of the three common types of ergometer have been discussed extensively

(Shephard *et al.*, 1968*a*, *b*; Shephard, 1969*a*, 1970, 1971*b*; Taylor *et al.*, 1969; Andersen *et al.*, 1971), and this chapter will do no more than stress issues relevant to human adaptability studies.

The IBP working party suggested that uphill treadmill running was the technique most likely to develop a true, centrally-limited maximum oxygen intake (Shephard *et al.*, 1968*a*). With this method, the oxygen cost of running per kilogram of body weight can be predicted with a precision of about 10% (Shephard, 1969*c*). However, the appropriateness of running depends on the likely maximum speed, and for many older subjects uphill walking is a preferable alternative. Moody, Kollias & Buskirk (1969) have shown a negligible difference in the aerobic power as measured by the two procedures.

The von Döbeln (1954) apparatus was commended as a relatively cheap and rugged portable form of bicycle ergometer. Several types of electrically-braked bicycle ergometer theoretically allowed maintenance of a constant work load in the face of minor variations in the speed of pedalling, but such machines were heavier than mechanical ergometers, and had less stable calibration characteristics when moved from one site to another (Stein, Rothstein & Clements, 1967). It was recommended that the pedalling rate normally be established in the range most efficient (Mellerowicz, 1966; Banister & Jackson, 1967; Hermansen & Saltin, 1969) for submaximal or maximal effort (50–60 pedal r.p.m.), but that in physically fit subjects (particularly endurance athletes) the rate be increased to 80–90 r.p.m.; the purpose of the faster rate was to avoid the muscle fatigue seen with a combination of heavy loads and slow pedalling speeds.

The proposed step test (Shephard, 1967*a*) was a modification of the widely known clinical Master's step. One face, for use in submaximum effort, had two 23 cm (9 inch) risers climbed at rates of ten to twenty-five ascents per minute (as paced by a metronome). The other face had a single 48 cm (18 inch) riser, for use in maximal testing. The purpose of this arrangement was to allow an optimum pace of stepping in both submaximal and maximal effort; in particular, the subject did not stand poised awaiting the beat of the metronome at very slow rates of climbing (Wyndham & Sluis-Cremer, 1968; Shephard & Olbrecht, 1970), and yet avoided the danger of tripping in maximum effort (Kasch *et al.*, 1965). While more complicated designs of step have been proposed by other authors (for example, Nagle, Balke & Naughton, 1965; Wyndham & Sluis-Cremer, 1968; Erb, 1970), these seem less suitable for field use.

Patient preparation

The IBP working party stressed that both the test room and the patient should be carefully prepared if valid test results were to be obtained. Rowell, Taylor & Wang (1964) had previously suggested 17 °C (62 °F) as an optimum laboratory temperature. The IBP team recognized the conflict between the ideal and the possible under field test conditions, and recommended a room temperature of no more than 20–22 °C, with the possibility of a further 1–2 °C increase over this figure if a fan was provided. In fact, some IBP investigators were unable to meet even these last standards, and at least one group had to conduct tests at an air temperature of 26 °C (Thinkaran *et al.*, 1974).

Equally, it was important for the subject to avoid exposure to extremes of climate and physical activity on the day preceding examination, to ensure adequate sleep on the preceding night, and to avoid the use of alcohol, drugs, and food immediately before the test (Jones & Haddon, 1973). Unfortunately, all these matters are easier to organize in a well-regulated urban community than when dealing with nomads who have their own tribal language.

The initial emotional increase of heart rate should be minimized, both during submaximal (Antel & Cumming, 1969) and maximal effort (Shephard *et al.*, 1974*a*), through an appropriate schedule of habituation or familiarization visits (Glaser, 1966; Shephard, 1966*b*, 1969*b*). Taylor *et al.* (1969) had previously remarked that about a quarter of their railroad workers were very anxious when first exercised on the treadmill; with repetition of the test at a constant and moderate work load, the average heart rate for the group decreased from 124 to 111 per minute. The IBP working group found that in young white Canadians, anxiety was greatest on the treadmill and least on the bicycle ergometer. The step test caused little anxiety at moderate rates of ascent, but in near maximum effort (presumably because of fear of stumbling), anxiety tachycardia was intermediate between that observed for the bicycle and the treadmill. It was recommended that, where possible, the oxygen consumption should be measured rather than estimated from the work performed. If it was necessary to rely on work measurements, the task should be fully learnt. Over a five-day period of observation, the IBP team traced minor gains in the mechanical efficiency of stepping (15.7 to 16.1%) and cycling (21.1 to 21.9%), with a 7% diminution in the cost of uphill treadmill running (Shephard *et al.*, 1968*a*, *b*). The main advantages of the bicycle ergometer relative to the step test seemed a consistent body position (thus facilitating ancillary measurements such as blood pressure), little evidence of habituation (as shown by the constant heart rate at a fixed oxygen consumption), and a somewhat more consistent mechanical efficiency.

Nevertheless, it was recognized that these last two features did not necessarily extend to groups other than fit young civilized adults; rates of habituation and test learning could well be quite different for the young, the elderly and primitive groups unfamiliar with modern laboratory equipment.

Directly measured maximum oxygen intake ($\dot{V}_{O_2(max)}$)

The directly measured maximum oxygen intake was conceived as the international reference standard of cardio-respiratory fitness (Shephard *et al.*, 1968*a*). It was recommended that tests used to measure $\dot{V}_{O_2(max)}$ should activate the majority of the body muscles and produce at exhaustion symptoms and signs of a central cardio-respiratory limitation of effort (nausea, severe breathlessness, cyanosis progressing to an ashen-grey pallor, incoordination, a confused response to questioning and impending loss of consciousness) rather than the picture of a peripheral limitation (muscular weakness, pain, and fatigue). On these criteria, the treadmill was judged superior to the bicycle ergometer, with the step test occupying an intermediate position (Shephard *et al.*, 1968*a*).

It was recommended that, where possible, objective criteria for attainment of the maximum oxygen intake should be met. These criteria would include an increase in oxygen consumption of less than 2 ml/kg min with an appropriate small increase of work load (Milic Emili *et al.*, 1959), a pulse rate close to the age-related maximum (Asmussen & Molbech, 1959), a respiratory gas exchange ratio of approximately 1.15, and a blood lactate reading $\geqslant$ 100 mg/dl two minutes after cessation of exercise (with lower limits of 80 mg/dl in children and 60 mg/dl in the elderly; G. R. Cumming & Borysyk, 1972). The original basis for definition of the oxygen plateau was an increment in oxygen consumption two standard deviations less than that anticipated for a 2.5% increase of treadmill slope (Taylor *et al.*, 1969). Wyndham *et al.* (1959) and Glassford *et al.* (1965) cautioned that a plateau could underestimate $\dot{V}_{O_2(max)}$; Wyndham suggested, as an alternative, statistical curve fitting to the work load/oxygen consumption relationship. However, this seems impractical with the number of observations obtained in the usual field experiment. A fixed criterion (such as an increase in oxygen consumption of less than 2 ml/kg min or 150 ml/min) becomes unsatisfactory when applied to subjects with a poor aerobic power. It is also quite difficult to define a plateau in children (P. O. Åstrand, 1952; G. R. Cumming & Friesen, 1967), and better results can sometimes be obtained in the young by requiring 2½–3 min of exhausting work at a 'supra-maximal' load calculated by extrapolating the linear portion of the heart rate/work load relationship to a theoretical heart rate of 247 per minute (G. R. Cumming & Friesen, 1967). The alternative, adopted by the IBP working group, was to require the children to make one or more

Table 5. *Comparison of maximum oxygen intake as estimated by bicycle ergometer, step, and treadmill tests* (*mean*±*SD; range of data*)

Exercise mode	$\dot{V}_{O_2(max)}$ (l/min STPD)	Heart rate (/min)	Blood lactate (mg/dl)
Treadmill	3.81±0.76	190±5	122±21
	(2.54–5.84)	(178–197)	(78–166)
Bicycle ergometer	3.56±0.71	187±9	112±15
	(2.57–5.23)	(167–207)	(89–143)
Step test	3.68±0.73	188±8	105±26
	(2.66–5.59)	(170–195)	(45–165)

Data for twenty-four healthy young white Canadians.
Reproduced from Shephard *et al.* (1968*a*), courtesy of World Health Organization.

preliminary familiarization visits to the laboratory (Shephard *et al.*, 1968*c*); however, even when much time was taken to gain the confidence of the child, plateaux were still only realized with difficulty.

A further group where extended anaerobic work is undesirable are older patients with suspected ischaemic heart disease; here, a useful alternative approach seems a very rapid increase of work load, with either a continuous record of oxygen consumption (Auchincloss & Gilbert, 1973 – a procedure only possible in the base laboratory) or collection of expired gas in the final fifteen seconds of effort (Niinimaa *et al.*, 1974).

The subjective impression from the experiments on young adults was that the treadmill gave a truer account of cardio-respiratory power than the bicycle ergometer; this was confirmed by objective data (Table 5). In keeping with reports from other laboratories (Table 6), the treadmill maximum oxygen intake was 3.4% larger than that found during stepping, and 6.6% larger than that during cycling. While it was recognized that the bulk of the treadmill and the need for three-phase electrical wiring precluded the field use of such equipment in most instances, the results obtained by step and bicycle ergometer exercise were sufficiently similar to treadmill data that results could be equated after application of appropriate scaling factors (at least in young adults, where the majority of comparisons have been made).

A preliminary warm-up at 70% of maximum oxygen intake was advised, partly to give an indication of appropriate loading for the definitive test, partly to reduce the likelihood of muscle tears, and partly because such a warm-up produced a small augmentation of $\dot{V}_{O_2(max)}$ (Taylor, Buskirk & Henschel, 1955; Pirnay *et al.*, 1966).

Other details of the test protocol were judged to be of secondary importance. Tests where the load was increased at two-minute intervals

Table 6. *A comparison between maximum oxygen intake data yielded by bicycle ergometer, step, and treadmill tests*

Sample	Treadmill (%)	Step (%)	Bicycle (%)	Author
Canada				
Young adults				
Miscellaneous	100	95.1	95.3	Bouchard *et al.* (1973)
Moderately fit	100	—	92	Glassford *et al.* (1965)
Moderately fit	100	96.6	93.4	Shephard *et al.* 1968*a*)
Sedentary	100[a]	93[a]	97[a]	Shephard (1966*a*)
Czechoslovakia				
Sedentary adults				
Aged 20–29 yr	—	82.2	100	Skranč, Havel & Barták (1970)
Aged 30–39 yr	—	88.4	100	Skranč *et al.* (1970)
South Africa				
Soldiers				
Trained	100	99	93.1	Wyndham *et al.* (1966*a*)
Untrained	100	—	91.2	Wyndham *et al.* (1966*a*)
Sweden				
Athletes	100	—	93	Saltin & Åstrand (1967)
United States				
Aged 5–57 yr	100	99	—	Kasch *et al.* (1966)
Young adults				
Japanese resident in US	100[c]	—	95.3[c]	Taguchi, Raven & Horvath (1971)
	100[d]	—	91.6[d]	Taguchi *et al.* (1971)
Miscellaneous	100	—	88.7	McArdle, Katch & Pechar (1973)
Young men				
Fit	100	93.3	88.5	Kamon & Pandolf (1972)
Moderately fit	100	98.8	89.2	Kamon & Pandolf (1972)
Unfit	100	96.2[b]	85.3	Kamon & Pandolf (1972)
Young women				
Fit	100	95.5	93.0	Kamon & Pandolf (1972)
Moderately fit	100	103.8	91.3	Kamon & Pandolf (1972)
Unfit	100	111.2	95.1	Kamon & Pandolf (1972)
Lean	100	—	98.5	Moody *et al.* (1969)
Obese	100	—	90.5	Moody *et al.* (1969)

Data from world literature.
[a] Predicted $\dot{V}_{O_2(max)}$. [b] Laddermill. [c] 50 r.p.m. [d] 60 r.p.m.

(continuous tests) reached a slightly higher $\dot{V}_{O_2(max)}$ with a slightly smaller oxygen debt than tests where differing loads were applied on separate days (discontinuous tests; Table 7). The continuous test was thus recommended as being more economical in time and better suited to a population survey. Other authors, also, have found little difference in results with changes in the pattern of exercise loading. P. O. Åstrand & Saltin (1961*a*) found

Table 7. *A comparison of maximum oxygen intake values as determined by continuous and discontinuous tests (mean±SD; range of data)*

	Continuous test			Discontinuous test		
Exercise mode[a]	$\dot{V}_{O_2(max)}$ (l/min STPD)	Heart rate (/min)	Lactate (mg/dl)	$\dot{V}_{O_2(max)}$ (l/min STPD)	Heart rate (/min)	Lactate (mg/dl)
Treadmill	3.69±1.14 (2.54–5.77)	185±10 (167–197)	79±19 (48–104)	3.59±1.32 (1.43–5.84)	186±5 (178–195)	123±15 (112–150)
Bicycle ergometer	3.82±0.77 (2.85–4.94)	187±5 (178–195)	76±15 (54–99)	3.79±0.61 (3.05–4.69)	187±6 (179–198)	111±7 (100–123)
Step test	3.97±0.71 (2.81–5.04)	185±6 (175–195)	81±21 (58–103)	3.81±0.60 (2.93–4.85)	189±4 (184–195)	99±19 (68–129)
All tests	3.84	186	79	3.74	187	110

From Shephard *et al.* (1968*a*), courtesy of World Health Organization.
STPD = standard temperature and pressure, dry gas.
[a] Different subjects used for the three test modes.

Table 8. *Difference between directly measured maximum oxygen intake (l/min STPD) and that predicted by Åstrand, Margaria and Maritz procedures*

Exercise mode	Day of test	Åstrand nomogram	Margaria nomogram	Maritz extrapolation
Step test	1	−0.18±0.46	−0.09±0.39	−0.03±0.36
	5	+0.16±0.25	+0.12±0.28	+0.12±0.28
Bicycle	1	+0.30±0.31	+0.18±0.27	+0.19±0.28
ergometer	5	+0.33±0.35	+0.13±0.35	+0.18±0.34
Treadmill	1	−0.27±0.37	−0.35±0.67	−0.17±0.57
	5	−0.04±0.39	−0.39±0.61	−0.22±0.44

Data for young Canadian men.
From Shephard *et al.* (1968*b*), courtesy of World Health Organization.

the same $\dot{V}_{O_2(max)}$ with two and eight minutes of exhausting exercise. Glassford *et al.* (1965) reported very similar results whether using the treadmill protocols advocated by Taylor (see Taylor *et al.*, 1969) or those of Mitchell, Sproule & Chapman (1958). Pirnay *et al.* (1966) commented on the slightly higher aerobic power attained by a continuously increasing load, results being similar whether the load was increased every second minute or every third minute (Bottin *et al.*, 1968). Bonjer (1966), working within the framework of the Dutch IBP, favoured a 'stepwise' or a

Table 9. *Some assessments of the systematic error in methods used to predict maximum oxygen intake, together with validity (coefficients of correlation) and the coefficient of variation of discrepancies between direct and predicted values. The prediction method is the Åstrand nomogram unless otherwise specified*

Sample	Exercise mode	Systematic error (%)	Validity (r)	Coefficient of variation (%)	Author
Belgium					
Trained students (30)	BI[a]	−4.4	—	4.3	Bottin *et al.* (1966)
Britain					
Miscellaneous (85)	BI	−18.3	0.79	11.1	C. T. M. Davies (1968)
Canada					
Children (M)	BI	—	0.51–0.62	—	Hyde (1967)
Children (F)	BI	—	0.27–0.51	—	Hyde (1967)
Students (24)	TM	+1	0.78	—	Glassford *et al.* (1965)
	BI	+6.1	0.65	—	Glassford *et al.* (1965)
Finland					
Military students (48)	BI	—	0.54	13.2	Oja, Partanen & Teräslinna (1970)
France					
Trained men } (68)	BI	−4.4	0.73	—	Flandrois *et al.* (1962)
Untrained men } (68)	BI	+2.2	0.48	—	Flandrois *et al.* (1962)
Germany					
Students (19)	BI	+10	0.78	13.6	Eichhorn *et al.* (1967)
Italy					
Aged 9–47 yr	ST[b]	+1.1	—	5.8	Margaria *et al.* (1965)

Table 9. (*cont.*)

Sample	Exercise mode	Systematic error (%)	Validity (*r*)	Coefficient of variation (%)	Author
South Africa					
Young men (26)	BI	+10	—	17.7	Maritz *et al.* (1961)
	BI	−5.6	—	8.7[c]	Maritz *et al.* (1961)
Sweden					
Building workers (84)	BI	—	—	8.4	von Döbeln, Åstrand & Bergström (1967)
	BI	—	—	12.7[d]	von Döbeln *et al.* (1967)
Well-trained students (50M; 62F)	BI	0	—	6.7	P. O. Åstrand & Ryhming (1954)
	ST	0	—	6.8	P. O. Åstrand & Ryhming (1954)
United States					
Students (10 athletes; 12 sedentary)	TM[e]	−5.6	—	4	Rowell *et al.* (1964)
University staff (31)	BI	+7.6	0.92	—	Teräslinna, Ismail & MacLeod (1966)

BI = bicycle ergometer; ST = step test; TM = treadmill. Number of subjects in parentheses.

[a] Need for high f_h stressed.

[b] Margaria nomogram.

[c] Maritz line.

[d] New formula Åstrand nomogram.

[e] Hot room (26 °C) and fear of catheterization.

'triangular' pattern of increasing load, completed in a single session. Moody *et al.* (1969) reported comparable values of $\dot{V}_{O_2(max)}$ for discontinuous and continuous progressive tests, where the load was augmented by several discrete stages. Several other groups (Horvath & Michael, 1970; Falls & Humphrey, 1973) have found little change of results with minor alterations of protocol, although Froelicher *et al.* (1974) have suggested that slightly higher results are obtained with a progressive test if five-minute rest intervals are allowed between stages, in the manner originally suggested by Taylor *et al.* (1969).

Submaximal tests

The IBP working group recognized that, in many field situations, circumstances such as the absence of a physician might preclude direct measurements of maximum oxygen intake on an entire population sample. It was then recommended that indirect predictions be made, exploiting the linear relationship between pulse rate and oxygen consumption during submaximal work.

Three commonly used prediction procedures were compared-linear extrapolation of the oxygen consumption/heart rate line to a theoretical age-related maximum heart rate (Maritz *et al.*, 1961), a nomogram based on the heart rates at two fixed rates of working (Margaria, Aghemo & Rovelli, 1965), and a nomogram based on the heart rate at a single oxygen consumption or equivalent work load (P. O. Åstrand & Ryhming, 1954; I. Åstrand, 1960). Theoretically, the precision of predictions might have been anticipated to vary according to the square root of the number of observations, but in practice this advantage was not realized. The explanation was thought to be a larger influence of anxiety on pulse rates at relatively low work loads (Shephard *et al.*, 1968*b*); thus, the slope of the Maritz line was systematically distorted.

As might be anticipated from the anxiety associated with treadmill running (Table 8), predictions based on submaximal loadings systematically underestimated the directly measured $\dot{V}_{O_2(max)}$ when using this apparatus. On the other hand, submaximum data from the bicycle ergometer slightly overestimated directly measured results for the same equipment, perhaps because of difficulties in reaching a true, centrally limited $\dot{V}_{O_2(max)}$ on the bicycle. The least systematic error between the two sets of results was observed with the step test. After subjects had opportunity to familiarize themselves with the stepping procedure, the submaximal data overestimated directly measured values by 3–4%, with a scatter of 6–7% about this average prediction. Data for the bicycle and the treadmill were a little less satisfactory, particularly with regard to the scatter about the mean discrepancy.

Table 10. *A comparison of continuous and discontinuous test modes in the prediction of maximum oxygen intake from submaximal test data (l/min STPD, mean±SD)*

Exercise mode[a]	Discontinuous test	Continuous test	Δ[b]
Step test	3.75±0.87	3.73±0.93	−0.013±0.34
Bicycle ergometer	3.57±0.81	3.64±0.86	+0.069±0.29
Treadmill	3.83±0.90	3.79±0.89	−0.043±0.61

Data for young Canadian men.
From Shephard *et al.* (1968*b*), courtesy of World Health Organization.
[a] Subjects differed for the three test modes.
[b] Δ = continuous test value−discontinuous test value.

Data from the world literature (Table 9) generally have substantiated the IBP findings. Large systematic errors were noted by only two groups (Rowell *et al.*, 1964; C. T. M. Davies, 1968); in the first of these two experiments, the poor results were due largely to a hot environment (78 °F; 25.6 °C) and anxiety concerning impending surgical procedures. C. T. M. Davies (1968) apparently gave his subjects rather little time to become familiar with the laboratory, but there is no other obvious explanation of his anomalous results. We may conclude that submaximum prediction procedures have a substantial variability, thus limiting their usefulness for the description of the individual, but that systematic errors are small enough that such data should give a reasonable indication of the work capacity of populations.

With regard to the exercise protocol (Table 10), the IBP team found no significant differences in results between experiments where subjects exercised for five minutes at each of four loads (discontinuous tests, with rest intervals increasing from seven to ten minutes) and other experiments where the same four loads were executed consecutively (continuous tests, with three minutes at each stage). The continuous format took a total of only fifteen minutes, and was thus better for field testing than a discontinuous test requiring one hour or longer. Other authors (Rutenfranz, 1964; Bonjer, 1968; Andersen *et al.*, 1971; S. M. Fox, 1974) have also discussed the question of test schedules. If the load is increased in a truly continuous fashion (as in the Müller Leistungs-Pulsindex, 1950), the cardiac response is somewhat less than where several minutes are allowed at each work load. Equally, in older subjects and those with cardio-vascular disease it may be desirable to allow four minutes at each exercise stage. However, with these exceptions, it seems possible to use quite a wide range of submaximal test protocols and yet to achieve very similar results.

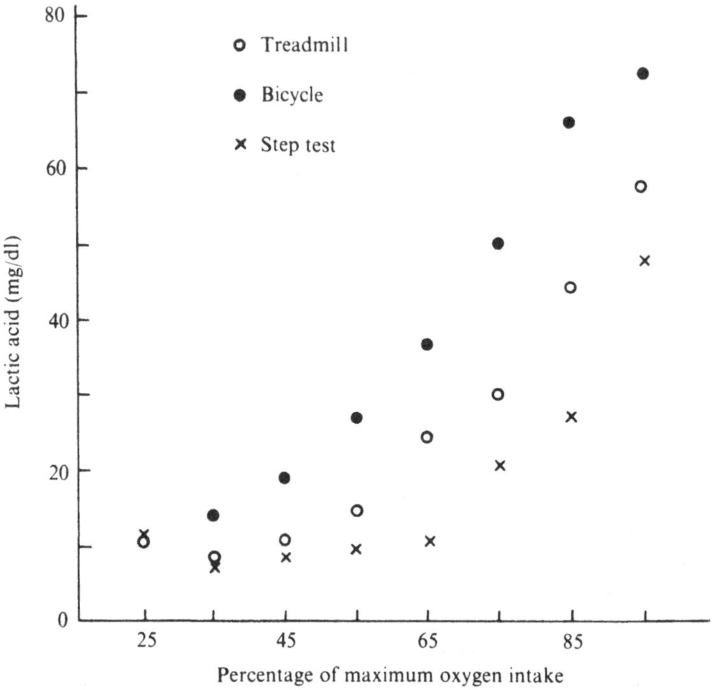

Fig. 2. The accumulation of lactic acid in the arterialized capillary blood two minutes after five minutes of exercise at varying percentages of maximum oxygen intake. (From Shephard *et al.*, 1968*b*, courtesy of World Health Organization.)

Bicycle ergometry and the quadriceps muscle

Many investigators have voiced the suspicion that performance on the bicycle ergometer is limited not by the cardio-respiratory system *per se*, but rather by the ability to sustain blood flow through over-vigorously contracting quadriceps muscles (Glasford *et al.*, 1965; Wyndham *et al.*, 1966*b*; G. R. Cumming & Friesen, 1967; Hoes *et al.*, 1968; Moody *et al.*, 1969; C. T. M. Davies, Tuxworth & Young, 1970; Katch, Girandola & Katch, 1971; Rowell, 1974). The practical consequences are (i) the commencement of anaerobic work at a low percentage of aerobic power (Fig. 2; Saltin & Åstrand, 1967; Dujardin *et al.*, 1967–8), thus delaying the attainment of a steady state and complicating estimates of mechanical efficiency in brief tests, and (ii) exhaustion from peripheral weakness before the maximum oxygen intake has been attained (Table 5).

In support of the IBP criticism of bicycle ergometry, Hoes *et al.* (1968) demonstrated that the forces developed during cycling reached a level where flow limitation was likely, and Kay & Shephard (1969) observed that the accumulated arterial lactate (mg/dl) two minutes after performance of a standardized ergometer load (80% of aerobic power) was negatively correlated with the maximum voluntary force developed by the quadriceps muscle. MacNab & Conger (1966) also commented on a positive correlation ($r = 0.45$) between the knee extension strength and the $\dot{V}_{O_2(max)}$ of female university students. On the other hand, Katch, McArdle & Pechar (1974) found no correlation between the peak isokinetic muscle force measured during a $\dot{V}_{O_2(max)}$ test and the discrepancy between treadmill and bicycle ergometer maxima, after allowance for the effects of body weight. This discordant finding may reflect the measurement of peak forces rather than the integrated force over the pedalling cycle; further, as Royce (1959) has stressed, the critical determinant of local blood flow restriction is not the absolute muscle force but rather the percentage of the maximum force exerted.

Some concessions in field methodology

Careful consideration of habituation and test learning (page 25) is necessary if the oxygen consumption is to be predicted from the work load rather than measured directly.

One important observation of the IBP working group was that 'constant' mechanical efficiency was by no means a unique feature of bicycle ergometer exercise. Although often quoted as a constant efficiency of 23%, P. O. Åstrand & Ryhming (1954) actually admitted a 6% coefficient of variation about their mean figure for the efficiency of a relatively homogeneous group of subjects exercising on the bicycle ergometer. The IBP team found a coefficient of variation of 4–5% for young Canadian men while on the bicycle ergometer and only 7 % while stepping; (Shephard, 1966*a*; Shephard *et al.*, 1968*b*); even the cost of uphill treadmill running could be predicted from a nomogram (Shephard, 1969*c*) with an accuracy of ±10%. In children, the IBP group found the mechanical efficiency of cycling increased with work load, from 20.1 to 22.2% in the boys, and from 19.4 to 22.8% in the girls (Shephard *et al.*, 1968*c*). Other studies of bicycle ergometry in children have quoted efficiencies as low as 14.5% at light work loads (Shephard, 1971*c*). Plainly, it is dangerous to assume a 'constant' mechanical efficiency of 23%, unless this has been verified for a given population under the proposed test conditions. G. R. Cumming & Alexander (1968) stressed the errors arising from traditional methods of calibrating mechanically-braked bicycle ergometers. Such methods neglect a substantial and variable loss of mechanical work in the chains and bearings, and for accurate determination of the work load it is necessary to clip a torque generator to the pedals.

With regard to measurements of heart rate, the IBP group recommended use of an electrocardiograph, since this added appreciably to the safety of testing. Nevertheless, it was demonstrated that a well-trained observer could measure the heart rate with reasonable accuracy during near-maximal effort using either auscultation (with a stethoscope strapped to the region of the apex beat) or carotid palpation. Post-exercise readings obtained within the first few seconds of recovery are closely correlated with the final exercise values (Cotton & Dill, 1935; Millahn & Helke, 1968; McArdle, Zwiren & Magel, 1969; Bailey *et al.*, 1974*a*). There is a substantial decrement of heart rate (10–30 beats/min) within the first thirty seconds of stopping exercise (Ryhming, 1954; Shephard, 1967*d*; Millahn & Helke, 1968), and the coefficient of correlation with exercise readings drops to about 0.8; nevertheless, in terms of the practical objective of sorting the fit from the unfit members of a population late pulse readings can be quite useful (Brouha, 1943; Shephard, 1966*c*).

Validity of IBP methodology

In concluding this chapter, it seems worthwhile to examine the validity of the IBP methodology in terms of its basic philosophy, the tools and exercise schedules suggested, and the extent to which systematic errors were eliminated.

Philosophy

Maximum oxygen intake

There has been considerable discussion regarding the interpretation of maximum oxygen intake over the past five years (Kaijser, 1970; Keul, 1973; Shephard, 1974*a*). The main areas of criticism are as follows.

(i) It has been suggested that a high percentage (94%) of the variance in aerobic power is determined by genetic rather than environmental factors (Klissouras, 1971). While this may represent an overstatement of the genetic contribution (Chapter 6), the measurement of $\dot{V}_{O_2(max)}$ remains of interest to the human adaptability project, whether environmentally or genetically determined.

(ii) Attention has been drawn to a lack of communality in the data when measurements of maximum oxygen intake are repeated on different types of ergometer (Bouchard *et al.*, 1973). It has been recognized for many years that arm ergometry yields only about 70% of the oxygen intake achieved in uphill treadmill running (Christensen, 1932; Bobbert, 1960; P. O. Åstrand & Saltin, 1961*b*; Stenberg *et al.*, 1967; Simmons & Shephard, 1971*a*). It is also known that peak levels of oxygen intake are lower with single-legged than with two-legged cycling (Düner, 1959; Freyschuss & Strandell, 1968; Gleser, 1973), and the IBP working group (Table 5) together with many other authors (Table 6) have shown that slightly lower results are obtained from stepping and cycling than from uphill treadmill running. However, discrepancies are not large enough to invalidate the IBP methodology. The lack of communality in Bouchard's data reflects largely the attempt to establish coefficients of correlation between tests using a population with a rather limited variance of aerobic power.

(iii) It has been argued that if uphill treadmill running yields a centrally-limited maximum oxygen intake, there should be no increase of aerobic power when a subject supplements leg work by arm work (P. O. Åstrand & Saltin, 1961*b*; Stenberg *et al.*, 1967). A recent report that aerobic power is 10% higher for combined work (Gleser, Horstman & Mello, 1974) seems attributable to use of a bicycle ergometer rather than a treadmill as the method of leg work.

(iv) Clausen, Trap-Jensen & Lassen (1970) published results suggesting that endurance training was limb specific; forearm training had relatively little effect on the heart rate response to subsequent leg exercise and vice versa. They thus reasoned that tests based on the use of the legs would have limited validity when applied to populations performing mainly arm work (for example, certain classes of athlete). Clausen's original experiments did not include direct measurements of maximum oxygen intake, and when this omission was corrected (Clausen *et al.*, 1971; Clausen, 1973) it was found that leg training did increase the maximum oxygen intake during arm ergometry, some 57% of the training effect being transferred.

There seems little possible criticism of measuring endurance fitness by a treadmill test when normal daily activities involve mainly use of the lower limbs. If arm work is required in daily life, a treadmill test indicates the power of the cardio-respiratory system, but does not demonstrate how far this can be realized during activities involving the arm muscles (Rowell, 1974). Fortunately, well-trained athletes such as rowers (Hagerman & Lee, 1971; Carey, Stensland & Hartley, 1974) and swimmers (Stenberg *et al.*, 1967; Magel, 1971; Dixon & Faulkner, 1971; Holmér, 1972; Shephard, 1975*b*) seem to achieve very similar maximum oxygen intakes with uphill treadmill running and with their chosen sport.

(V) Some authors – for example Cureton (1972) – have complained that there is a poor correlation between maximum oxygen intake and athletic performance. Many athletic events are completed within a few seconds, and one would then hardly anticipate a strong relationship between competition results and oxygen transport. Even if long-distance competitors are distinguished, the average team of athletes shows a rather limited distribution of performances, and this inevitably reduces the significance of correlation coefficients between performance and maximum oxygen intake; nevertheless we have demonstrated coefficients amounting to

0.84 in university-class swimmers (Shephard *et al.*, 1973*a*) and 0.67 in provincial white-water paddlers (Sidney & Shephard, 1973).

(vi) Kaijser (1970) reasoned that if circulatory transport of oxygen limited endurance, then performance should be improved by exposure to oxygen at a pressure of three atmospheres. In fact, he did not find any change of endurance in a hyperbaric chamber. However, he did not measure maximum oxygen intake directly, but relied on the rather subjective criterion of endurance time; it seems likely that voluntary performance at the high oxygen pressure was affected adversely by oxygen toxicity. Certainly, there have been many experiments both before and subsequent to the work of Kaijser that have shown gains of oxygen transport with more modest increments of oxygen pressure, and decrements of aerobic power with a decrease of oxygen pressure (see Shephard, 1974*a*).

(vii) Kaijser (1970) also argued that if maximum oxygen intake were dependent on cardio-vascular transport, it would change in parallel with changes of maximum cardiac output. In fact, the cardiac output was greater in long (8–10 min) than in brief (3–5 min) periods of maximum effort, although maximum oxygen intake remained unchanged (P. O. Åstrand & Saltin, 1961 *a*). Equally, administration of propranolol reduced maximum cardiac output without influencing maximum oxygen intake (P. O. Åstrand, Ekblöm & Goldberg, 1971).

In general, maximum oxygen intake does parallel maximum cardiac output. However, a logic that insists on such a relationship neglects both the major fraction of the cardiac output directed to the skin in sustained effort (Simmons & Shephard, 1971*b*) and possible alterations in blood flow distribution between nutritional and shunt vessels within muscle; inevitably, maximum oxygen intake depends upon cardio-respiratory transport to active muscle fibres, and in a very hot environment as much as a quarter of the maximum cardiac output may be directed to the subcutaneous vessels.

(viii) Lastly, there is good evidence that endurance training increases enzyme concentrations within the muscle fibres (Gollnick & Hermansen, 1973; Holloszy, 1973; Howald, 1975). Holloszy (1973) has reasoned that since endurance training also widens arterio-venous oxygen differences in maximal effort, this implies a peripheral limitation of oxygen transport in unfit individuals, with increased oxygen extraction after development of tissue enzyme systems. However, blood leaving active muscles is usually almost completely depleted of oxygen (content < 6 ml/l in the experiments of Hartley & Saltin, 1969). Little increase of oxygen extraction could thus arise from enzymic changes; Doll, Keul & Maiwald (1968), for example, found no difference of femoral venous oxygen tension between athletes and sedentary subjects. A more reasonable hypothesis seems that training, like heat acclimatization, facilitates sweating, thereby allowing a diversion of blood flow from the skin to the active muscles (Simmons & Shephard, 1971*b*). Teliologists seeking 'reasons' for the increase of muscle enzyme concentrations with training may consider some alternative suggestions: (1) there is a need to compensate for the longer diffusion path from capillary to mitochondrion as a muscle fibre hypertrophies; (2) if enzyme activity is increased, less phosphagen depletion is needed to activate tissue mechanisms of oxygen utilization (Saltin & Karlsson, 1973), thereby reducing oxygen debt at the commencement of exercise; (3) in steady-state exercise, greater enzyme activity increases the relative utilization of fat, conserving glycogen; this is particularly helpful to the performance of those muscles that contract too strongly or too long per cycle to permit adequate perfusion (Holloszy, 1973).

In sum, despite five years of vigorous debate, the weight of evidence still supports the IBP premise (Shephard *et al.*, 1968*a*) that the maximum oxygen intake provides the best single measure of man's fitness for endurance-type activities.

Predictions of maximum oxygen intake

Procedures for the prediction of maximum oxygen intake from submaximum test data remain generally well-accepted (see, for example, Kozlowski *et al.*, 1968); nevertheless, they have been the subject of substantial criticism (for example, Rowell *et al.*, 1964; C. T. M. Davies, 1968). It has been argued that the information content of submaximum data is not increased by extrapolation to a supposed maximum. This is a valid criticism of some 'indirect' physiological tests such as the 'indirect maximum voluntary ventilation' (where the value reported is the one-second forced expiratory volume multiplied by a constant of 40). However, each of the three common methods for the prediction of maximum oxygen intake (I. Åstrand, 1960; Maritz *et al.*, 1961; Margaria *et al.*, 1965) introduces an allowance for the age-related decrease in maximum heart rate. Predicted results are thus preferable to such alternative procedures as reporting the work rate at a heart rate of 170/min (Wahlund, 1948), the heart rate at an oxygen consumption of 1.5 l/min (Cotes, 1966), or the $\dot{V}_{O_2(max)}$ as estimated from a linear regression equation based on measurements of heart rate at a fixed work load (E. L. Fox, 1973). Prediction methods can be so arranged that all subjects are carried to a comparable submaximal stress (for example, 75% or 85% of maximum oxygen intake as judged from a target heart rate; Table 11); on the other hand, if submaximal test data are recorded at a constant pulse rate, oxygen consumption or work load, this places a severe stress on the older members of a population, while providing an inadequate stimulus to younger and more active subjects. Prediction procedures have both a small systematic error and a

Table 11 *Target heart rates corresponding to approximately 75% of aerobic power*

Age (yr)	Heart rate (/min)
20–29	160
30–39	150
40–49	140
50–59	130

From Shephard (1971*b*), courtesy of C. C. Thomas, Publisher, Springfield, Illinois.

substantial coefficient of variation about this systematic error (Tables 8, 9). The predicted maximum oxygen intake thus has rather limited value in interpreting the fitness and working capacity of the individual. However, the human adaptability project is concerned primarily with the status of populations, and there is general agreement that submaximum predictions yield relatively unbiassed estimates of maximum oxygen intake when applied to substantial groups of subjects.

Choice of ergometer

With the exception of laboratory tests on city-dwellers and some US studies of Alaskan Eskimos, the treadmill proved impractical for IBP investigations. The choice of ergometer was thus distributed rather evenly between mechanically-braked bicycles, usually of the von Döbeln type, and various forms of step.

Most authors took the precaution of allowing their subjects one or more familiarization runs, but no formal studies of habituation and test learning were made. Ratings of perceived exertion (Borg, 1971) have shown a remarkably consistent relationship to heart rate in Swedish (Borg & Linderholm, 1967), Canadian (Kay & Shephard, 1969), Israeli (Bar-Or *et al.*, 1972) and US (Skinner *et al.*, 1973) populations. Nevertheless, rather different rates of habituation might be anticipated in a sophisticated city-dweller and a primitive tribesman, since the latter has little previous experience of civilized man and no knowledge of equipment such as bicycles and ergometers.

The assumption that a bicycle ergometer can be operated with a constant mechanical efficiency of 23% (doubtful even in the civilized population – see page 35) becomes yet more questionable when extended to tribes with no previous experience of cycling. Andersen (1971) described the difficulties he encountered in Ethiopia: 'Many subjects were unable to pedal the ergometer with sufficient skill and as time could not be spared

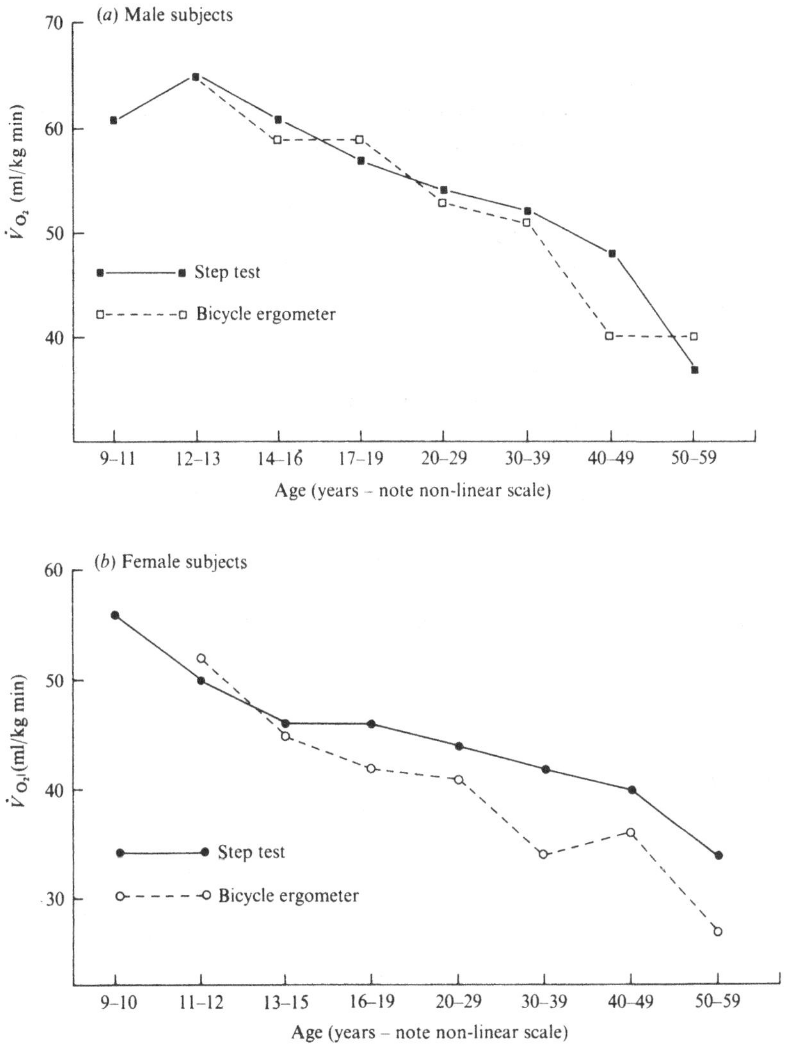

Fig. 3. The discrepancy between the predicted maximum oxygen intake for stepping and bicycle ergometer exercise. Data for (*a*) male and (*b*) female Igloolik Eskimos. (From Rode & Shephard, 1973*b*, courtesy of the editor and publishers, *Int. Z. angew. Physiol.*)

for practice and training, stepping was used as a supplementary procedure.' Mechanical efficiency is also low in the Eskimo (Andersen & Hart, 1963; Shephard, 1975*a*) and the Ainu (Ikai *et al.*, 1971), but not in the more acculturated Kautokeino Lapps (Andersen *et al.*, 1962). One paper sometimes quoted as demonstrating that the mechanical efficiency of cycling remains constant in primitive groups (van Graan & Greyson, 1970) records oxygen consumptions equivalent to efficiencies of 17.3% for Bantu and 21.0% for 'Kalahari Bushmen' who were no longer nomadic. Fortunately, ethnic differences in the efficiency of cycling and stepping are relatively unimportant from the practical point of view, since most IBP investigators related fitness to oxygen consumption rather than the work performed.

The IBP step test (Weiner & Lourie, 1969) was learnt quickly even by those unfamiliar with western languages. Two very effective expedients were a practical demonstration of the test by the investigator, and the chanting of the native words for 'up' and 'down' to a six-beat rhythm. The musical pacing of the exercise has since been exploited further in the Canadian 'home fitness test'; with only minor modifications, a gramophone recording will soon make the IBP stepping protocol available to the entire North American population (Bailey *et al.*, 1974*a*). The main practical difficulty encountered in field use of the step was that the eighteen-inch riser (46 cm) proposed for maximum testing was rather tall for prepubescent children.

One disturbing point that emerged from studies of the Canadian Eskimo was that whereas discrepancies between step test and bicycle ergometer predictions of maximum oxygen intake were small in the young adult, the bicycle ergometer gave substantially lower readings in older people (Rode & Shephard, 1973*b*; Fig. 3). A similar large discrepancy (Table 40) was subsequently reported for older white Canadians (Bailey *et al.*, 1974*a*). The explanation may be that the quadriceps muscles of older subjects are weak and sustain greater relative loading when exercise is performed on the bicycle ergometer. Certainly, step test data for the Saskatoon group (Bailey *et al.*, 1974*a*) agree better than the bicycle ergometer results with direct treadmill measurements of maximum oxygen intake, and since the objective of testing is to specify the centrally-limited aerobic power, this seems a significant argument in favour of using the step rather than the bicycle ergometer for the field testing of physical fitness and working capacity.

Practicality of testing schedules

Maximum testing

Quite a number of investigators took primitive peoples to their directly measured maximum oxygen intake, and as far as can be ascertained there were no resultant complications. The Canadian IBP team provided a physician equipped with oxygen and a direct-current defibrillator for their maximum tests, arguing that Eskimo subjects merited the same safety precautions that would be provided for citizens carrying out a comparable procedure in a major urban laboratory. Nevertheless, such a requirement immobilized a physician and specialized items of equipment in a remote area, placing a severe restriction on the practicality of widespread maximum testing. Not all IBP groups took these stringent precautions, nor is there unanimous agreement that medical coverage is essential (compare, for example, the 'Viewpoint' of Faulkner (1973) and the response of G. R. Cumming (1973)). McDonough & Bruce (1969) have suggested that in the coronary-prone middle-aged man the risk of ventricular fibrillation or cardiac arrest is about 1 in 3000 for 'symptom-limited' maximum effort tests, but drops to about 1 in 15000 for submaximum test procedures. The hazard of maximum effort in the unselected adult population is certainly less than this (Shephard, 1973*a*), and may be no more than four times the risk of relaxing in an armchair (Vuori, 1974). Nevertheless, in an era of instant communication, one accident could have disastrous repercussions for applied physiologists throughout the world. Thus, on both ethical and pragmatic grounds, the present author would argue that if maximum testing is to be permitted in primitive societies, the associated medical precautions should equal those anticipated in a modern hospital clinic.

A second more practical difficulty arises when maximum tests are initiated in a primitive group. A civilized subject understands only too well the concept of exceeding the performance of his contemporaries. However, the majority of primitive societies are built on the principle of cooperation rather than ruthless competiton, and for this reason, subjects see litte point in continuing with an exercise test once they become tired and breathless. Even in a competitive population, the attainment of a good plateau of oxygen consumption depends very much upon the persuasive power of the investigator. Given a limited vocabulary of phrases in the primitive dialect, it becomes very difficult to sustain the necessary motivational drive. Some authors (for example, C. T. M. Davies & Van Haaren, 1974) have contented themselves with extrapolations from 'near-maximal' data, while others have admitted their results could underestimate true maximum values by as much as 20% (Andersen & Hart, 1963). One useful tactic is to subdivide subjects into three categories, based on either observer ratings of the attained effort or objective data such as blood lactate levels and respiratory gas exchange ratios (Shephard *et al.*, 1968*c*; Rode & Shephard, 1971). The results of such an analysis support the view that those rated as making a 'good' effort, and many of those making a 'fair' effort attain a true maximum oxygen intake; however, the directly measured value is an underestimate of $\dot{V}_{O_2(max)}$ in the third of the population making only a 'poor' effort (Table 12).

Table 12. *A comparison between the directly measured maximum oxygen intake and that predicted from measurements of heart rate and oxygen consumption in submaximum effort using the Åstrand nomogram*

		Direct measurements				
Subjects	Age (yr)	Max. heart rate (/min)	Resp. gas exchange ratio	Arterial lactate (mg/dl)	Aerobic power (l/min STPD)	Δ, prediction (l/min STPD)
White boys	11–13					
Good effort		195±7	1.09±0.16	105.2±18.4	1.79±0.32	0.00±0.32
Moderate effort		191±8	1.13±0.28	73.0±2.6	1.91±0.29	−0.02±0.41
Poor effort		192±10	1.03±0.18	47.2±8.0	1.80±0.46	+0.17±0.21
White girls	11–13					
Good effort		199±7	1.28±0.22	109.4±31.1	1.43±0.26	−0.05±0.12
Moderate effort		196±8	0.99±0.16	71.9±4.2	1.64±0.34	−0.14±0.27
Poor effort		195±9	1.07±0.15	51.6±2.6	1.50±0.25	+0.02±0.21
Eskimo men						
Good effort	23.0±7.8	185±14	1.14±0.04	—	3.46±0.55	+0.28±0.43
Moderate effort	26.8±12.5	179±15	1.05±0.03	—	3.10±0.82	+0.29±0.54
Poor effort	23.7±8.3	176±3	0.97±0.03	—	3.27±0.72	+0.61±0.47

Data on white schoolchildren (Shephard *et al.*, 1968*c*) and Eskimos (Rode & Shephard, 1971) classified according to observer ratings of the quality of effort.

Δ = difference (predicted aerobic power − direct measured aerobic power).

Submaximum testing

Despite such problems, it remains important that direct maximum tests be performed on at least a sub-sample of the various primitive communities to resolve nagging doubts regarding the relationship between maximum and submaximum test results in such groups. To date, fears of atypical maximum heart rates and unusual oxygen consumption/heart rate lines have not been realized (Table 13). Data have generally been as in civilized communities of comparable age and physical fitness, exercised to comparable levels of exhaustion (Wyndham *et al.*, 1963; Andersen, 1971; Rode & Shephard, 1971). It thus seems possible to use submaximum prediction procedures on primitive populations with about the same accuracy as that anticipated in civilized populations (Rode & Shephard, 1971; Joseph *et al.*, 1973).

Test schedule

One of the main areas of departure from standard IBP methodology has been with respect to the duration of individual stages of the exercise test. Fortunately, neither the plateau value of maximal oxygen intake nor the

Table 13. *Maximum heart rates of young adult men in selected populations*

Population	Max. heart rate (/min)	Author
Canada		
Igloolik Eskimos	185±13	Rode & Shephard (1971)
Easter Island	180–200	Ekblöm & Gjessing (1968)
Greenland		
Eskimos	191	Lammert (1972)
Israel		
Kurds	193±6	C. T. M. Davies *et al.* (1972)
Yemenites	187±7	C. T. M. Davies *et al.* (1972)
Jamaica	185±10	Miller *et al.* (1972)
Japan		
Ainu	189±8	Ikai *et al.* (1971)
Malaya		
Asians	182±2	Duncan (1972)
Nigeria		
Yoruba		
Active	190±8	C. T. M. Davies *et al.* (1972)
Inactive	192±4	C. T. M. Davies *et al.* (1972)
Scandinavia		
Kautokeino Lapps	197	Andersen *et al.* (1962)
South Africa		
Bantu (aged 20–40)	180	Wyndham *et al.* (1966*c*)
Kalahari Bushmen	180	Wyndham *et al.* (1966*c*)
Trinidad	188±13	Miller *et al.* (1972)

maximum oxygen intake predicted from submaximum test data seem changed to any great extent with minor modifications of the required test schedule. Small increments of work at one-minute intervals and larger increments at intervals of two, three, four or, with submaximum effort, six minutes of work all yield essentially the same answer, as do submaximum tests with recovery intervals of five to fifteen minutes and maximum tests repeated on subsequent days (page 33).

Other tests of physical fitness and working capacity

The main practical difficulty limiting the use of the anaerobic power measurement was the absence of a suitable long staircase in many primitive communities; results also depended on persuading the peoples concerned to engage in the unusual activity of all-out competition.

The use of skinfold calipers and soft-tissue radiographs to measure body fat was complicated in many primitive populations by the extreme thinness of subcutaneous tissue. Standard equations proposed for the conversion of skinfold readings to percentages of body fat had little validity, since most of the body reserves of fat were carried internally (Shephard, Hatcher & Rode, 1973*b*). More complex procedures such as the use of deuterated water (Shephard *et al.*, 1973*b*), measurements of naturally radioactive $^{40}K^{+}$ (Cotes *et al.*, 1969), and underwater weighing (Brožek *et al.*, 1963*a*; Shephard *et al.*, 1968*c*) also remain open to suspicion, since tissue hydration, cellular potassium and the density of lean tissues are not known precisely for primitive groups.

Activity measurements and performance tests have been attempted on a few primitive groups; however, cultural factors such as lack of experience in gymnasium manoeuvres and the non-competitive nature of society severely restrict the interpretation of findings.

Validation of methodology

Unfortunately, despite preliminary warnings of inconsistencies in the data reported from different laboratories (page 22), very few of the investigators associated with the human adaptability project made formal checks to eliminate systematic errors from their data. This issue is particularly critical when assessing results from field laboratories. Gas samples may change in composition because they have been transported over long distances, chemical methods of gas analysis may be plagued by unusual laboratory temperatures, and electronic gas analysis may be invalidated by variations in line voltage or frequency. Voltage stabilizers are mandatory in small communities, and where possible an independent power line should be established from the local generating station.

One possible expedient, used by at least two groups (Ikai *et al.*, 1971; Shephard, 1975*a*) is a biological calibration, using identical methods on both the primitive and the civilized subjects resident in the same community. A second guide is provided by the estimated mechanical efficiency; apparently unusual values may reflect an error of gas analysis. Unfortunately, this approach seems ruled out, because the mechanical efficiency is abnormal in many primitive groups; the low efficiency of the Eskimos during use of the bicycle ergometer (Shephard, 1975*a*) is established with a fair degree of confidence, since (1) biological calibration of the equipment against white subjects has been completed, and (2) the Eskimos show a relatively normal mechanical efficiency during performance of the IBP step test.

4 Climate, season and local geography

In this chapter, we shall look first at some general effects of climate, season and local geography upon patterns of physical activity and associated fitness levels. We shall then consider in more detail the problems encountered by both indigenous peoples and recent immigrants when working in extremes of heat and cold, and finally will examine briefly the influence of circadian rhythms on working capacity and human activity.

Influences upon physical activity patterns

Climate

Extremes of heat, cold, heavy rain and drought can all modify physical fitness and working capacity by causing changes in the range of habitual activity.

A harsh climate, whether desert heat (Roberts, 1970) or arctic cold (Shephard, 1975*a*), may demand vigorous physical activity for mere survival. Game and other sources of food may be in very short supply, dictating the need for a nomadic life-style; there are then added energy costs involved in the daily transporting of personal possessions (including children) and the regular building of temporary nocturnal shelters. Lee (1972*a*) has estimated that African families of the !Kung tribe walk 2400 km each year. Mothers carry their children over this distance until they reach the age of four – a significant factor limiting potential birth spacing. In the Botswana region, rainfall varies very widely from season to season and from year to year (Lee, 1972*b*). Precipitation governs the spatial organization of the community, particularly population density and the area of territory over which one family must wander. In some seasons, sufficient berries and other basic foods can be found within a half-day's journey; under less favourable conditions, the forays become so extended that personal possessions must be carried from one site to another.

In other cultures, the transition to nomadic life follows a regular annual cycle. Thus, the traditional Eskimo will journey on his sledge for several weeks at a time during the winter months (page 20), but in the summer may establish a more permanent camp near the seal hunting grounds. The Tarahumara Indian of central Mexico moves to the highlands in the spring, to plant, tend and harvest his crops. But in the fall and winter he retreats to the warmth of the canyons (Balke & Snow, 1965). Even in richer geographic regions such as the Indian settlements of coastal British Columbia, there is a tradition of summer dispersal to prime salmon runs, with all the modifications of activity patterns demanded by camp life (Drucker, 1955).

Some observers have considered the indigenous populations of hot and humid areas as lazy. Thus, Bates (1884) wrote of the Amazon peoples: 'Our neighbours, the Indian and Mulatto inhabitants of the open palm thatched huts, as we returned home fatigued with our ramble, were either asleep in their hammocks or seated on mats in the shade, too languid even to talk.'

Where there is a low level of physical activity, this may represent an important cultural adaptation to an extreme climate. As Ladell (1964) explains: 'there is no advantage in developing a high sweat rate in a climate where evaporative cooling is difficult, and the Africans instinctively avoid this by taking frequent rests after bouts of heavy work, which allows the body to cool'.

However, there are also problems of reporting. Wyndham (1966) commented that the traditional Bantu arose early and had most of his work done before the sun reached its peak; the European observer, seeing him asleep under a tree in the afternoon, gained a misleading picture of his daily caloric expenditure. Ladell (1964) similarly observed that West Africans completed most of their paid work before 'thermal midday' (two hours after the sun had passed its zenith); a prolonged meal break and afternoon siesta was followed by work on his own farm or by dancing. In the words of Pales (1950): 'C'est que cette épargne physique diurne soit dilapidée le soir venu dans les activites très intenses et non-utilitaires à nos yeux; la danse, par exemple.'

A formal study by two agricultural economists (Clark & Haswell, 1964) estimated that African and Indian peasants spent from seventeen to thirty-four hours each week working in the field; however, in addition to this, much energy was expended in building and repairing houses, making shoes and furniture, spinning and making clothes, and grinding corn. P. T. Baker (1966) noted that Peruvian Indians living in hot wet conditions worked mainly in the morning and late afternoon. Nevertheless, if occasion demanded they were capable of sustained physical activity in extreme heat; possibly, in the past this facility had some survival value in permitting escape from marauding beasts. Other indigenous groups also perform feats that are beyond the tolerance of an acclimatized white person. MacPherson (1966) reported that Aborigines living in the wet heat of north Australia undertook exhausting day-long exercise under adverse conditions, with a surprising economy of water and salt. The Kalahari Bushmen are reputed to show prodigious endurance when tracking game wounded by their poisoned arrows; on occasion, they are said to have run horses to a standstill. Wyndham (1966) joined a group of Kalahari hunters in pursuit of an antelope, but in the midday heat he was forced to abandon the chase due to 'thirst and exhaustion'. He suggested that while the very light Bushmen had some inherent advantage over Caucasians in terms of the energy expenditures needed to traverse soft sand, much of their success was due to 'know-how' rather than peculiarities of anatomy and physiology.

In cold climates, physical activity may be undertaken deliberately to increase body temperature. Our group have noted that, in the winter months, Eskimos interrupt sledge journeys for a period of running about in the snow until thermal comfort is restored.

Climate has other more general effects upon the culture of a population, including inevitable modifications of clothing and housing. Adaptations to a cold environment are considered in a subsequent section (page 70). It seems logical that a man working in humid heat should wear as little as possible. Ladell (1964) had difficulty demonstrating this point on British soldiers serving in Africa, partly because 'tropical' kit still covered most of the skin surface, and partly because a 'funnel effect' was created by briskly moving and loose-fitting long trousers. However, the majority of indigenous peoples, including Bushmen (Wyndham, 1966), Peruvian Indians (P. T. Baker, 1966) and Australian Aborigines (MacPherson, 1966) have accepted the principle of minimal clothing, continuing in this sensible habit despite strong pressure from missionary groups (P. T. Baker, 1966); only persistent attacks from insects such as the dim-dam fly (Malhotra, 1966) have induced greater modesty.

With regard to housing, tropical man has to choose between daytime comfort and night-time warmth. Naked Africans cool quite rapidly at an environmental temperature of 27 °C (Ladell, 1964), and the nights are often colder than this. Malayans (Webb, 1959) erect well-shaded houses with wide eaves, large unglazed windows, and lightly thatched roofs. Such structures are well ventilated, and have a low thermal capacity; although comfortable by day, they are cold at night. The Peruvian Indians also tend to build their villages on a rise of ground; the houses have no sides, and the forest is cleared to further increase ventilation (P. T. Baker, 1966). The West Africans, on the other hand, prefer a solid mud hut with small windows and almost no ventilation. This has a high thermal capacity, and remains comfortably warm at night; however, during the daytime it becomes so hot that all activities must be carried on outdoors, usually under the shade of nearby trees (Ladell, 1964). In some areas (for example, north-eastern India), dampness may dictate the support of the house on piles (Malhotra, 1966).

Season

Seasonal differences of activity are more marked at moderate and high latitudes than in the tropics, where conditions may change relatively little from winter to summer.

Detailed comparisons of summer and winter activities have been made for the Canadian Eskimo community of Igloolik (Godin & Shephard, 1973). The hunting pursuits of traditional Eskimos are dictated by ground conditions and the availability of game (Fig. 1). In the winter, long sledge journeys are made across the frozen sea, in search of caribou, bear, arctic fox and other mammals. During periods of intense cold, petrol-driven snow-mobiles are distrusted, and more energetic travel is undertaken by dog sleigh. As the snows begin to melt, seal hunting at the floe edge becomes the prime activity, and fishing begins from small boats, often with nets dragged under the ice. Then there is walrus hunting in a larger petrol-driven Peterhead boat, with harpoon trips in satellite vessels and much handling of meat carcasses. The late summer sees long journeys on foot over the rocky hills of Baffin Island, interspersed with searches for caribou, small mammals, arctic birds and their eggs. Each of these pursuits has a characteristic daily energy cost (Table 14), and each produces corresponding changes in the diet available to the villagers.

Acculturated Eskimos who have accepted salaried employment within the settlement also show very different patterns of activity in summer and winter months (Fig. 4). In the summer, there is much heavy work cleaning up the year's accumulation of garbage and unloading the supply ship, with lighter tasks in the construction and maintenance of property. In the winter, heavy work is encountered in cutting and distributing ice (the source of domestic water), while operation of mechanical snow-clearing equipment provides more moderate activity; the energy cost of moving about the village is also increased by the need to walk through quite deep snow (Goldman, Haisman & Pandolf, 1976).

Despite substantial seasonal differences in activity patterns and diet, changes of the Eskimo's physical fitness and working capacity from summer to winter are slight (Rode & Shephard, 1973*c*). Indeed, with the exception of a small winter increase of skinfold thickness previously noted in white circumpolar residents (Lewis, Masterton & Rosenbaum, 1960; Edholm & Lewis, 1964), the changes are statistically insignificant. Studies of the Ainu (Ikai *et al.*, 1971), equally, show little difference of fitness levels between the cold winter and more temperate summer months.

Edholm and his associates (Edholm *et al.*, 1973; Samueloff, Davies & Schvartz, 1973) have documented the summer and winter activities of farming communities in a hot desert region (Kurdish and Yemenite migrants to Israel). During the summer, the wet bulb globe thermometer

Table 14. *The estimated daily energy cost of several forms of hunting. Average data for the Canadian Eskimo community of Igloolik*

Hunt type	24-hour energy expenditure (kcal)	(kJ)
Caribou		
Winter	3590	15000
Summer	3600	15000
Fishing		
Ice	4040	16900
Summer	4350	18200
Seal		
Boat	3435	14400
Floe edge	2440	10216
Ice hole	3310	13900
Walrus	3610	15100
Average, 8 hunts	3550	14900

From Godin & Shephard (1973), courtesy of William Heinemann Medical Books Ltd.

index (Minard, Belding & Kingston, 1957) averages 30 °C throughout the working day, but in winter it does not exceed 20 °C, even if the weather is clear and sunny. Over this temperature range, seasonal differences in daily energy expenditures and physical working capacity are small. In the summer, both Kurdish and Yemenite men expend a daily average of 3050 kcal. In the winter, the Yemenite men use 3000 kcal, and the Kurds 3110 kcal. Among the women, summer activities are slightly less demanding (Yemenites, 2280 kcal; Kurds, 2250 kcal) than those of the winter season (Yemenites, 2400 kcal; Kurds, 2390 kcal). Predicted aerobic powers average 3.6% higher in the men and 12.3% higher in the women during the winter, but these differences remain statistically insignificant since a relatively small sample (31M, 8F) has been examined. Activities include the growing of grass, carrots, potatoes, and oranges. Rather larger seasonal effects might be found in other agricultural communities, for the Israelis are somewhat atypical in claiming three or more consecutive harvests over the course of a year.

With regard to tolerance of work in the heat, Wyndham (1966) noted that the Bantu show slightly more natural acclimatization in the summer than in the winter months; thus, at any given work load, both rectal temperature and heart rate are somewhat lower. There are also seasonal differences in the energy cost of activities; thus Gold, Zornitzer & Samueloff (1969) established the following data for a mixed population of Israelis living in the Negev.

	Percentage energy cost			
	Summer		Winter	
	Thermo-neutral conditions	Hot conditions	Thermo-neutral conditions	Hot conditions
Lying	100	104	114	108
Sitting	100	109	105	106
13 cm stepping	100	106	107	108
26 cm stepping	100	106	109	108
Lying	100	111	115	119

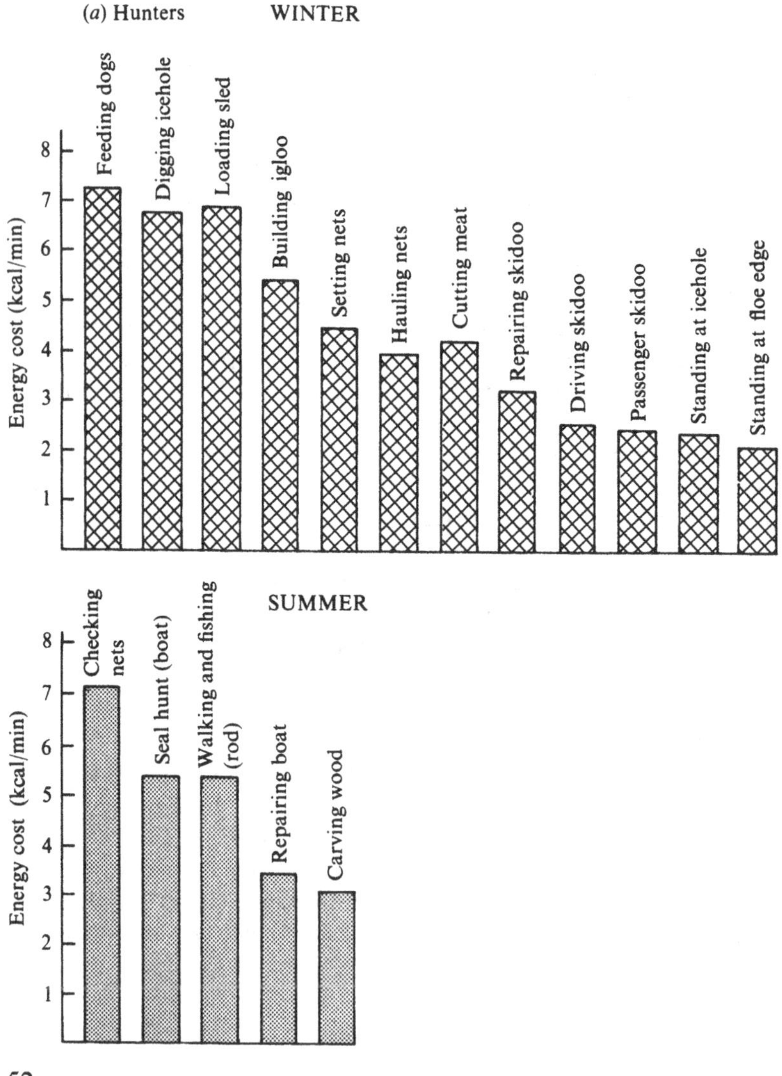

Expressing the results for summer thermo-neutral conditions as 100%, winter values were higher for all levels of activity. In the summer period, immediate hot conditions increased energy expenditures (as noted previously by Consolazio *et al.*, 1963); however, in the winter months, perhaps because of the higher energy expenditures under thermo-neutral conditions, there was no further increase of cost under hot conditions.

Summer seems associated with a reduction of resting blood pressure (Ladell, Waterlow & Hudson, 1944), and perhaps for this reason those who prefer the summer months have a higher mean level of blood pressure than

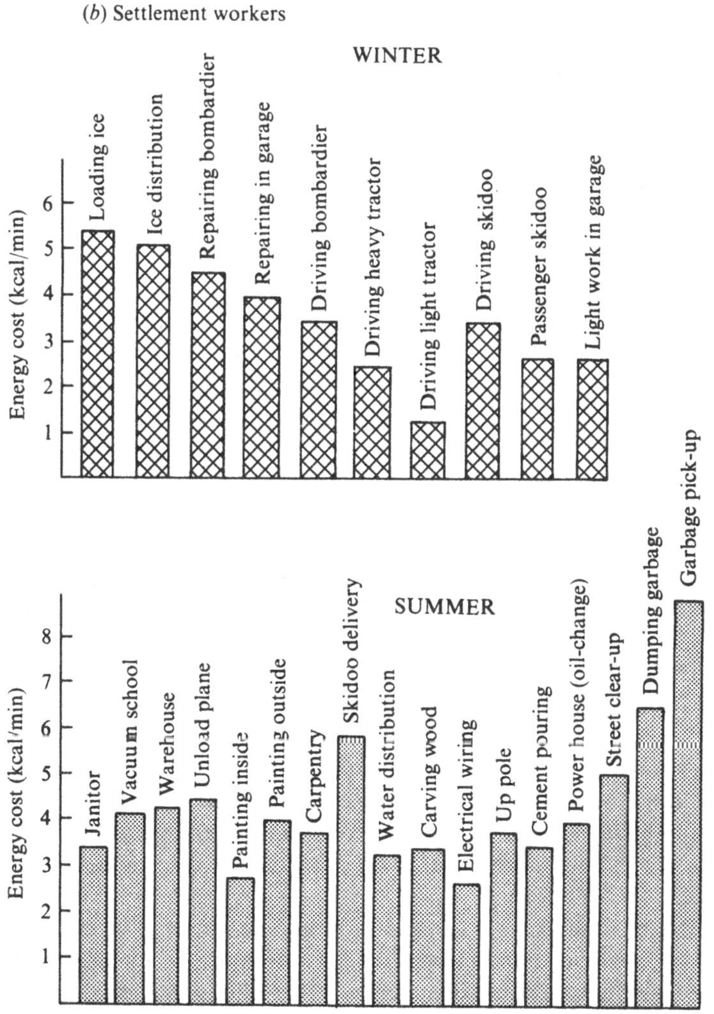

Fig. 4. The caloric cost of some typical daily activities of (*a*) hunters and (*b*) settlement workers of Igloolik. Based on data of Godin & Shephard (1973).

those who prefer the winter (Driver, 1958); in the Chinese, at least, body temperatures are also higher in the summer than in winter months.

Local geography

We have already noted the influence of soft sand and of snow upon the energy cost of daily outdoor activities. An even more important consideration is the steepness of the terrain. An Eskimo who carries a caribou carcass twenty kilometers over the rocky hills of Baffin Island, for example, inevitably has a high caloric expenditure, and is likely to develop a substantial working capacity.

Durnin & Passmore (1967) provided a dramatic example of the importance of local topography when they reported the maximum daily caloric expenditure of Swiss peasants as 5000 kcal in men and 3860 kcal in women.

A second group living in rugged territory are the Tarahumara Indians of Mexico. 'Tarahumara' is a Spanish corruption of the Indian word 'raramuri', which means literally 'fleet foot'. Lumholtz (1902) cited the case of a Tarahumaran who made a round trip mail delivery of 800 km (500 miles) in five days, and many of the runners in the tribe can cover 160 km at 10–13 km/hr, sustaining an estimated oxygen consumption of 43 ml/kg min over a total energy expenditure of 10000 kcal and more (Balke & Snow, 1965). A writer for *Sports Illustrated* described his experience thus: 'For five miles we climbed that trail, which seemed designed only for goats. At one point, as we toiled upwards, the Indians passed us, each carrying a 60 pound pack of our gear. Suddenly, I realized it was their third trip of the day.' (*Sports Illustrated*, 21 October 1963, p. 33.)

Balke & Snow (1965) estimated aerobic power from the distance covered in a fifteen-minute run. Values were high for unacculturated Tarahumara boys, but in a group who had been attending a mission boarding school the mean estimate (41.5 ml/kg min) was no better than in their Mestizo classmates (41.2 ml/kg min). More recent step-test predictions (Aghemo, Limas & Sassi, 1971) have confirmed this view, runners having an aerobic power of 63 ml/kg min, and their civilized counterparts' values averaging only 39 ml/kg min (Table 15). It thus seems that the unusual endurance capacity of this tribe stems from the conditioning effect of jogging long distances over mountainous territory, rather than from any genetic peculiarity.

Parts of New Guinea, also, are very mountainous. MacPherson (1966) described one highland population (the Chimbu) thus: 'When met casually, he [the Chimbu] may appear leisurely or even idle in his behaviour, but, of necessity, he usually proves to be extraordinarily fit. . . he has, in the highlands, to ascend and descend hills of incredible steepness by seemingly impossible paths. His total energy expenditure per day might not be greater than that of the active European, but the pattern might be very different...although malnutrition has been amply demonstrated...both men and women perform physical feats and show sustained stamina that few of us could match.' Sinnett & Solomon (1968) studied 152 of the 2000 Enga-speaking people in the western highlands. The average altitude of residence was 2485 m (8150 ft). Activity comprised mainly cultivation and harvesting of sweet potatoes, collection of firewood, and occasional house-building. A Harvard pack test yielded a mean score of 83, compared to 73 and 89 in previously studied groups of New Guineans (Hipsley & Kirk, 1965), 89 for Nigerian infantrymen (Ladell & Kenney, 1955), 80 for members of the British Commonwealth division stationed in Korea (Ladell & Kenney, 1955), 88 for British aircrew stationed in Poona (610 m; 2000 ft), and 80 for Indian aircrew at the same station (Munro, 1950). The aerobic power of the New Guinea highlanders, as predicted by bicycle ergometer was 45.1 ml/kg min, and, although this does not at first sight seem particularly high, allowance must be made for (*a*) the inclusion of some older subjects (age range 16–37 years) and (*b*) the effect of altitude itself in diminishing maximum oxygen intake (at least 10% at 2440 m (8000 ft)). Patrick & Cotes (1971) provided more striking evidence of the fitness of some New Guinea highlanders (Table 15). Maximum oxygen intake was expressed per kilogram of lean body mass, and in such terms the people of Lufa (an eastern highland settlement, altitude 2000 m) were much superior to the inhabitants of Kaul, a sea-level village; however, only a part of their advantage was retained when the highlanders migrated to the coastal area. For comparative purposes, the data has been recalculated as approximate values per kilogram of body weight; on this basis, the Lufa people seem superior even to the Tarahumara Indians. Measurements of cardiac output by the CO_2 rebreathing method show a parallel development of stroke volume.

The International Biological Programme has stimulated a number of studies of working capacity in groups residing at even higher altitudes. Donoso *et al.* (1974*b*) examined forty Aymara Indians living in the Chilean Andes (altitude 3680 m). All were working as farmers or shepherds. The maximum oxygen intake, measured directly on a bicycle ergometer, was a high normal value (Table 15), particularly if corrected for the altitude of measurement; the apparent rate of ageing of aerobic power was also less than in sea-level populations. Andersen (1971) compared two groups of Ethiopians, living respectively at 1500 m and 3000 m; the people, Amharas, were a mixture of Mediterranean and Negro stocks, and the climate dry heat at the lower altitude, but somewhat cooler at the higher altitude. The state of nutrition was not specified, but both populations were

Table 15. *Aerobic power, skinfold thickness and/or estimated percentages of body fat. Male populations living and/or working in rugged mountainous areas*

Group	Age (yr)	Period at altitude[a]	Test altitude (m)	Aerobic power (ml/kg min STPD)	Skinfold (mm) or estimated body fat (%)	Author
Canada						
Igloolik Eskimos						
Active hunters	29.3	Intermittent	0	56.4	5.6	Shephard (1974*c*)
Settlement dwellers	30.2	None	0	50.7	7.3	Shephard (1974*c*)
Chile						
Aymara Indians	20–29	Life	3680	49.1	—	Donoso *et al.* (1974*a*)
	40–49	Life	3680	44.0	—	Donoso *et al.* (1974*a*)
Ethiopia						
Amhara Indians	20–29	Life	1500	39.9	6.0	Andersen (1971)
	20–29	Life	3000	34.4	6.8	Andersen (1971)
	50+	Life	1500	27.8	6.5	Andersen (1971)
	50+	Life	3000	27.4	7.9	Andersen (1971)
India						
Army personnel	18–30	> 6 months (3505 m)[b]	0	42.3[c]	—	Dua, Ramaswamy & Sen Gupta (1966)
			3500[b]	32.7[c]	—	Dua *et al.* (1966)
	18–30	> 6 months (3100 m)[d]	0	44.3[c]	—	Dua *et al.* (1966)
			3100[d]	39.6[c]	—	Dua *et al.* (1966)
Sea-level residents	18–30	> 6 months	3100[e]	38.4[c]	—	Ramaswamy *et al.* (1966)
Gorkhas	18–30	> 6 months	3100[e]	41.5[c]	—	Ramaswamy *et al.* (1966)
Altitude natives	18–30	Life	3100[e]	43.5[c]	—	Ramaswamy *et al.* (1966)
Sea-level residents	31–40	> 6 months	3100[e]	32.8[c]	—	Ramaswamy *et al.* (1966)
Gorkhas	31–40	> 6 months	3100[e]	38.2[c]	—	Ramaswamy *et al.* (1966)
Altitude natives	31–40	Life	3100[e]	42.0[c]	—	Ramaswamy *et al.* (1966)
Tamilians	18–30	None	0	45.5	11.9%	Bharadway *et al.* (1973)
	18–30	10 months	3960	35.0	10.5%	Bharadway *et al.* (1973)
	18–30	Life	3960	40.3	10.6%	Bharadway *et al.* (1973)
	?	Life	2900	40.0	—	Lahiri (personal communication)
	?	Life	3800	40.0	—	Lahiri (personal communication)

Table 15. (*cont.*)

Group	Age (yr)	Period at altitude[a]	Test altitude (m)	Aerobic power (ml/kg min STPD)	Skinfold (mm) or estimated body fat (%)	Author
Mexico						
Tarahumara Indians						
Runners	29.8	Life	2000	63.0	11.1%	Aghemo *et al.* (1971)
Former runners	52.3	Life	2000	50.0	14.7%	Aghemo *et al.* (1971)
'Civilized' or non-runners	28.0	Life	2000	38.9	17.9%	Aghemo *et al.* (1971)
New Guinea						
Tukisenta-Lagaip	16–37	Life	2485[f]	45.1	~ 5	Sinnett & Solomon (1968)
Coastal Kauls	25	None	0	53.2[g]	—	Patrick & Cotes (1971)
Highland Lufas	25	Life	2000	67.0[g]	—	Patrick & Cotes (1971)
Migrant Lufas	25	Migration (2000 m to 0 m)	0	63.2[g]	—	Patrick & Cotes (1971)
Peru						
Sea-level Quechua	22	None	0	49.3	6.3[h]	Buskirk (1974)
	22	4 weeks	3992	44.5	—	Buskirk (1974)
Nunoa Quechua	22	Life	3992	49.1	—	Buskirk (1974)
Russia						
Kirghiz	18–25	Life	940	33.9	—	Mirrakhimov (personal communication)
	18–25	Life	2500	45.7	—	Mirrakhimov (personal communication)
	18–25	Life	3600	34.8	—	Mirrakhimov (personal communication)

[a] This altitude is given where it differs from the test altitude.
[b] 11500 ft in original paper.
[c] Data collected using Kofranyi–Michaelis respirometer.
[d] 10500 ft in original paper. Region more rugged than that at 3500 m (11500 ft).
[e] 10500 ft in original paper.
[f] 8150 ft in original paper.
[g] Original values expressed relative to lean body mass; body weights taken from Cotes *et al.* (1973*a*).
[h] Cited by Elsner (1963).

extremely light in relation to their height (56.2 kg for 168.3 cm, 53.5 kg for 169.9 cm). A 'hyper-maximal' bicycle ergometer or step test to exhaustion showed a relatively poor aerobic power at both altitudes; however, taking account of the respective altitudes of testing, those living at 3000 m were somewhat fitter and again had a somewhat slower apparent rate of ageing than those living at 1500 m. Buskirk (1974) summarized the results of IBP experiments on Peruvian Quechua living at sea level and at 3992 m. The high-altitude residents (Nunoa) were not particularly active, three estimates of caloric intake for groups of adult men being 2125, 2719 and 2833 kilocalories per day (P. T. Baker, 1974); nevertheless, their aerobic power was 10% superior to that of sea-level Quechua allowed to acclimatize to altitude for four weeks. Malhotra (1966) described the primitive tribespeople of north eastern India. Many were huntsmen, while some engaged in terrace cultivation, burning the native forests and planting seeds in the ashes. The population 'spend their entire lives on the slopes of mountains, and so have grown very strong with well-developed calf-muscles'. While the physical fitness of many high-altitude peoples is partly an expression of the enforced activity of mountain climbing, the difference between sea-level and altitude natives cannot be fully abolished even if sea-level populations are allowed an extended period of acclimatization. Thus, Bharadway *et al.* (1973) found Ladakhis native to 3960 m had a 12% higher aerobic power than sea-level Tamilians even after the latter group had spent ten months at altitude.

Heat and working capacity

Climatic regions

Most climatic physiologists distinguish two main types of hot climate – hot dry areas (deserts) and hot wet areas (jungles). In the desert, summer daytime temperatures commonly exceed skin temperatures, so that heat is gained by convection; physical activity increases air movement over the body surface and thus augments convective heat gain. There is little cloud or moisture to screen the sun's rays and 'sunshine hours' can rise to 95% of potential; the radiant heat burden in direct sunlight is thus high, and most indigenous peoples seek shade during the hottest part of the day. In the desert sand more than in the savannah, reflected radiation from the ground further restricts the possibility of daytime work. Heat loss may be entirely through the evaporation of sweat, making activity dependent on the availability of water and salt. In continental deserts, the vapour pressure of the air is generally low, and sweat is readily evaporated; however in regions bordering the sea (such as the Persian Gulf), the air may have a fair moisture content, particularly if air movement is from the sea (D. H. K. Lee, 1964). Wind speeds are very variable, and on a calm day limb movement can facilitate the evaporation of sweat.

During the night, the clear skies of the desert permit a rapid reversal of radiant heat loss, and dramatic drops of temperature may occur. Primitive groups with limited shelter and clothing must thus resist not only daytime heat but also nocturnal chilling (see page 49). Fortunately, it appears that man has the potential to develop heat and cold acclimatization simultaneously (Glaser & Shephard, 1963).

In hot and wet areas, cloud cover, water vapour, a blue haze believed to be derived in part from plant hydrocarbons, and dense foliage all reduce the burden of radiant heat. However, these same factors serve to sustain nocturnal temperatures and block cooling breezes. Higher altitudes usually bring the relief of cooler temperatures, but in some regions such as Hawaii this may be offset by a rainfall as large as 25 cm per day (F. N. Young, 1964). If the air is 97% saturated with moisture, almost no heat loss can occur by sweating, and the working man must attempt to dissipate heat by the slow process of convection during rest pauses.

Physiological constraints

If submaximal work is performed in a hot climate, the pulse rate tends to be higher than in a temperate region, even in the early stages of activity; this reflects a greater skin blood flow. Predictions of aerobic power based on the pulse response to submaximum effort are erroneously low if based on data collected in an over-heated test facility (page 25). On the other hand, sudden exposure to severe heat has little influence upon the directly measured maximum oxygen intake (Saltin, 1964); very similar values are observed in a brief test (< 3 min) with minimal cutaneous vaso-dilation and longer experiments (10–20 min of progressive effort) where skin vessels are widely patent (Saltin, 1973).

Nevertheless, heat exposure reduces the endurance for sustained work (Saltin, 1964). Men who can work an eight-hour shift at 40–50% of aerobic power in a cool climate (I. Åstrand, 1967; Bonjer, 1968) become fatigued when subjected to the same load under hot conditions, and if maximum oxygen intake is measured at the end of the shift it may be reduced by as much as 20–30% (Pirnay, Petit & Deroanne, 1969); presumably, there is not only a diversion of blood flow to the skin, but also a progressive reduction of central blood volume through dilatation of venous reservoirs and peripheral exudation of fluid.

All the general risks of heat exposure (Ladell, 1964; Wyndham & Strydom, 1972; Shephard, 1972*b*, 1976) – heat collapse, heat exhaustion and neurasthenia, heat stroke, anhydrotic heat exhaustion, heat shock, heat cramps and salt deficiency exhaustion – and heat rashes are increased when work is performed in the heat. If the activity is out of doors, as is likely in primitive society, there are also risks of sunburn, keratosis and skin carcinomas.

Heat acclimatization

With repeated exposure to heat, it becomes progressively easier for a man to carry out physical tasks in a hot environment. Part of the change in performance is a matter of know-how. The hours of working are modified to avoid the heat of the day (page 48), shade and a free access of air are sought, and necessary tasks are performed with an economy of movement. Such practical details are readily learnt in the tropics, but may be acquired less easily with artificial exposures to heat in an environmental chamber, a point not appreciated by scientists who have later used the same environmental chamber to test acclimatization. Thus, a study of British Royal Naval personnel found no differences of reactions between those artificially acclimatized and other personnel who had lived in Singapore for eighteen months (Hellon *et al.*, 1956); again, British soldiers who had spent several months in Nigeria were said to react in the same way as those artificially acclimatized (Ladell, 1957). However, a few years later Edholm and his associates (1963) used field exercises to assess the adjustment of military personnel to their environment. Such an approach demonstrated that troops who had already passed several weeks in the desert were capable of more efficient performance and sustained fewer heat casualties than did reinforcements who had spent an equal period of preparation in an environmental chamber in England.

Although some primitive indigenous peoples have had difficulty in mastering the use of the bicycle ergometer, many show an economy of effort for other more familiar simple tasks such as stair-climbing and walking. Thus, Robinson *et al.* (1941) found Negroes from the southern United States mechanically more efficient than those living in northern states, Phillips (1954) noted a lower oxygen cost of such activities in Nigerians than in Europeans, and Wyndham (1966) reported that the Bantu and Kalahari Bushmen were both slightly more efficient than Caucasians when performing the Johannesburg type of step test. This may be partly a long-term effect of heat acclimatization (see seasonal effects, page 50). With some types of activity such as Wyndham's design of step test, a lightweight indigenous subject may also have an advantage over his Caucasian counterpart, independent of any inherent or acquired skill (Wyndham & Sluis-Cremer, 1968; Shephard & Olbrecht, 1970). When more vigorous work is performed, it becomes difficult to demonstrate any difference in metabolic cost between Europeans, Bushmen, and West African Negroes (Phillips, 1954; Wyndham & Morrison, 1958).

In any event, changes of behaviour, improvements of physical efficiency, and habituation to subjective discomforts account for only a part of the gains in performance with acclimatization. The physiological strains induced by a given combination of heat and exercise stress are also lessened. In particular, sweating occurs earlier, at a lower rectal temperature, and heat can be dissipated with a lesser cutaneous blood flow. Salt, also, is conserved by a reduction in the sodium ion content of the sweat. Some of the physiological marks of acclimatization are seen if a resting man is exposed periodically to heat. Equally, some changes of response can occur if body heating is produced by vigorous exercise (Piwonka *et al.*, 1965; Marcus, 1972). Nevertheless, if full acclimatization is to occur, it seems necessary for the body to experience exercise in a hot environment (Ladell, 1957; Wyndham & Strydom, 1972).

Work tolerance of indigenous peoples

Stigler (1952) concluded from studies of his manservant that the body temperature of East Africans was lower than that of Caucasians, and that the rise of body temperature for a given amount of physical work was less. However, subsequent more detailed studies have not altogether confirmed this advantage.

Ladell (1964) suggested that the comfortable working temperature was higher for tropical people than for Europeans, the former group preferring a vaso-dilated to a vaso-constricted skin; nevertheless, he cautioned that at least a part of their preference was due to differences in clothing and mode of life, rather than any inherently unusual feature of their physiology. Indeed, when vigorous physical work was required of Nigerians under hot and wet conditions, sweat production was only slightly more than that of unacclimatized Europeans (Ladell, 1957), the alteration of response being equivalent to no more than three of the fourteen days required for 'full' acclimatization of a European. The inherent heat tolerance of other indigenous groups such as the Bantu (Weiner, 1950) and the Indian (Caplan, 1944) is somewhat better than that of unacclimatized whites (Wyndham, 1966), but nevertheless the indigenous peoples need a deliberate programme of acclimatization prior to work in deep mines.

When fully acclimatized by a course of physical activity in the heat, both the Nigerian (Ladell, 1951) and the Bantu (Wyndham *et al.*, 1953) apparently have a similar heat tolerance to the acclimatized European, and there would seem much to support Robinson's view (1952) that there are no real differences between races that cannot be explained in terms of nutrition, acclimatization and training.

There is some evidence that the linear regression relating rectal temperature (T_R, °C) and pulse rate (f_h) is flatter for tropical workers than for Europeans. Thus, Ladell (1955*a*) established the following relationships for well-hydrated subjects:

For whites, $f_h = 33.2\ (T_R) - 1142.8$
For Nigerians, $f_h = 29.6\ (T_R) - 842.0$

Similarly, Asians living in Singapore (Whittow, 1961) had a rather shallow heart rate/body temperature regression. However, both the Nigerians and the Asians had high resting heart rates, and it could thus be argued that the shallow regression arose from difficulties in habituating the more primitive populations to the laboratory situation.

Some authors have reported racial differences in sweat secretion. Wyndham (1966) suggested that curtailment of sweating might be advantageous in a desert climate. He estimated that if a European were exposed to the same climate as a Bushman, he would sustain a sweat loss of four to five litres each day, and cumulative exposure would lead to a dangerous dehydration. Ladell (1964) had previously disputed the validity of the predicted four-hour sweat rate in calculations such as those made by Wyndham. Nevertheless, a Medical Research Council report (1960) described three tropical subjects (two Malayans and one Indian) who produced less sweat than a European at a comparable stage of acclimatization. Wyndham (1966) also observed that after acclimatization the Bantu sweated less than a white person. On the other hand, Yoshimura (1960) found 20% more active sweat glands in Filipinos and 6% more in Siamese than in Japanese. Pales (1950) also reported more active sweat glands in Africans than in Europeans, a claim subsequently disputed by Thomson (1954). Wyndham (1966) compared the heat responses of river and desert Bushmen, and found that despite their similar genetic background the desert dwellers had a higher sweat rate at a given rectal temperature and heart rate. He suggested this was due to differences in the habitual activity of the two groups; the sweat rate reflected mainly the history of heat stress, while changes of rectal temperature and heart rate during effort were strongly influenced by the physical fitness of the individual. Various authors have commented on the low sodium ion content of the sweat of tropical peoples including West Africans (Ladell, 1947), Asians (Medical Research Council, 1960), Indians (Malhotra, 1966), New Guinea highlanders (MacPherson, 1966), and Queensland Aborigines (D. H. K. Lee, 1964). Theoretically, this could indicate an altered functioning of the adrenal cortex, but at least in India and in New Guinea the main factor responsible seems a low dietary salt intake.

Despite a low salt intake, heat cramps are said to be rare in West Africans (Ladell, 1955*b*). Heat syncope is less common than in comparably adapted Europeans (Ladell, 1964), and salt deficiency exhaustion is rare. However, the hazards of hyperpyrexia seem to be the same as in equally acclimatized Caucasians, and heat deaths have been reported both in the South African mines (Wyndham & Strydom, 1972) and among the Tarahumara Indians (Balke & Snow, 1965).

Adaptations of body build and skin colour

Physical anthropologists have made much of the Bergman and Allen 'laws' in explaining adaptations of man to life in the tropical habitat. Bergman (1847) wrote: 'Within a polytypic warm-blooded species, the body size of the sub-species usually increases with decreasing temperature of its habitat.'

J. A. Allen (1877) suggested an increased linearity of body form as the environmental temperature rose, with an increase in the ratio limb to trunk length, and a decrease of limb diameter.

Both authors were in essence postulating a maximizing of body surface to facilitate heat dissipation during heavy work. Cases can be cited in support of these 'laws'; for example, the West African has very long arms (Schreider, 1957), and the same seems true of the Kalahari Bushman and the Australian Aborigine (Wyndham, 1966; P. Baker, 1958*a*; Newman, 1961). However, there are several arguments against accepting the hypothesis that these peculiarities of body build have adaptive value.

(a) Groups with widely differing stature and relative limb length thrive under apparently similar geographic and environmental conditions (Hiernaux, 1966).

(b) Both stature and limb circumferences change quite rapidly with acculturation and associated changes of diet. Thus, Malhotra (1966) noted what he thought a nutritional gradient of stature among Indian men from the Punjab (average 168.4 cm) to Madras (163.7 cm); further, Canadian-born children of Indian parentage were both taller and heavier than those of the same generation who had remained in India. Wyndham (1966) reported that the indigenous Bantu lived mainly on the mealie, eating little fat or animal protein; however, on migrating to a mining camp he was given a 4000 kcal diet that included 60 g of animal or fish protein and 60 g of vegetable protein. Improved feeding plus six days of hard physical work per week increased his average body weight by 3.5 kg (7.7 pounds) over a four-month period. Hiernaux (1966) commented on the accelerated growth of the urban Hutu in Rwanda, again attributing this to better nutrition and hygiene. Tobias (1966) noted that while the traditional Bushman of southern Angola was still quite short, his height had increased by about 3 cm over the last three generations. Even the young Twa now achieves an adult height of 160 cm and more (Ghesquière, 1971). Such improvements of physique are not confined to tropical regions; the circumpolar peoples, also, are showing a rapid secular trend to increase of standing height (Lewin, Jurgens & Louekari, 1970; Shephard, 1974*c*), tall individuals being particularly common among hybrids (Jamison, 1970; Jamison & Zegura, 1970).

(c) Linearity increases convective heat exchange. However, in the desert where the sun is bright and mid-day air temperatures exceed skin temperatures this has doubtful adaptive value; the added surface increases radiant and convective heat gain, and homeostasis becomes dependent on an evaporation of fluid that the body can ill-afford. Furthermore, at night the large surface area augments radiant and convective heat losses, imposing a nocturnal cold stress on such groups as the Kalahari Bushmen and Australian Aborigines.

In dense jungle, a small body has been considered of adaptive value (Roberts, 1953; Newman, 1960). Thus, Hiernaux (1966) described a gradation of standing heights for indigenous African men, ranging from 171 cm for those living in arid zones through 169 cm for the savannah dwellers to 164 cm for those inhabiting wet forests; the ultimate expression of this gradient was seen in the Twa pygmies, with an average height of 144 cm. Other small forest dwellers have been described in South India (the Kadars and the Paniyans; Malhotra, 1966), in the Andaman Islands (the Onge; Malhotra, 1966), in Papua (MacPherson, 1966) and in South America (P. T. Baker, 1966). Hiernaux (1966) suggested the compact body form might be a genetic adaptation to a limited food supply, while Malhotra (1966) speculated that both a short stature and short hair would be advantageous in crawling through densely meshed forests. In European children, the hot wet conditions of Kenya have been suggested as favouring growth (MacKinnon, 1923); however, until recently, the European settlers occupied a very privileged socio-economic position in Kenya, and as with indigenous populations it is likely that nutrition was the main factor modifying adult stature.

A thin layer of subcutaneous fat might be considered an adaptation of body form to assist heat elimination. At a first glance, the published measurements (Table 16) would seem to indicate such a development. However, equally thin folds have been reported for those working in mountainous and circumpolar regions (Tables 15 and 20). Further, several of the tropical populations face problems of heat conservation at night. Lastly, game is scarce and difficult to preserve in the desert heat. A well-adapted individual thus has the capacity to gorge himself and then to live on his reserves of fat and protein for an extended period; thin skinfolds are compatible with this life-style only if the main reserves of body fat are held elsewhere than in the subcutaneous tissues (Shephard *et al.*, 1973*b*).

The black skin of tropical man has been suggested as protecting the outdoor worker against sunburn and malignant cutaneous tumours (Ladell, 1964). However, many tropical peoples spend their time in the shade of forests, where sunburn and keratoses can hardly be considered serious risks. Further, the black skin absorbs an additional 20–30% of solar radiation, and perhaps for this reason Negroes marching in the Arizona

Table 16. *Average skinfold thicknesses of young men living in tropical areas*

Population	Average skinfold thickness (mm)	Author
Arabs		
Chaamba	6.15	Wyndham (1966)
Australia		
Aborigines	6.63	Wyndham (1966)
Ethiopia		
Adi Arkai	6.03	Andersen (1971)
New Guinea	~ 5	Sinnett & Whyte (1973*b*)
South Africa		
Bantu	5.25–5.95	Wyndham (1966)
Kalahari Bushmen	4.47–4.73	Wyndham (1966)
Tanzania		
Active indigenous	6.35	C. T. M. Davies & Van Haaren (1973)
Inactive indigenous	6.52	C. T. M. Davies & Van Haaren (1973)
Trinidad		
Negroes	7.90	Edwards *et al.* (1972)
East Indians	10.62	Edwards *et al.* (1972)
Venezuela		
Warao Indians	5.87	Gardner (1971)

sun have been less tolerant of the desert heat than white men (P. Baker, 1958*b*).

Physical working capacity

Reported values for the aerobic power of tribes living in tropical regions are summarized in Tables 17 and 18. Many of the values are as poor or poorer than those found in white communities. Exceptions to this generalization are Tanzanian and Yoruba men who preserve an active life-style (C. T. M. Davies *et al.*, 1972; C. T. M. Davies & Van Haaren, 1973), Warao Indians (Gardner, 1971), Yemenite Jews (Samueloff *et al.*, 1973), Malayan Temiars (Chan *et al.*, 1974) and the Kaul lowlanders of New Guinea (Patrick & Cotes, 1971). In some instances, genetic factors could contribute to the low $\dot{V}_{O_2(max)}$; however, in most populations there are other simpler explanations of the poor showing. Where body dimensions are unusual, as in the Twa and the Australian Aborigine, standardization per unit of body weight may not provide a fair basis of comparison. The age of some groups has not been ascertained and may not even have been known; it is notoriously difficult to guess the age of

Table 17. *Aerobic power of young men living in tropical regions*

Population	Age	Aerobic power (ml/kg min STPD)	Author
East Africa			
Dorobo, Turkana	25	46	Di Prampero & Cerretelli (1969)
Easter Island[a]	25	42	Ekblom & Gjessing (1968)
Ethiopia			
Addis Ababa			
Workers	18–25	37.7[b]	Areskog *et al.* (1969)
Airforce cadets	18–22	40.0[b]	Areskog *et al.* (1969)
Adi Arkai	20–29	39.9	Andersen (1971)
India			
Nepalese (Biratnagar)	?	35	Lahiri (personal communication)
Israel			
Kurds	26.4	48.4 (observed)	Samueloff *et al.* (1973)
		44.5 (predicted)	Samueloff *et al.* (1973)
Yemenites	25.3	52.4 (observed)	Samueloff *et al.* (1973)
		46.9 (predicted)	Samueloff *et al.* (1973)
Jamaica	26.0	47.0	Miller *et al.* (1972)
Malaya			
Asian medical students	20–23	34.6	Duncan (1972)
Temiars	?	53.2	Chan *et al.* (1974)
New Guinea			
Kaul	25	53.2[c]	Patrick & Cotes (1971)
Nigeria			
Yoruba			
Active	25.1	55.5	C. T. M. Davies *et al.* (1972)
Inactive	25.8	45.9	C. T. M. Davies *et al.* (1972)
Villagers	25.3	48.5	C. T. M. Davies *et al.* (1972)
South Africa			
Bantu	?	47.7–48.0	Wyndham *et al.* (1963)
Kalahari Bushmen	?	47.1	Wyndham *et al.* (1963)
Venda (Bantu)			
Rural	32	37.6	Wyndham (1973)
Urban	34	41.9	Wyndham (1973)
Ten Bantu tribes	?	41.1–47.8	Wyndham *et al.* (1966*c*)
Tanzania			
Active indigenous	25.4	57.2	C. T. M. Davies & Van Haaren (1973)
Inactive indigenous	21.8	47.2	C. T. M. Davies & Van Haaren (1973)
Trinidad			
Negroes	25.5	38.3	Edwards *et al.* (1972)
East Indians	26.0	39.4	Edwards *et al.* (1972)
United States			
Navajo Indians	25	44	Gardner (1971)

Table 17 (*cont.*)

Population	Age	Aerobic power (ml/kg min STPD)	Author
Venezuela			
Warao Indians	20–24	51.2	Gardner (1971)
	25–29	44.6	Gardner (1971)
Zaïre			
Hoto	?	42.7	Ghesquière (1971)
Twa	?	47.5	Ghesquière (1971)
Male workers	?	42	Ghesquière (1971)
Students	?	44	Ghesquière (1971)

[a] Strictly speaking, Easter Island is sub-tropical, with a well-nigh constant summer and winter temperature of 24 °C.
[b] Expressed by authors as PWC_{170}/kg.
[c] Estimate based on data for body weight published by Cotes *et al.* (1973*a*).

Table 18. *Aerobic power of young women living in tropical regions*

Population	Age	Aerobic power (ml/kg min STPD)	Author
East Africa			
Dorobo, Turkena	25	41	Di Prampero & Cerretelli (1969)
Easter Island	25	31.0	Ekblom & Gjessing (1968)
Israel			
Kurds	25.5	29.0	Samueloff *et al.* (1973)
Yemenites	24.3	35.4	Samueloff *et al.* (1973)
Jamaica	26.5	27.3 (direct)	Miller *et al.* (1972)
		32.7 (predicted)	Miller *et al.* (1972)
Nigeria			
Yoruba	21.5	31.6	C. T. M. Davies *et al.* (1972)
Tanzania	27.9	40.2	Davies & Van Haaren (1973)

primitive groups, and samples of supposed 'young' subjects may have been diluted by the inclusion of older individuals. Levels of daily energy expenditure certainly play a role, and at least among Tanzanians (C. T. M. Davies & Van Haaren, 1973) the aerobic power is substantially augmented if the sample is restricted to active members of the community. Responses to submaximal work have sometimes been distorted by extremely hot laboratory conditions (for example, 24–31 °C in Trinidad and 24–26 °C in Jamaica). Lastly many tropical populations suffer from poor nutrition and anaemia due to malarial and hookworm infestations. Pre-existent malnutrition is suggested by the sudden weight gain of the Bantu on migrating to mining camps (page 63); Wyndham *et al.*

Table 19. *Excess weight of tropical peoples relative to standards proposed by Society of Actuaries (1959), as modified by Shephard (1972b)*

Population	Excess weight (kg)	Author
Easter Island	0	Ekblom & Gjessing (1968)
Ethiopia		
Addis Ababa		
Workers	−10.7	Areskog *et al.* (1969)
Airforce cadets	−7.1	Areskog *et al.* (1969)
Adi Arkai	−13.0	Anderson (1971)
Israel		
Kurds	−0.4	Samueloff *et al.* (1973)
Yemenites	+2.5	Samueloff *et al.* (1973)
Jamaica	−5.3	Miller *et al.* (1972)
Malaya		
Medical students	−9.1	Duncan (1972)
Nigeria		
Yoruba		
Active	−3.7	C. T. M. Davies *et al.* (1972)
Inactive	−4.0	C. T. M. Davies *et al.* (1972)
South Africa		
Bantu	−4.7	Wyndham (1966)
Tanzania		
Active indigenous	−1.2	C. T. M. Davies & Van Haaren (1973)
Inactive indigenous	−4.1	C. T. M. Davies & Van Haaren (1973)
Trinidad		
Negroes	−1.3	Edwards *et al.* (1972)
East Indians	−3.4	Edwards *et al.* (1972)
Zaïre		
Hoto	−6.7	Ghesquière (1971)
Twa	−7.7	Ghesquière (1971)

(1963) note that gross body weight provides a simple and relatively reliable guide to the working capacity of new recruits. This point is further emphasized if excess weights are calculated relative to a modification (Shephard, 1972*b*) of the standards proposed by the Society of Actuaries (Table 19). The Canadian Eskimo, who is well nourished, has a substantial excess weight due to well-developed musculature. However, many tropical groups show a substantial deficit. Interestingly, the deficit is not shown by fitter groups (active Tanzanians and Yemenites), nor is it seen in the Easter Islanders who are known to be both well nourished and relatively inactive. C. T. M. Davies & Van Haaren (1974) encountered haemoglobin levels of less than 8.5 g/dl in many of their Tanzanian subjects, and in Venezuela hookworm infestation may lead to haemoglobin levels in the range 2.4–10.7 g/dl (Gardner, personal communication to author). Such anaemia has the effect not only of greatly reducing maximum oxygen intake, but also of causing cardiac dilation. Thus, whereas the $\dot{V}_{O_2(max)}$/heart volume line was independent of ethnic origin in normal subjects, it was displaced substantially to the right in Tanzanians who were anaemic (C. T. M. Davies & Van Haaren, 1974). On the other hand, a more modest haemoglobin deficit (1.4–2.6 g/dl) had no influence upon the PWC_{170} of Burmese men, relative to controls with a normal haemoglobin level (Yin Thu, Mya Tu & Aung-Than-Batu, personal communication).

In young Africans, maximum oxygen intake is fairly closely correlated ($r = 0.38$ to 0.71) with various indices of body composition such as gross body weight, estimated lean body mass and calf volume (C. T. M. Davies & Van Haaren, 1973); however, older Africans follow the European pattern in no longer showing such a relationship. Although there is a substantial (21%) difference of $\dot{V}_{O_2(max)}$ between active and inactive Tanzanians, differences of body weight (4.7%), lean body mass (5.1%) and leg volume (10.7%) are a good deal smaller.

Di Prampero & Cerretelli (1969) commented on the low anaerobic power of the East Africans. The maximum vertical speed of running was only about 75% that found in Italians. While it is conceivable that the skill or motivation of the Africans is less than that of the Italians, the test is relatively simple, and it is thus tempting to attribute the discrepancy to differences of anatomy and geometry of the limb muscles, together with possible deficiencies of high-energy phosphates secondary to poor nutrition. Somewhat surprisingly, Di Prampero & Cerretelli found that after correction for subcutaneous fat, the maximum diameters of the limb and calf were as in European subjects; however, because of shorter limb length, the total volume of muscle was smaller in the East Africans.

Miller *et al.* (1972) found that the lean body mass of Jamaican and Trinidadian men was lower than that of English subjects (−12.2%, −9.2% respectively), with correspondingly smaller thigh muscle widths (−10.4%, −7.5%).

Measurements of muscle strength were made in Addis Ababa (Areskog *et al.*, 1969) and Venezuela (Gardner, 1971); these figures also confirm the poor level of muscular development in many tropical peoples.

Population	Age (yr)	Right hand-grip strength (kg)	Right knee extension strength (kg)
Workers, Addis Ababa	18–25	34.6	29.4
Airforce cadets, Addis Ababa	18–22	41.4	33.6
Warao, Venezuela	20–24	42.5	37.1
Warao, Venezuela	40+	37.0	37.0

Cold and working capacity

Macro- and micro-climate

The extremely low effective winter temperatures of the circumpolar regions apparently place a severe stress on the range of human adaptability. However, a simple statement of meteorological conditions can be misleading. Various studies of white circumpolar expeditions have shown no more than 6–15% of a total sojourn is spent out of doors (Norman, 1960; Milan & Rodahl, 1961; A. Cumming, 1961). In summer, the highest recorded temperatures are usually experienced by the subjects, since this is when people are outside. In winter, outdoor activities tend to occur when the dry-bulb temperature is relatively low but there is little wind. Indigenous peoples typically encounter a wider range of climate than Caucasians living in the same areas; depending on their degree of acculturation, they spend more time out of doors, and their activities are inhibited less by what a white person would consider adverse weather conditions.

Physiological constraints

While resting out of doors, there may be difficulty in conserving body temperature, and this could explain the tendency to increase of subcutaneous fat during the winter months (page 50). It is particularly necessary to conserve manual dexterity and avoid cold damage to the extremities, and there are circumstances where a high rate of peripheral circulation protects against loss of function and cold injury (Rennie, 1963). Nevertheless, the usual problem during hard work is to eliminate sufficient body heat. Because very thick clothing is worn, skin temperatures are high; measurements on white workers using a wire vest resistance thermometer have shown values of 20–34 °C during sledding and other forms of outdoor work (Norman, 1962). Although formal observations on indigenous peoples have been limited (Milan, 1960), the insulation of Eskimo caribou and seal-skin garments and Mongolian quilted outer-wear is extremely high; Renbourn (1972) speaks of a 'tropical' micro-climate. Unfortunately, excessive sweating rapidly degrades the insulating properties of clothing, and eventually causes them to rot. Nansen (1892, 1898) graphically described the problem of the white explorer: 'During the course of the day, the damp exhalations of the body had little by little become condensed in our outer garments which were now a mass of ice and transformed into complete suits of armour.' Renbourn (1972) details the care taken by the Eskimo to avoid accumulation of sweat: 'With work, the belt is first loosened, the neck opened up, the gloves taken off, and if necessary the top garment may even be taken off.'

Physiological adaptations

Several adaptations have developed to meet these thermal stresses. Local dexterity of the fingers is conserved by a diminution of cold pressor responses (LeBlanc, personal communication), and the Eskimo can carry out repairs to his snowmobile under conditions when a white person's fingers would be completely numbed; the problem of the recent immigrant in sustaining manual dexterity is apparently even more severe in the Negro than in the white person (Rennie & Adams, 1957; Iampietro *et al.*, 1959). On the other hand, the blood flow to the superficial tissues of the face during a standard cold exposure seems the same in city-dwelling Norwegians and Skolt Lapps who habitually work out of doors (Krog, personal communication).

With regard to the regulation of core temperatures, the Eskimo and other circumpolar groups choose to rely on the insulation of removable layers of animal clothing rather than irremovable layers of subcutaneous fat (Table 20). This is important not only in allowing heat dissipation during heavy work, but also in maintaining the temperature of the extremities and ensuring a continuation of necessary insulation if a negative caloric balance develops during a long hunting trip (Renbourn, 1972; C. Allen, O'Hara & Shephard, 1976). There seems an interesting redistribution of body fat in the Eskimo; the overall percentage is much as in the white person (Shephard *et al.*, 1973*b*), but the skinfolds are very thin, and one must presume there are substantial internal depots to meet the demands of long and arduous hunting trips. Renbourn (1972) draws attention to a redistribution of subcutaneous fat towards the hands and feet, where insulation is most vital. Some authors have considered the pyknic build of the traditional Eskimo a useful adaptation to cold, but this is hard to substantiate since (a) many primitive groups from a wide range of habitats have a small size, and (b) the Eskimo (in common with other circumpolar peoples) is now showing a rapid secular trend to increase of stature (page 190; Shephard, 1974*c*). A stronger case could perhaps be made for accepting the square nose as an adaptation. In many Eskimos, the shape of the nose is such as to make it extremely difficult to use a standard respiratory noseclip, and it is obviously an advantage if this organ does not protrude beyond the protective micro-environment of the parka hood. With regard to sweating, Rennie *et al.* (1962) commented that in the comfort zone (28–32 °C) the seated Eskimo sweated as much or more than his white counterpart, and that at low temperatures he conducted heat more easily to the skin surface. Nevertheless, there seem regional differences in sweat gland activity. During vigorous work, sweat accumulates on the face (Shephard, 1975*a*), rather than on the trunk as in a white person, and formal counts of active glands after pharmacological stimulation show

Table 20. *Average skinfold thicknesses of young men living in cold environments*

Population	Season	Skinfold thickness (mm)
Argentina		
Alcaluf Indians	Winter 1960	7.90
Australia		
Central Aborigines	Summer 1959	7.04
Tropical Aborigines	Summer 1959	9.10
Canada		
Arctic Indians	Fall 1960	5.79
	Spring 1960	6.65
Eskimos	Summer 1963	6.0
Fort Chimo Eskimos	Summer 1970	5.8
Igloolik Eskimos	Summer 1970	5.5
	Winter 1970	6.5
Japan		
Ainu (Hokkaido)	Summer 1971	5.3
Korea		
Diving women	Summer 1962	7.98
Non-divers	Summer 1962	8.42
Scandinavia		
Lapps	Summer 1963	7.7
United States		
Alaskan Eskimos	Summer 1963	6.0
	Summer 1969, 1970	11.0

For sources of data, see Hammel (1964), Shephard (1975*a*), and Auger (1975).

increased activity over exposed areas such as the nose and cheeks, but a progressive diminution over heavily clothed regions (50% reduction over the trunk, 80% reduction on the feet; Schaefer *et al.*, 1976).

Cold and work tolerance

One immediate effect of cold exposure that has adverse consequences for the performance of physical work is a bronchospasm. However, this seems a transient phenomenon even in recent white immigrants (O'Hara & Shephard, in preparation). Dr O. Schaefer has speculated that the oral inhalation of large quantities of arctic air during vigorous work might have more permanent effects upon cardio-respiratory function. Our data for healthy Igloolik Eskimos (Rode & Shephard, 1973*a*) certainly show relatively low ratios of one-second forced expiratory volume to forced vital capacity, but, as in some classes of athlete, this reflects the large forced vital capacity rather than a true impairment of dynamic lung volumes. A high incidence of resting right bundle branch block has also been reported for the Igloolik Eskimos (10% in younger men, 33% in older men, and 18% in older women; Hildes *et al.*, 1976). If there were no associated clinical abnormalities, a partial block might merely reflect a well-developed cardio-respiratory fitness (Mitrevski, 1969); however, Hildes *et al.* (1976) stress that the block was complete, and in many instances was associated with enlargement of the main pulmonary arteries. While many facets of Eskimo life (acute and chronic respiratory disease, cigarette smoking, and incomplete fuel combustion in igloos) could contribute to these findings, Hildes *et al.* (1976) suggest that one possible factor is 'breathing very cold air at high flow rates during the heavy work involved in hunting and winter travel'. Exercise electrocardiograms were obtained (Shephard, 1975*a*) on thirteen of the Igloolik Eskimos where the medical team had found a resting right bundle branch block (complete in six, incomplete in seven). Unfortunately, the anomaly was particularly apparent in leads AVR, V4R, V1 and V2, while exercise electrocardiograms were uniformly taken in the CM5 position. Two of the thirteen Eskimos showed definite bundle block in exercise, and in three of the remaining eleven there was a suspicion of QRS broadening; nevertheless, all thirteen had a good effort tolerance.

Lack of fuel and other difficulties in melting snow and ice can compound the problems of heavy water loss in sweat and expired air, leading to a progressive dehydration of the Arctic traveller. Nansen (1892) thus described his discovery of a fresh-water lake towards the end of his journey across the Greenland ice cap: 'We lay down and let the water fairly flow down our throats. It was truly a divine pleasure to be able once more to drink to the very end of one's thirst.' Some more recent circumpolar expeditions have succeeded in maintaining a good urine flow (Edholm & Lewis, 1964), but other less heavily equipped patrols of white soldiers have shown dehydration (C. Allen *et al.*, 1976), with a poor working capacity that persisted until fluid volumes were restored. Some authors (G. M. Brown *et al.*, 1954; Baugh *et al.*, 1958) have reported 'large' blood volumes for Eskimo groups, but in fact the data do not seem outstanding relative to working capacity. In the Igloolik studies (Shephard, 1975*a*), it was necessary to test some hunters within a few hours of their return to camp, and, under these circumstances, blood volumes were appreciably lower for the nomads (84 ml/kg) than for acculturated Eskimos permanently resident within the Igloolik settlement (95 ml/kg). Confirmation that this difference was likely an expression of acute cold dehydration came from the correlation between cold adaptation and low values for blood volume and total body water. We developed a cold tolerance index from the summation of nine measurements of pulse rate and systemic blood pressure obtained by Dr J. LeBlanc during hand immersion, face immersion, and combined hand and face immersion; significant coefficients of correlation related this index to blood volume (ml/cm standing height, $r = 0.86$, $P \sim 0.07$) and body water (ml/cm, $r = 0.91$, $P \sim 0.05$).

The energy cost of work in the cold should be increased by the weight and hobbling effect of clothing (Renbourn, 1972), shivering, changes of basal metabolism (Hart, 1967), clumsiness (Pugh, 1971), increased muscle viscosity and decreased muscle endurance (R. S. J. Clarke, Hellon & Lind, 1958), but in practice this is hard to demonstrate in Eskimo subjects (Godin & Shephard, 1973); possibly, the reduction of body temperature leads to a simultaneous diminution of energy expenditures in inactive regions of the body (the Q_{10} effect of Arrhenius).

Acute exposure to cold can induce small improvements in working capacity through an increase in tone of venous reservoirs and a redirection of cardiac output from skin to active muscles; however, there is little evidence that cold in itself has any more long-term effect upon physical fitness (Andersen, 1967*a*).

Other primitive groups

Other primitive groups who encounter cold conditions include the Alacaluf Indians who live in the wet cold region bordering the Magellan Straits, Australian Aborigines and Kalahari Bushmen who face the dry cold of the desert night, Peruvian Indians who experience dry or wet cold at high altitudes, and the Korean Ama, who are perpetually diving into very cold water.

As with the circumpolar groups, there is considerable debate as to how far cultural adaptations avoid exposure to cold. H. T. Hammel (1964) and P. T. Baker (1966) concluded that the Alacaluf Indians derived considerable protection from fires, skin garments and their shelters of bent saplings and skins; only when they were working in small boats with minimal clothing was there likely to be a sustained cold stress. The same seemed true of the Peruvian Indians living at altitudes of 4000 m and above. Although night-time temperatures dropped below freezing point every month of the year, good shelter was provided by solid adobe housing; indoor temperatures were 7 °C higher, and blankets preserved an even more comfortable micro-climate. In wetter regions, such as Lake Titicaca, better construction of housing and clothing compensated for the adverse weather conditions. Cold stress was only likely in the nomadic Hama herders living at the highest altitudes; these groups had recourse to temporary rock shelters, and in the absence of wood were obliged to use animal dung as a rather inefficient source of heat.

The Aborigines use almost no clothing and build no formal shelters despite night temperatures that can drop below 0 °C (Hicks, 1964). While cold adaptations are shown by this group (page 49), the main basis of survival is cultural, and again white visitors can share their windbreak of small bushes and open camp fire with a fair degree of comfort (MacPherson, 1966). Equally, the Kalahari Bushmen make skillful use of skin cloaks and fires, and their skin temperatures throughout a cold night remain close to thermo-neutral values (Wyndham, 1966).

In view of doubts as to how far these various groups are exposed to cold, it is not surprising that there is little information on interactions of the supposed cold exposure and work stress. Body build hardly favours heat conservation; Wyndham (1966) cites a surface area/weight ratio of 2.60 cm^2/kg for Caucasians, 2.80 cm^2/kg for Bantu, and 3.11 Cm^2/kg for Kalahari Bushmen. Average skinfold thicknesses, while in some instances greater than in the Eskimo (Table 20), are still much lower than in the white community.

Physical working capacity

Early authors measured the fitness of the Eskimo in terms of respiratory recovery (Erikson, 1958) or the Harvard step test (Rodahl, 1958). In such terms, the Eskimo had a 'better cardio-respiratory function than white service volunteers', scores being 3.5 times those for untrained airmen, and 2.5 times those for trained Arctic soldiers. More recent experiments, including those sponsored by the IBP, have measured or predicted the maximum aerobic power (Fig. 5, Table 21) but in some groups such as the Nellim Lapps the maximum heart rate (180/min) and the blood lactate immediately following maximum effort (7 mmol/l) have been below maximum values anticipated in the white community. The authors of one study (Andersen & Hart, 1963) freely admitted that the reported aerobic power (44 ml/kg min STPD) could underestimate the true value by as much as 20%; the subjects examined in this experiment were also affected by both tuberculosis and episodic malnutrition. Having regard to such technical problems, the aerobic power of some of the more active groups (such as the Igloolik Eskimos and the Nellim and Kautokeino Lapps) is higher than in many of the tropical communities (Table 17). However, the main factor responsible for the difference seems a higher level of physical activity rather than cold exposure *per se*; even within Igloolik, sedentary Eskimos have a poorer aerobic power than the active hunters (page 107) and more highly acculturated groups such as the Alaskan Eskimos have quite low maximum oxygen intakes.

Paradoxically, malnutrition seems less of a factor limiting performance in the Arctic than in the tropics. Skinfold thicknesses remain quite low in many groups (Table 20), but in most instances body weight reaches or exceeds actuarial standards (Igloolik +5.7 kg (M), +5.2 kg (F); Wain-

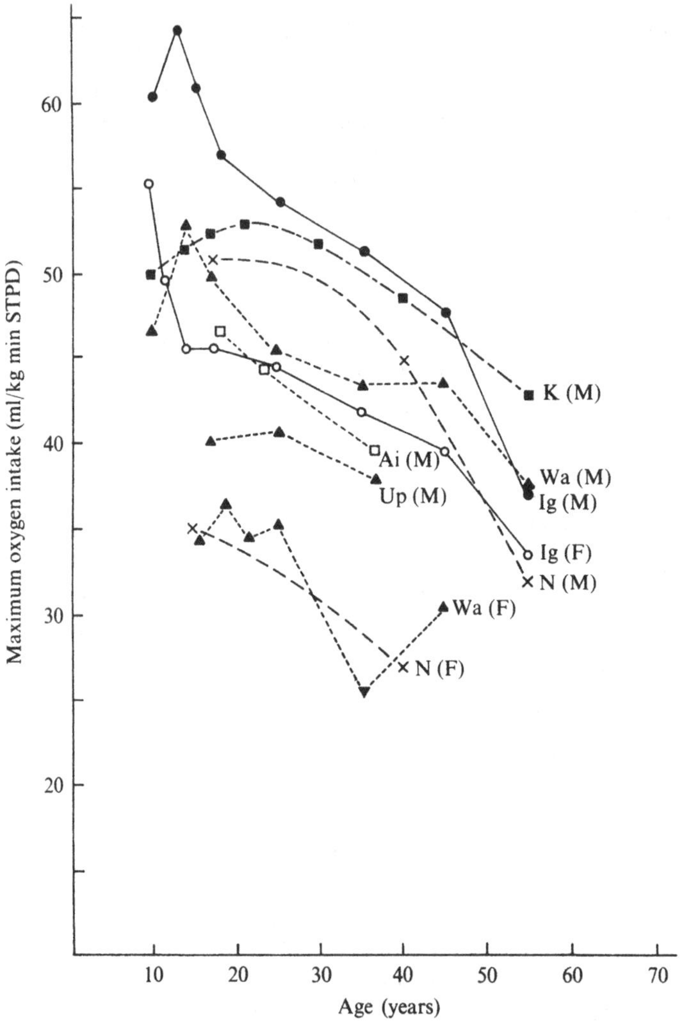

Fig. 5. The relationship between age and maximum oxygen intake. Direct measurements using bicycle ergometer on Ainu (Ai), Upernavik Eskimos (Up), and Kautokeino (K) and Nellim (N) Lapps; linear extrapolation of step test data on Wainwright Eskimos (Wa); Åstrand nomogram applied to step test data and corrected downwards by 8% on Igloolik Eskimos. M = male; F = female. (From Shephard, 1975*a*, by permission of the editor, *Circumpolar Peoples*, and Cambridge University Press.)

wright +5.1 kg (M), +8.4 kg (F); Hokkaido, +2.4 kg (M)). During the 1950s, a substantial incidence of anaemia was reported among the adults of Southampton Island (G. M. Brown, 1954) and among infants and children of the central Arctic (Sellers, Wood & Hildes, 1959). This seemed related to a transition from a nomadic to a more settled pattern of existence; thus

Table 21. *Aerobic power of populations living in cold climates – direct measurements of maximum oxygen intake*

Population	Age (yr)	Sex	Aerobic power (ml/kg min STPD)	Author
Canada				
Igloolik Eskimos	23	M	52.3	Shephard (1974*c*)
Arctic Indians	17–39	M	49.1	Anderson *et al.* (1960)
Greenland				
Upernavik Eskimos	14–19	M	40.1	Lammert (1972)
	20–29	M	40.7	Lammert (1972)
	30–45	M	37.2	Lammert (1972)
Japan				
Ainu (Hokkaido)	16–19	M	46.5	Ikai *et al.* (1971)
	20–29	M	44.4	Ikai *et al.* (1971)
	30–39	M	39.5	Ikai *et al.* (1971)
Ama	33.1	F	31.6	Hirota *et al.* (1969)
Scandinavia				
Nellim Lapps	15–19	M	52	Karlsson (1970)
	20–29	M	49	Karlsson (1970)
	10–14	F	35	Karlsson (1970)
	20–29	F	33	Karlsson (1970)
Kautokeino Lapps	18–30	M	53	Andersen *et al.* (1962)
	50	M	45	Andersen *et al.* (1962)
	15–38	M	57.4 (Skiing)	Andersen *et al.* (1962)
United States				
Alaskan Eskimos[a]	11–18	M	44.9	Rennie *et al.* (1970)
(Point Hope)	24	M	42.3	Rennie *et al.* (1970)
	35	M	41.6	Rennie *et al.* (1970)
	13–18	F	33.5	Rennie *et al.* (1970)
	24.7	F	32.0	Rennie *et al.* (1970)

[a] Data selected from larger sample on basis of terminal heart rates.

13% of the men and 17% of the women living in the permanent settlement of Cambridge Bay were clinically anaemic in 1959–60, but in the more isolated areas of Bathurst Inlet and Boothia Peninsula (where native diet predominated), anaemia was virtually absent (L. E. C. Davies & Hanson, 1965). Such problems of transition now seem resolved. Our haemoglobin values for Igloolik, measured against international cyanmethaemoglobin standards, show high normal values throughout childhood and adult life. Parallel studies from Wainwright, Alaska, confirm the current absence of anaemia in adults and children older than six (Sauberlich *et al.*, 1970).

Measurements of muscle strength and anaerobic power (Table 22) also show the superior fitness of the circumpolar peoples. Grip and leg extension strengths at Igloolik far exceed the figures for Venezuela and Ethiopia; the leg strength results are also substantially higher than would

Haemoglobin levels for Igloolik Eskimos

Male		Female	
Age (yr)	Haemoglobin (g/dl)	Age (yr)	Haemoglobin (g/dl)
11–13	13.6±0.8	11–12	13.6±1.0
14–16	14.4±0.8	13–15	13.5±0.8
17–19	15.2±0.9	16–19	13.4±0.9
20–29	16.0±0.9	20–29	14.2±0.8
30–39	15.7±0.9	30–39	14.4±0.7
40–49	15.4±0.7	40–49	14.7±1.5
50–56	14.7±0.6	50–59	14.0±0.6

be anticipated in a white community, perhaps because of the effort involved in walking through deep snow, while the anaerobic power of the Wainwright Eskimos is 20% greater than that of Italian subjects and 70% greater than that of East Africans. On the other hand, Hirota *et al.* (1969) found that the maximum oxygen debt of the Ama per kilogram of body weight was the same as in normal women.

Table 22. *Strength and anaerobic power of circumpolar peoples*

	Igloolik					
	Grip strength (kg)		Leg strength (kg)		Wainwright: anaerobic power (M)	
Age (yr)	M	F	M	F	(ml/s)	(ml/kg s)
20–29	52.0±5.3	29.8±4.0	88.3±17.4	68.9±17.8	1.67±0.20	24.2±2.9
30–39	46.1±8.1	28.4±4.9	88.9±24.1	66.9±16.5	1.49±0.13	21.3±2.6
40–49	42.4±5.0	29.0	76.5±15.6	68.1	1.22±0.31	18.6±5.4
50–59	38.5	27.3±2.0	76.9	69.5±2.7	1.17	17.0
60–69	—	—	—	—	0.99	14.8

Data of Canadian and US teams as collected by Shephard (1974*c*). From Shephard (1975*a*), by permission of the editor, *Circumpolar Peoples*, and Cambridge University Press.

Formal studies of body composition, using deuterated water, confirm that the Igloolik Eskimo has an above average proportion of lean tissue (Shephard, 1975*a*). Expressing total body water in litres per centimetre of standing height, we have: Igloolik Eskimo (male), 0.267; (female), 0.207; US white (male), 0.225; (female), 0.179.

Since the Eskimo is also somewhat shorter than the average US white, we may conclude he is carrying approximately the same quantity of muscle as his American counterpart on shorter legs.

Circadian rhythms

The white population living in moderate latitudes shows a number of quite marked circadian rhythms (Guberan *et al.*, 1969; Mills, 1973). Some, such as body temperature (Glaser & Shephard, 1963; Mills, 1973), seem endogenous in character. Others such as variations in the resting (Kleitman & Kleitman, 1953; Lewis & Lobban, 1957; Howitt *et al.*, 1966) and exercise (Voigt, Engel & Klein, 1967; Klein, Wegmann & Bruner, 1968; Crockford & Davies, 1969) pulse rates are probably secondary to rhythms of body temperature and physical activity. Nevertheless, they are sufficient to cause a 25% difference in predictions of working capacity, the lowest pulse rates and highest predictions occurring at 02.00 to 04.00 hours, and the highest pulse rates and poorest predictions at 16.00 to 18.00 hours of each day. Direct measurements of $\dot{V}_{O_2(max)}$ are 5% larger in the morning (09.00 to 13.00 hours) than at night (01.00 to 05.00 hours), according to the experiments of Wojtczak-Jaroszowa & Banaszkiewicz (1974).

Sudden exposure to a warm or a cold climate, or displacement of the normal diurnal cycle of temperatures can modify the body temperature rhythm (Glaser & Shephard, 1963). There are probably corresponding changes of working capacity, although this point has not yet been examined experimentally.

In the circumpolar regions, the problem is complicated by winter darkness and twenty-four hours of sunshine in the summer. The Eskimo has a natural tendency to make the most of long and warmer days, and the newly arrived investigator may be surprised to find the main 'street' of a settlement busy with men, women, and children at 2 or 3 a.m. Formal studies of hunting trips (Shephard & Godin, 1973) show that vigorous physical activity sometimes continues through several 24-hour periods. While no formal studies of the effect of such trips have been made, it is hard to imagine that any other than the most robust endogenous circadian rhythms would withstand such treatment. Historically, Eskimos have shown almost a complete absence of circadian rhythms, but with acculturation, 24-hour cycles have begun to appear (Lobban, 1976). These seem related more to the availability of social synchronizers than to natural cycles of light and darkness. The implications of this developing rhythmicity for working capacity have yet to be explored.

While previous work may have underestimated the importance of social synchronizers (Lobban, 1976), in many regions natural environmental cycles are of prime importance. Some synchronization may result from phasic changes in environmental temperature, although the magnitude of such signals tends to be regional rather than latitudinal (large in Siberia, Alaska and some deserts, small over the polar pack-ice and in rain forests). Nevertheless, there is widespread agreement that the hours of daylight remain the major synchronizer of circadian activities, not only for man, but also for other mammals (Simpson & Bohlen, 1973). Equatorial regions are marked by a maximum 24-hour periodicity of light, and a minimum annual periodicity; indigenous residents of such regions (for example, Galapagos islanders; Lobban, 1969) show a well-developed circadian rhythm. Conversely, the polar regions have a minimum 24-hour periodicity and a maximum annual periodicity; daily rhythms are depressed more in the Eskimos than in Arctic Indians or temporary white residents but even the last group find that their normal circadian rhythms are depressed during the winter periods of darkness (Lobban, 1965, 1967; Simpson & Bohlen, 1973).

In a subsequent section (Chapter 7) we shall be examining data on 'athletes' as illustrating the potential for development of human working capacity through a combination of selection and training. Where athletes are tested at international competitions, it is important to note that they may be encountering problems in adapting their circadian rhythm of performance to the constraints imposed by international travel and unaccustomed cycles of light and temperature.

5. Socio-economic status and working capacity

Socio-economic status can have an important influence upon the potential and the realized working capacity of a community through the interaction of such factors as the quantity and quality of available nutrients, family size and habits of child care, the prevalence of disease, and patterns of habitual activity.

Nutrition and physical working capacity

A low total caloric intake and/or a deficiency of first-class protein can retard or permanently stunt growth, thus restricting physical working capacity. Hyponutrition at the time of testing may limit glycogen stores and thus immediate performance. Chronic malnutrition may also sap will-power and the desire to develop potential working capacity.

At the turn of the twentieth century, malnutrition was a significant problem even in countries such as Britain and the United States. Partly for this reason, Dreyer (1920) included a table of weight for height standards in his book *The assessment of physical fitness*. During World War II, the German authorities discovered that a certain minimum quantity and quality of nutrients were helpful in sustaining the output of slave-labour employed in coal mining.

The working capacity of many indigenous peoples still seems limited by malnutrition, particularly in tropical regions. The activities necessary to subsistence are often completed in the face of severe constitutional handicaps. Thus, MacPherson (1966) describes the highland populations of New Guinea: 'Although malnutrition has been amply demonstrated in children and protein deficiency is frequent in pregnant and lactating women, and skinfold thickness measurements are low and there is a decline in the weight–height ratio with age and the serum albumin is low, both men and women perform physical feats and show sustained stamina that few of us could match.'

In this section, we shall look at several simple indices of nutritional status (skinfold thicknesses, body weight, haemoglobin level and serum lipids), and will then discuss more complex measures of body composition and their relation to working capacity. The impact of nutrition on habitual activity patterns will be discussed later (page 113).

Table 23. *Typical skinfold readings and excess weights for a white metropolitan population*

	Men		Women	
Age (yr)	Average skinfold (mm)	Excess weight (kg)	Average skinfold (mm)	Excess weight (kg)
10–12	8.0	3.2	10.3	2.1
16–19	11.3	0.8	—	—
20–29	11.2	1.7	15.2	8.3
30–39	16.1	6.4	13.5	1.4
40–49	14.0	9.3	17.3	6.8
50–59	15.2	8.8	18.2	4.9
60–69	15.4	5.1	22.5	4.5

Author's data for adults resident in metropolitan Toronto.
From Shephard (1969*a*), by permission of the University of Toronto Press.

Simple indices of nutrition

Skinfold thicknesses

Skinfold thicknesses have already been reported for those living in tropical (Table 16), circumpolar (Table 20) and mountainous (Table 15) regions. All indigenous groups except the highly acculturated Alaskan Eskimos of Point Barrow show extremely low readings relative to those encountered in white city-dwellers (Table 23). Indeed, on the reasonable assumption that a double fold of skin in itself has a thickness of at least 2 mm, we would conclude that many primitive tribes have almost no subcutaneous fat. The increase of skinfold readings that the white population shows over the adult life-span is plainly an indication of over-nutrition. However, it is far less certain that thin folds signify malnutrition. Our values for the young male hunters of Igloolik, for example, are extremely low (Table 20), yet by several other criteria (general observation, dietary records, haemoglobin levels, lean body mass, and specific biochemical assays of nutrients) the group are well nourished, with an excellent aerobic power and leg strength (Tables 21 and 22).

There are several other problems in interpreting skinfold readings. Novak (1970) examined American and Filipino nurses resident in Rochester, Minnesota, and stressed the difficulty that he encountered in comparing skinfolds and other linear anthropometric data on populations differing markedly in body size. Although his sample of American nurses carried 4.9 kg more fat than the Filipinos, the percentages of body fat (34.4 and 33.7% respectively) were very similar for the two groups. Among indigenous populations, another problem arises from the small proportion

Table 24. *The effect of the choice of prediction formulae on estimates of body density as derived from typical measurements of skinfold thickness on a white person and an Eskimo*

	Calculated body density		Age and sex of population originally tested		
Formula	White	Eskimo			Author
$1.1533-0.0643\ (\log_{10} \Sigma S)$	1.048	1.065	15	M	Durnin & Ramahan (1967)
$1.1610-0.0632\ (\log_{10} \Sigma S)$	1.057	1.074	22	M	Durnin & Ramahan (1967)
$1.1369-0.0598\ (\log_{10} \Sigma S)$	1.032	1.054	15	F	Durnin & Ramahan (1967)
$1.1581-0.0720\ (\log_{10} \Sigma S)$	1.032	1.059	22	F	Durnin & Ramahan (1967)
$1.1447-0.0612\ (\log_{10} \Sigma S)$	1.050	1.068		M	Durnin (personal communication)
$1.1309-0.0587\ (\log_{10} \Sigma S)$	1.036	1.057		F	Durnin (personal communication)
$1.130-0.055\ (\log_{10} S_t)$ $-0.026\ (\log_{10} S_s)$	1.049	1.065	13–16	M	Pařízková (1961)
$1.114-0.031\ (\log_{10} S_t)$ $-0.041\ (\log_{10} S_s)$	1.035	1.055	13–16	F	Pařízková (1961)
$1.0923-0.0202\ (S_t)$	1.074	1.080	22	M	Pascale *et al.* (1956)
$1.0896-0.0179\ (S_s)$	1.068	1.077	22	M	Pascale *et al.* (1956)
$1.0764-0.00088\ (S_t)$ $-0.00081\ (S_s)$	1.051	1.065	20	F	Sloan, Burt & Blyth (1962)

Reproduced from Shephard, Hatcher & Rode (1973*b*), by permission of the editor and publishers, *European Journal of Applied Physiology*.
ΣS = sum of skinfolds; S_t = triceps skinfold; S_s = subscapular skinfold.

of body fat carried in the skinfolds. In consequence, there is a poor correlation between skinfold readings and other indices of body fat (Shephard, Hatcher & Rode, 1973*b*). In all populations, attempts to predict the percentage of body fat from skinfold readings give widely differing answers, depending on the assumptions that are made (Tables 24 and 25).

Body weight

Does gross body weight provide a more reliable criterion of nutritional status than skinfold thicknesses? We have already noted the interest of Dreyer (1920) in body weight, and have remarked that Wyndham and his associates (1963) used weighing as a rapid method of screening the work tolerance of Bantu mining recruits.

The Bantu gained weight rapidly with a combination of vigorous physical effort and improved diet. Areskog (1971), equally, noted that Ethiopian military recruits gained 5 kg over their first year of training. While the low initial body weight was associated with a poor absolute physical working

Table 25. *Percentage body fat if density (D) equals 1.060*

Formula	Body fat (%)	Author
$\left(\frac{5.548}{D}-5.044\right)100$	19.0	Rathbun & Pace (1945)
$\left(\frac{4.971}{D}-4.519\right)100$	17.1	Brožek *et al.* (1963*a*)
$\left(\frac{4.570}{D}-4.142\right)100$[a]	16.9	Brožek *et al.* (1963*a*)
$\left(\frac{4.0439}{D}-3.6266\right)100$	18.8	Grande (1961)
$(1.10-D)\,500$	20.0	MacMillan *et al.* (1965)

Reproduced from Shephard, Hatcher & Rode (1973*b*), by permission of the editor and publishers, *European Journal of Applied Physiology*.
[a] Formula adopted by IBP.

capacity, if expressed per kilogram of body weight the PWC_{170} of groups receiving less than 75% of the WHO recommended caloric intake was no poorer than that of much better nourished subjects. On these grounds, Areskog (1971) queried whether the UN Food and Agricultural Organization and the WHO were not recommending excessively large minimum dietary allowances. However, from the viewpoint of man's adaptation to life in a country where hard physical work is still required, the absolute PWC_{170} seems of greater significance than the relative value. Among some more developed populations such as the industrial Yugoslavs studied by Buzina (1972), there is still a relationship between a low body weight and a poor physical working capacity. On the other hand in Canada (Table 23) and the United States, excess weight has become a sign of over-nutrition. Heavy subjects do poorly on such tasks as the Harvard fitness test, and there is a highly significant negative correlation ($r = -0.46$) between the estimated percentage of body fat and fitness test scores.

As with the skinfold readings (above), problems of data interpretation can arise in populations with an unusual body build. Thus, figures for the Oto and Twa populations of central Zaïre (Ghesquière & Andersen, 1971) show respective weight deficits of 6.7 and 7.9 kg relative to actuarial standards. However, the protein intake of both populations is reported as 'sufficient', haemoglobin levels are normal (averages of 15.5 and 16.0 g/dl respectively) and kwashiorkor is 'virtually unknown'. By way of contrast, the weight deficits of many tropical groups almost certainly reflect malnutrition. Some circumpolar groups show quite substantial excess weights (page 75), and from the skinfold readings it can be presumed that the majority of this excess is muscle.

Haemoglobin

Low haemoglobin readings may reflect parasitic infections rather than a deficiency of minerals, vitamins, and first-quality protein. Thus Cotes *et al.* (1973*a*) reported the following figures for the indigenous peoples of New Guinea:

	Coastal plain (g/dl)	Coastal migrants (g/dl)	Highlanders (g/dl)
Men	11.0	11.9	14.4
Women	9.9	—	13.2

The highlands and islands of New Guinea are relatively free of malaria (Vines, 1967), but on the coastal plains infection is sufficiently prevalent to cause a 25% drop in average haemoglobin levels, both in lifetime residents and in those who have migrated to the coast from highland settlements. Hookworm infections in Venezuela may give rise to even lower haemoglobin readings (Vetencourt, personal communication). Ethiopians, on the other hand, present the interesting paradox of poor nutrition and relatively normal haemoglobin concentrations (13.0 g/dl in men living at an altitude of 1500 m; Andersen, 1971); this can be traced to a contamination of the staple food (teff) with iron-rich soil (Hofvander, 1968).

Despite these various complicating factors, Donoso *et al.* (1974*b*) established a positive correlation ($r = 0.55$) between the haemoglobin level and the aerobic power of ninety-six poorly nourished foundry workers in Santiago, Chile. Interestingly, the working capacity of this sample was unrelated to total caloric intake or to other stigmata of possible vitamin deficiencies. C. T. M. Davies & Van Haaren (1974) found that the relationship of aerobic power to limb volume was distorted by anaemia, but that a normal relationship was restored if the anaemia was corrected.

Serum lipids

Serum cholesterol is a further simple quantitative index of nutritional status. In civilized communities, average values are much higher than among most primitive groups, and readings seem to rise yet further with access to a particularly rich diet.

The nomadic peoples of Somalia, Kenya, and Tanzania adopt a diet that consists largely of milk, meat, and blood, yet succeed in preserving low blood cholesterol readings – possibly because their level of habitual activity is high in relation to total caloric intake (Lapiccirella *et al.*, 1962; Shaper *et al.*, 1963; Mann *et al.*, 1964). One exception from this region is provided by the camel-herding Rendille nomads of northern Kenya, who reputedly have quite high cholesterol levels. Wealthy, city-dwelling East Africans also commonly reach the blood lipid figures found among North American whites (Scott *et al.*, 1963). In the Pacific, blood lipid levels apparently reflect the degree of acculturation of the population studied. Among the indigenous New Guinea highlanders, cholesterol readings average less than 150 mg/dl throughout adult life (Goldrick, Sinnett & Whyte, 1970). On Pukapuka (a low atoll), the indigenous population present values of 160–180 mg/dl but among the Maoris of New Zealand, figures of 200–220 mg/dl are encountered (Prior & Evans, 1970). The Eskimos of Greenland show low figures not only for serum cholesterol (Sagild, 1972), but also for other measures of fat metabolism, including total lipids, triglycerides, β-lipoproteins and pre-β-lipoproteins (Bang & Dyerberg, 1972). A genetic explanation can be ruled out for this group, since Eskimos who have moved to Denmark show the same lipid profile as typical Danes. In the younger Greenland Eskimos, physical activity may help in controlling blood lipids, but Band & Dyerberg (1972) comment that the older members of their population were no longer active hunters and fishermen; possibly, a diet rich in unsaturated fat from fish and whale oil helps to hold down the cholesterol readings in this group (Keys, Anderson & Grande, 1959). Canadian Eskimos from the northern Foxe Basin have low cholesterol levels (Sayed *et al.*, 1976*a*); however, more acculturated Eskimos in northern Quebec show figures of 182–254 mg/dl (Carrier *et al.*, 1972), and very high cholesterol readings have recently been noted among Alaskan Eskimos (Maynard, 1976).

The nutrition of the average Indian is poor, and despite substantial ethnic differences within the Indian subcontinent, cholesterol levels (152–174 mg/dl) are consistently low (Pinto *et al.*, 1970). Among Israelis, the lowest cholesterol levels are found in nomadic Bedouins. Immigrants to Israel show a gradation from an average of 195 mg/dl for those born in North Africa to 219 mg/dl for European Jews. While more of the North African immigrants have manual work, the main basis for the persistent difference of cholesterol readings is probably the preservation of eating habits acquired in the country of birth (Medalie, 1970).

Other simple indices of nutrition

Sinnett (1972) has suggested several other possible criteria of good nutrition, including a high birth rate, a satisfactory population growth, and a high level of physical fitness. In such terms, New Guineans make a good adaptation to a diet (94.6% carbohydrate, 3.0% protein and 2.4% fat) that is based largely on the sweet potato. However, other stigmata of deficient feeding remain, particularly a 25% drop of body weight between the ages of twenty and sixty years. Measurements of 24-hour creatinine excretion indicate that at least a part of the weight loss is protein, and Sinnett has suggested that the age-related decrease of electrocardiographic voltage may reflect a decreasing myocardial mass.

Anthropometry, body composition and human performance

Early studies by Cullumbine (1949–50) showed significant correlations between anthropometric measures of leg musculature and the Harvard fitness test scores of Sinhalese, Tamils and Moors living in Ceylon. The importance of quadriceps strength to bicycle ergometer performance has also been noted in our discussion of IBP methodology (page 34).

Analysis of data obtained by the IBP working group (Shephard, Weese & Merriman, 1971) showed that the aerobic power of Toronto schoolchildren could be predicted quite accurately from anthropometric measurements. Lean body mass (*LBM*), as estimated from underwater weighing or skinfold readings, accounted for 39.1% of the variance in the directly measured aerobic power ($\dot{V}_{O_2(max)}$) of the boys, and 75.5% of the variance in aerobic power for the girls. A three-term multiple regression equation based on lean body mass, thickness of the medial knee skinfold (S_k) and age (A) increased descriptions to 57.7 and 85.3% of the variance respectively:

$$\dot{V}_{O_2(max)} \text{ boys} = 0.0216\,(LBM) + 0.0117\,(A) + 0.105\,(S_k) - 1.19$$
$$\dot{V}_{O_2(max)} \text{ girls} = 0.0480\,(LBM) + 0.0050\,(A) + 0.043\,(S_k) - 0.89$$

In girls, inclusion of a term describing summated muscle strengths (ΣF) reduced the coefficient of variation of the prediction equation to 7.9%, and an even more accurate prediction (CV 7.3%) was obtained by introducing body surface area (*BSA*):

$$\dot{V}_{O_2(max)} \text{ girls} = 1.127\,(BSA) + 0.0058\,(A) + 0.0327\,(S_k) - 0.753$$
$$\text{or } = 0.0384\,(LBM) + 0.0035\,(A) + 0.043\,(S_k) + 0.0020\,(\Sigma F) - 0.63$$

The British contribution to the IBP human adaptability project included a somewhat analogous study of forty-six English workers aged eighteen to twenty-eight years (Cotes *et al.*, 1969). A soft-tissue radiographic measurement of thigh muscle width (D_m) predicted maximum oxygen intake with a coefficient of variation of 11.4% in men and 11.2% in women, while body potassium estimation ($\Sigma\,^{40}K^+$) yielded predictions with a CV of 10.7% in men and 10.9% in women. Multiple regression analysis was attempted, despite the rather small sample size. This showed that the incorporation of anthropometric data together with the forced vital capacity (*FVC*) yielded small improvements in the prediction of maximum oxygen intake relative to statistics based simply on the heart rate at an oxygen consumption of 1.5 l/min ($f_{h,1.5}$):

$$\dot{V}_{O_2(max)} = 0.015\ (\Sigma\ {}^{40}K^+) - 0.0114\ (f_{h,1.5}) + 2.70$$
(CV = 9.5%, men; 8.1% women)

$$\dot{V}_{O_2(max)} = 0.19\ (D_m) + 0.16\ (FVC) - 0.011\ (f_{h,1.5}) + 1.38$$
(CV = 9.7%, men; 7.4% women)

In young British adults (C. T. M. Davies, 1972*a*), quite good predictions could be obtained from anthropometric data alone:

$$\dot{V}_{O_2(max)} = 0.001\ (H) + 0.26\ (D_m) + 2.71\ (CV = 10\%)$$

where *H* was the standing height. However, in older subjects (36–50 yr), the only useful anthropometric prediction (CV = 13%) was given by summated skinfold thickness (ΣS) and body weight (W):

$$\dot{V}_{O_2(max)} = -0.015\ (\Sigma S) + 0.042\ (W) + 0.68$$

Among British children, C. T. M. Davies (1971) found a variable and sex-dependent relationship between bicycle ergometer estimates of $\dot{V}_{O_2(max)}$ and either gross body weight or lean body mass. However, if account was taken of leg muscle volume, the variability was reduced and the sex difference disappeared. C. T. M. Davies (1971) thus suggested that the bicycle ergometer $\dot{V}_{O_2(max)}$ for these children depended essentially on the amount of muscle that could be brought into use.

In various populations of young adults, Cotes *et al.* (1973*b*) suggested that the heart rate at a given submaximal work load was inversely related to the estimated muscle mass (*M*) according to equations of the type

$$f_h = A + (B/M)(1 + C\dot{V}_{O_2})$$

where *A*, *B* and *C* are constants.

Unfortunately, each of the currently used indices of muscle mass has serious limitations (page 46). The whole-body counters required for $\Sigma\ {}^{40}K^+$ determinations are only available at a few specialized centres, and interpretation of the results assumes (i) a constancy of intracellular potassium concentrations between populations, and (ii) a constant proportional contribution of radioactive emissions from sources other than muscle (such as the viscera and the central nervous system). Nevertheless, the ${}^{40}K^+$ method has been used recently to show a close correlation between body cell mass and PWC_{170} in German schoolchildren (Burmeister *et al.*, 1972). Estimations of total body water by the deuterium dilution method again require complicated analytical procedures, and assumptions regarding the hydration of body tissues are necessary if lean mass is to be calculated. The deuterium method has been applied to one Eskimo population (Shephard, Hatcher & Rode, 1973*b*), urine samples being transported to Toronto for analysis; as noted above, a well-developed working capacity was associated with a greater lean mass per unit height than the usually accepted figures for the white population. Anthropometric (Jones & Pearson, 1969) and radiographic estimates of leg musculature have been applied more widely in studies of primitive groups; one problem with such techniques is that no account can be taken of 'dilution' of muscle by interstitial fat secondary to ageing or wasting (Durnin & Womersley, 1971). C. T. M. Davies (1973) used the method of Jones & Pearson (1969) to show that, in malnourished East African children, $\dot{V}_{O_2(max)}$ decreased *pari passu* with the reduced leg (muscle plus bone) volume. On the other hand, the slope of the $\dot{V}_{O_2(max)}$/leg volume relationship (established originally on normal sedentary adults) was modified by both anaemia and vigorous habitual activity. Cotes *et al.* (1973*b*) have suggested that the effect of regular vigorous physical activity is brought out particularly clearly by plotting heart rate against the reciprocal of lean body mass. Thus, sedentary Jamaican adults have higher heart rates for a given lean mass than Europeans; however, hill farmers from the Lawrence Tavern region (page 99) have a high daily caloric expenditure, and in this group heart rates are low relative to lean mass (Miller *et al.*, 1972). Other populations with lower heart rates than would be anticipated from lean mass include British distance cyclists and New Guinea highlanders who take 'much exercise' (Cotes *et al.*, 1973*a*).

Family size, child care and working capacity

Family size and habits of child-rearing have many practical implications for the fitness and survival of both the child and his parents. We shall look briefly at lactation as a method of contraception, child-rearing and the fitness of the mother, family size and population stability, and the influence of socio-economic conditions upon the working capacity of the developing child.

Lactation and contraception

Until recently, many indigenous communities from the Canadian Eskimo (Schaefer, personal communication) to the Kalahari Bushmen (Wyndham, 1966) breast-fed children to the age of three. Often, tribal taboos prohibited sexual intercourse during the period of lactation. But even if this was not the case, lactation itself served as a rather effective natural form of contraception. Acculturation is now causing rapid changes of breast-feeding practices. A study of Canadian Eskimos conducted at Igloolik in

Table 26. *The influence of child-rearing upon working capacity. Mean± SD of 15–29 years*

No. of children	Sample size	Age (yr)	Predicted aerobic power (l/min STPD)	Predicted aerobic power (ml/kg min)	Skin-folds (Σ3, mm)	Leg strength (kg)	Hand-grip (kg)
0	11	16.5±1.5	2.49±0.38	47.4±7.3	32.5±10.4	56.3±20.6	28.1±3.0
1	5	20.4±3.6	2.99±0.29	49.7±4.0	32.6±8.3	65.4±18.3	30.4±2.9
2	1	25.0	2.56	47.9	20.0	60.0	24.0
3 or more	11	25.7±2.3	2.80±0.50	48.0±8.3	23.9±7.6	67.6±18.6	29.4±4.8

From Shephard (1975*a*), by permission of the editor, *Circumpolar Peoples*, and Cambridge University Press.

1972 found forty-five infants under the age of thirty-six months; fifteen of these children had been breast-fed for a month or less, and the average period of lactation was only 6.7 months (Sayed *et al.*, 1976*b*). Solid food was introduced between the first and fourteenth months of life, but only 40% of the infants took their first solid food from traditional sources (caribou, seal or fish).

Child-rearing and maternal working capacity

In many indigenous populations, children under the age of three or four have traditionally been carried on the mother's back. We have already noted (page 47) the work load this imposes upon !Kung women. Equally, the Eskimo amauti and the American Indian papoose make a useful contribution to the development of physical fitness among young mothers of northern communities. Table 26 classifies Igloolik women aged 15–29 years according to parity. The relative aerobic power was similar in all groups, but the nulliparous had a lower absolute aerobic power and a lower leg strength than those with babies. Skinfold thicknesses also were larger in those who had no children or only one child.

While these several physiological differences probably reflect the added energy expenditures involved in raising a young family and transporting them through deep snow, it must also be stressed that women with two or more children were on average older and less acculturated than those with no family or but one child.

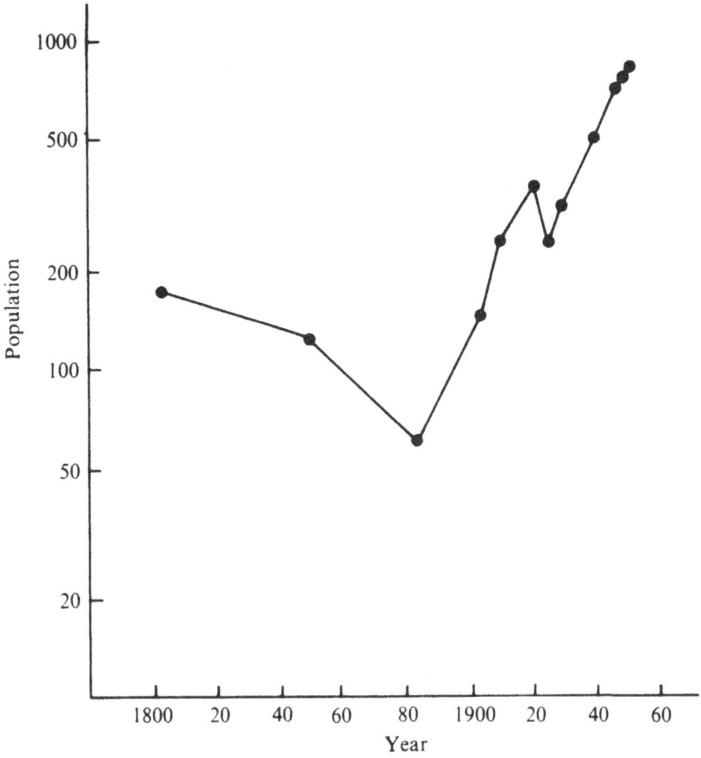

Fig. 6. Population of Igloolik region from the early nineteenth century to the mid 1970s. Probable factors contributing to the population explosion include improved nutrition, control of major diseases such as tuberculosis, availability of medical and nursing care, improved domestic and community hygiene, and altered breast-feeding habits.

Family size and population stability

An increase in the number of small children in a nomadic family restricts its ability to wander in search of food, and thus diminishes its survival potential. A simple change of community breast-feeding habits in the absence of alternative forms of contraception thus imposes a double burden, compounding other factors that are currently producing a 'population explosion' (Fig. 6), while at the same time limiting the mobility of the group.

Socio-economic factors and working capacity

Within the individual family, the course of physical development is strongly influenced by birth order and family size, with all that the latter implies for parental attention, living space, unit income, and standards of nutrition.

Table 27. *Social class and PWC_{170} of schoolchildren*

Sample	School type	PWC_{170} (kg m/min)	PWC_{170}/kg	Author
Canada				
11–12 yr boys	Public (lower class)	468	12.6	G. R. Cumming & Cumming (1963)
	Public (non-academic stream)	472	12.8	G. R. Cumming & Cumming (1963)
	Public (academic stream)	432	11.1	G. R. Cumming & Cumming (1963)
	Private	531	14.4	G. R. Cumming & Cumming (1963)
11–12 yr girls	Public (lower class)	379	8.2	G. R. Cumming & Cumming (1963)
	Public (non-academic stream)	393	11.6	G. R. Cumming & Cumming (1963)
	Public (academic stream)	323	8.3	G. R. Cumming & Cumming (1963)
Ethiopia				
10 yr boys	Public	315	12.0	Areskog *et al.* (1969)
	Private	252	9.9	Areskog *et al.* (1969)
13 yr boys	Public	379	11.9	Areskog *et al.* (1969)
	Private	357	10.2	Areskog *et al.* (1969)

Examples of poor working capacity due to malnutrition have already been cited. Among the urban populations of more-developed nations, social class may operate in an opposite sense, with over-protective middle-class parents unduly restricting the physical activity of their children. Thus Durnin (1966) found that British boys of the highest social class spent 94 minutes a day in moderate, heavy or very heavy activity, but that the period for boys of the lowest social class was 129 minutes; comparable figures for girls were 92 and 111 minutes respectively. The resultant gradient of physical working capacity (Table 27) is particularly evident in the data of G. R. Cumming & Cumming (1963); these authors found that children from the academic stream of a middle-class municipal school in Winnipeg had lower PWC_{170} scores than very wealthy children attending a private school, children at the same municipal school in a non-academic stream, and children from a municipal school in a poor neighbourhood.

Disease and physical working capacity

An epidemic of a lethal disease could theoretically improve the working capacity of an isolated community through the selective elimination of weaker members of the population. For such a process of natural selection to operate effectively, it would be necessary for the disease process either to kill the affected individual before the period of procreation was attained, or at least to reduce the likelihood of marriage and the successful rearing of progeny.

Table 28. *Infant deaths in Igloolik, 1959 to 1968*

Year	Actual deaths, infants < 1 yr: At camp	In village	Total[a]	Infant mortality rate: Igloolik	All Canadian Eskimos
1959	1	1	3	120	206
1960	8	1	9	360	211
1961	4	1	5	200	185
1962	9	1	11	440	194
1963	4	1	6	240	157
1964	2	2	4	160	92.1
1965	4	6	10	400	95.4
1966	1	3	4	160	92.3
1967	2	1	4	160	—
1968	1	1	2	80	—
Mean per annum	3.6	1.8	5.4	232	

From Shephard & Rode (1973), by permission of William Heinemann Medical Books, Ltd. There were about 25 live births per year during 1959–68, and an approximate infant mortality rate has been calculated on this assumption.

[a] A small number of infants died outside the region.

In a more immediate sense, disease usually has a negative effect on the effort tolerance of the individual. Restrictions of habitual activity may be imposed by the disease itself or by an attending physician, systemic effects of the disease may restrict growth and development, and acute impairments of cardio-respiratory function may reduce the potential for oxygen transport.

Disease and natural selection

Imported diseases of the white man, such as measles and tuberculosis, have had profound effects on the health and evolution of small and isolated populations (Rabinowitch, 1936; Peart & Nagler, 1954; Motulsky, 1960). Regular nursing care, improved health control measures and aerial evacuation of the severely ill is now improving the life expectancy of many primitive communities. Nevertheless, even in wealthy nations such as Canada, indigenous populations are still subject to a substantial evolutionary pressure, as can be gauged from the infant mortality statistics for the Igloolik region and for Canadian Eskimos in general (Table 28). Almost inevitably, such mortality leads to a selective elimination of the weaker children. Infant deaths remain particularly frequent at the summer camps. Here, housing is primitive, access to a district nurse is rare, and conditions generally are comparable with those encountered by less acculturated populations.

Episodic starvation is no longer a feature of the Canadian Arctic. However, older Eskimos speak of periods of hunger 'before the coming of the white man'. Again, it seems likely that natural selection would have operated when hunting conditions were poor, the children with physically weaker parents being the first to succumb to starvation.

Disease and physical activity

We have noted the impact of hookworm and malaria-induced anaemia upon physical activity, the will to work, and the aerobic power of populations living in humid tropical regions (page 85). Rest has been a traditional component of therapy for tuberculosis, and such enforced inactivity can modify physical fitness (see below). Deafness and blindness are other common disabilities of primitive groups. G. T. Cumming, Goulding & Baggley (1971) hypothesized that blindness would restrict habitual activity and thus working capacity, while deaf children would attempt to compensate for their disability through an increase of physical activity. Measurements of working capacity in blind, deaf and retarded Canadian children generally supported these hypotheses (Table 29). Deaf boys had a greater standing height, a lesser percentage of body fat, and a greater PWC_{170} per kilogram of body weight than those who were visually handicapped. However, perhaps because of difficulties in urging the deaf to maximum effort, the $\dot{V}_{O_2(max)}$ per kilogram was comparable in the deaf and the blind boys. The deaf girls also were taller and had a greater PWC_{170} per kilogram than girls who were blind; however, they had no advantage with respect to the percentage of body fat or the $\dot{V}_{O_2(max)}$ per kilogram.

Disease and oxygen transport

Tuberculosis and other forms of chronic chest disease have yet to be controlled in many primitive populations. Partly for this reason, the IBP handbook of methodology (Weiner & Lourie, 1969) recommended inclusion of vital capacity measurements in the assessment of physical working capacity.

A low vital capacity in specific members of a given population may reflect advanced 'restrictive' chest disease. However, the interpretation of systematic differences of vital capacity between populations (Table 30) is a more complex question. Sir Joseph Barcroft (1914) first drew attention to the unusual shape of the chest in the indigenous population of the Andes. More recent radiographic studies of Rhodesians (Paul, Fletcher &

Table 29. *Effects of deafness, visual handicap and mental retardation on physical working capacity of Canadian children*

Sample	Age (yr)	Height (cm)	Weight (kg)	Body fat (%)	PWC_{170}/kg	$\dot{V}_{O_2(max)}$/kg
Boys						
Deaf	8–12	144±10	37.3±6.7	16.0±3.5	14.3±2.6	43.8±4.6
	13–17	164±9	57.2±11.8	18.0±4.9	15.0±3.0	45.2±7.0
'Blind'	8–12	139±17	37.7±14.8	21.2±6.3	9.6±2.2	44.6±11.8
	13–17	159±2	52.6±18.2	21.1±8.8	10.4±3.3	44.6±11.8
Retarded	8–12	138±6	35.2±5.3	17.5±2.6	12.6±4.4	—
	13–17	159±14	60.1±15.2	21.5±5.7	12.8±3.5	—
Girls						
Deaf	8–12	149±7	46.3±11.0	25.5±4.9	11.4±2.5	37.0±6.1
	13–17	161±5	52.3±7.6	23.4±3.3	11.7±1.6	37.6±6.1
'Blind'	8–12	136±9	32.0±6.2	22.3±4.4	9.1±3.4	37.4±5.1
	13–17	152±2	48.5±1.3	26.7±2.7	9.4±0.5	33.6±4.2
Retarded	8–12	142±8	41.9±11.6	26.0±6.2	9.3±2.2	—
	13–17	153±7	59.8±17.6	30.0±5.0	7.9±2.8	—

Based on data of G. R. Cumming *et al.* (1971).

Addison, 1960) have shown that, even after allowance for differences of standing height, Europeans have wider and taller chests than Africans, the estimated volumetric discrepancy being 13.2%. Cotes (1974) has noted small vital capacities in Africans, Indians, and Chinese, and he has speculated that genetic differences between populations may be demonstrated more readily in 'static' measurements such as vital capacity than in 'dynamic' data such as aerobic power. However, considerable caution is necessary in attributing differences of body size and form to genes rather than to environment. We have noted elsewhere (page 190) that the secular trend is currently bringing about a rapid reduction of differences in standing height and in trunk/leg-length ratio between Caucasians and a variety of primitive groups. Again, in many high-altitude communities, such factors as vigorous daily activity, an increased respiratory work load and the absence of air pollution combine to yield lung volumes that (under BTPS conditions) are larger than would be predicted from height and age. In some groups, the cigarette habit is only recently acquired, but this advantage may be offset by the frequent inhalation of smoke from primitive heating systems. Lastly, results can differ both from one settlement to another and also within a given settlement, depending on the sampling techniques that are used; a medical team, for example, may encounter a high proportion of diseased subjects, with additional bias towards physically inactive, poorly-nourished subjects who are heavy smokers. Thus,

Table 30. *Lung volumes of selected ethnic groups, expressed as percentage of age and height standards of Andersen et al. (1968)*

Population	FVC (l BTPS % normal)	$FEV_{1.0}$ (l BTPS % normal)	$FEV_{1.0}$/FVC (%)	Author
African women	92	—	80.0	Cotes (1974)
Canada				
Igloolik Eskimos				
Men	110	100	76.0	Rode & Shephard (1973*a*)
Women	118	112	70.7	Rode & Shephard (1973*a*)
Chile				
Aymara Indians[a]				
Aged 20–29	129	125	83.3	Donoso *et al.* (1974*a*)
Aged 40–49	124	123	77.9	Donoso *et al.* (1974*a*)
Chinese				
Men (Singapore)	81	79	—	Da Costa (1971)
Women (Singapore)	95	96	—	Da Costa (1971)
Women	95	—	—	Cotes (1974)
Guyana				
Africans				
Men	78	82	80.9	Miller *et al.* (1970)
Women	85	85	81.1	Miller *et al.* (1970)
Indian				
Men	74	77	80.5	Miller *et al.* (1970)
Women	82	80	78.7	Miller *et al.* (1970)
India				
Indians				
Men	83	82	83.4	Cotes & Malhotra (1965)
Women	88	—	77.1	Cotes (1974)
North Indians	87	78	76.9	Joshi, Madan & Eggleston (1973)
South Indians	77	80	87.6	Singh, Abraham & Antony (1970)
	78	—	—	Bharadway *et al.* (1973)
	82[b]	—	—	Bharadway *et al.* (1973)
Ladakh	93	—	—	Bharadway *et al.* (1973)
Bhutanese				
Men	107	99	74.1	Cotes & Ward (1966)
Women	114	111	80.2	Cotes & Ward (1966)
Jamaica				
Men	78	78	78.0	Miller *et al.* (1972)
Women	82	81	79.2	Miller *et al.* (1972)
Japan				
Ama women	116	—	—	Song *et al.* (1963)
Control women	101	—	—	Song *et al.* (1963)
New Guinea				
Coastal men	82	76	80.1	Cotes *et al.* (1973*a*)
Coastal women	90	85	81.8	Cotes *et al.* (1973*a*)
Coastal migrant men	93	81	76.9	Cotes *et al.* (1973*a*)
Highland men	94	88	81.5	Cotes *et al.* (1973*a*)
Highland women	106	99	82.1	Cotes *et al.* (1973*a*)

Table 30. (*cont.*)

Population	FVC (l BTPS % normal)	$FEV_{1.0}$ (l BTPS % normal)	$FEV_{1.0}$/FVC (%)	Author
Peru				
American Indians at 3830 m	104	—	—	Mazess (1969)
Caucasians at 3830 m	97	—	—	Mazess (1969)
Trained students (mixed Indian/ Caucasian)				
Highland	109	—	—	Mazess (1969)
Lowland	104	—	—	Mazess (1969)
Trinidad				
Negroes	80	81	85.4	Edwards *et al.* (1972)
Indians	76	77	84.5	Edwards *et al.* (1972)
Turkey				
Boys	96	—	—	Akgün & Özgönül (1969)
Girls	103	—	—	Akgün & Özgönül (1969)
United States				
Alaskan Eskimos	110	110	—	Rennie *et al.* (1970)
Irish	100	100	—	Sidor & Peters (1973)
Italian	105	104	—	Sidor & Peters (1973)
Negro	88	87	—	Sidor & Peters (1973)
Other	102	102	—	Sidor & Peters (1973)
Negro	88	91	84.8	Abramowitz *et al.* (1965)
White	97	105	85.3	Abramowitz *et al.* (1965)
Negro	87	87	—	Rossiter & Weill (1974)
White	100	100	—	Rossiter & Weill (1974)
Negro	85	86	79.3	Oscherwitz *et al.* (1971)
Asian	89	90	77.8	Oscherwitz *et al.* (1971)
Yugoslavia				
Male smokers				
Control	—	100	—	Šaric & Palaić (1971)
Miners > 20 yr	—	96	—	Šaric & Palaić (1971)
Male non-smokers				
Control	—	99	—	Šaric & Palaić (1971)
Miners > 20 yr	—	91	—	Šaric & Palaić (1971)
Women				
Control	111	101	80.4	Šaric & Palaić (1971)
Miner's wives	110	90	79.2	Šaric & Palaić (1971)
Zaïre				
Hoto	78	—	—	Ghesquière (1971)
Twa	73	—	—	Ghesquière (1971)

FVC = forced vital capacity; $FEV_{1.0}$ = one-second forced expiratory volume.
Male subjects unless otherwise indicated.
[a] Altitude 3680 m.
[b] After 10 months at an altitude of 3962 m.

IBP physiologists reported high normal vital capacities for both Alaskan and Canadian Eskimos, but as recently as 1968 a medical team led by Dr Beaudry concluded that the lung volumes of Eskimos living in the eastern Arctic were lower than in white subjects.

In view of reports that chronic chest disease was rampant in the North-West Territories, we made a careful study of the impact of respiratory disease on the physical working capacity of the Canadian Eskimo (Rode & Shephard, 1971, 1973*a*). We found that Igloolik Eskimos with a previous history of pulmonary disease were on average 0.7% shorter than healthy villagers of the same age, presumably reflecting a systemic effect of tuberculosis upon growth and development. The diseased group also weighed 8.5% less than healthy subjects. In men, the weight loss was partly muscle and partly fat, reflecting a systemic effect of the disease. However, the younger women had greater skinfold thicknesses than those who were healthy, probably reflecting the inactivity imposed during treatment of their disease. Both sexes showed an association between a history of chest disease and poor muscular strength, the deficit amounting to 2.6% for hand-grip and 5.8% for leg strength. Aerobic power was also reduced by disease, the deficit averaging 12.1% in absolute units (l/min) and 9.0% when related to body weight (ml/kg min).

Twenty-eight of the 224 Igloolik villagers seen by the IBP physiology team had a history of primary tuberculosis and/or hilar calcification with minimal fibrosis. Cardio-respiratory function in this group was close to 100% of the age-related values for normal healthy Eskimos. A further seventeen of the subjects visiting our field laboratory had a history of secondary or advanced tuberculosis, and three had emphysema and/or chronic bronchitis with extensive fibrosis. These last twenty cases, 9% of our sample, accounted for most of the functional loss associated with respiratory disease. Subsequent review of nursing station records showed that the medical team saw almost all Eskimos with chest disease, whereas the physiology team saw only 40% of those affected. Such considerations of sampling may well reconcile previous impressions of a high incidence of respiratory disease in the Arctic (Beaudry, 1968) and the excellent lung volumes reported by IBP physiologists (Rennie *et al.*, 1970; Rode & Shephard, 1973*a*).

Working capacity and acculturation

There is good evidence that the habitual activity of many primitive groups has declined in recent years, leading to associated changes of physical working capacity. Sometimes, changes of working capacity have been associated with migration from rural areas to the city, and in other instances they have reflected the transition to a wage or welfare-based economy within what was initially a traditional primitive community. In this section, we shall look at urban/rural differences of working capacity for populations of less- and more-developed nations, and will examine correlations between physical fitness and indices of acculturation.

Urban/rural differences in less-developed nations

Perhaps the most dramatic example of loss of working capacity with urban migration was provided by Aghemo *et al.* (1971). They compared nomadic Tarahumara Indians, living in caves under very primitive conditions, with other genetically similar groups who had reached a higher level of civilization and were living in relatively comfortable huts. The latter had lost their 'primitive' habit of distance running, this being associated with a drop of aerobic power from 63.0 ml/kg min to 38.9 ml/kg min. Although the average body weight increased from 55.3 to 63.0 kg in the more-civilized group, changes in skinfold readings suggested that 5.2 kg of the 7.7 kg additional weight was fat.

Miller *et al.* (1972) compared Jamaicans living in suburban August Town with the rural population of Lawrence Tavern, some twenty-four kilometres distant. The rural group lived in rugged hill-country. The older men were mostly farmers who cleared, planted and reaped the steep hillsides by hand. Often, they could be seen carrying their produce (yams, bananas, plantains and potatoes) in head baskets with a laden weight of around forty-five kilograms. The younger men of Lawrence Tavern were beginning to look to Kingston for work, but there was much unemployment. The rural women, in addition to their domestic duties, engaged in straw-weaving and sometimes helped with the lighter tasks on the land. In August Town, both occupations and living standards varied from one household to another, but the people generally had a less strenuous life than those in the rural districts. Eight of the Lawrence Tavern men wore the SAMI heart rate integrator (Wolff, 1966) for one-week periods. Estimates of daily energy expenditures varied widely from 2000 to 4700 kcal per day, with a mean daily average of 3250 kcal; however, five of the eight subjects used over 4000 kcal on at least one day of the week. Unfortunately, the published reports did not distinguish aerobic power data for the August Town and Lawrence Tavern populations; however, the rural group were capable of expelling a larger percentage of their forced vital capacity in one second (respective values 82.6 and 88.0%). Ashcroft *et al.* (1966*a*) compared a larger sample of 288 male and 295 female African residents of Lawrence Tavern with the Kingston population. At this time, urban and rural women were of the same height, but urban men were slightly taller than their rural counterparts. The rural men weighed 5.8% more than weight for height standards at the age of twenty, and 7.7%

Table 31. *A comparison of working capacity in rural and urban Bantu*

	Rural sample		Urban sample	
Tribe	Body weight (kg)	Aerobic power (ml/kg min)	Body weight (kg)	Aerobic power (ml/kg min)
Venda	56.7	39.9	64.1	40.5
Pedi	56.2	37.6	60.6	41.9

Based on data of Wyndham (1973)

more than the height standards at age fifty. In contrast, the urban sample weighed only 2.4% above predictions at age twenty, increasing to 12.2% at age fifty. Rural women reached a peak of excess weight (11.3%) at age thirty, and were still of comparable build (excess weight 10.8%) at age fifty. Over the same time, the urban group of women advanced from an excess weight of 3.9% to 8.1–23.7%. The authors thus hypothesized that the early excess weight of rural men and women was due to muscular development, but that the later gain of weight in the urban population represented a gain of fat secondary to a much lower level of physical activity.

Among some primitive groups, factors such as malnutrition have led to poorer standards of working capacity in rural than in urban groups. The main contribution of South Africa to the human adaptability project of the IBP was a study of the changes that occur in Bantu tribesmen on moving from rural peasant life to the urban industrial society of the Johannesburg gold mines (Wyndham, 1973). Comparisons of body weight and predicted aerobic power were made between 241 rural and 240 urban men from the Venda tribe, and between 202 rural and 223 urban men from the Pedi tribe (Table 31).

Both tribes increased body weight and aerobic power in response to the better nutrition and vigorous physical activity required in the mining camps. From the industrial point of view, particular importance was attached to the proportion of the sample capable of sustaining hard work (aerobic power greater than 45 ml/kg min). This increased from 26.9 to 36.4% of the Venda sample, and from 17.8 to 34.0% of the Pedi sample. Wyndham speculated that the poor initial status of the Pedi recruits might reflect the arid nature of their 'homeland' (Sekhukhuniland).

In Nigeria, Davies (personal communication) noted that rural villagers and workers in light industry had a similar aerobic power, but that factory workers in heavy industry had a higher physical capacity.

Studies of African children living in Zaïre (Ghesquière, 1971) and Jamaica (Ashcroft & Lovell, 1964) have confirmed observations made on

Table 32. *Working capacity of selected groups of Zaïrean men*

Population	Height (cm)	Weight (kg)	Aerobic power (ml/kg min STPD)
Hoto	169	58.0	42.7
Twa	160	51.2	47.5
Workers	169	58.0	42.0
Physically active students	172	62.0	44.0
Lumberjacks	167	56.0	48.0
Soccer players	172	66.0	53.0

Based on data of Ghesquière (1971).

Table 33. *A comparison of working capacity for urban and rural children living in Sweden*

Sample	Height (cm)	Weight (kg)	PWC_{170} (kg m/min)	PWC_{170}/kg	Heart volume (ml)	Heart volume/kg (ml/kg)
Boys						
Urban	147.7	38.0	560	14.7	434	11.4
Rural	144.6	37.0	582	15.7	431	11.7
Girls						
Urban	146.6	36.0	434	12.1	383	10.7
Rural	148.6	39.0	471	12.0	427	11.0

Based on data of Adams *et al.* (1961*a*).

European children in showing that rural residents at any given age are shorter and lighter than those living in urban areas. Even within the cities, Zaïrean children of wealthy parents mature faster than those whose parents are poor (Ghesquière, 1971). Among the adults, cross-sectional comparisons suggest that acculturation leads to little drop of aerobic power, providing that physical activity is sustained (Table 32).

Urban/rural differences in more-developed nations

Children

Many of the urban/rural comparisons in more-developed nations have been restricted to children. Adams *et al.* (1961*a*) compared the PWC_{170} of two populations of Swedish children, one living within the city of Stockholm, the other in a farming district a few kilometres to the north. Both groups were aged ten to twelve years. Contrary to expectation, differences of working capacity with the area of residence were quite small; indeed, when plotted as a graph of PWC_{170} versus body surface area, the apparent advantage of the rural boys was shown to be statistically insignificant (Table 33).

In Canada, Green (1967) compared 809 urban and 108 rural students from the province of Alberta. Åstrand predictions of maximal oxygen intake from bicycle ergometer data showed a small but statistically significant advantage to the rural sample:

Sample	Boys (l/min STPD)	Girls (l/min STPD)
Urban	3.20	2.16
Rural	2.90	2.03

However, these differences became insignificant when aerobic power was related to body weight. In a later comparison of French Canadian schoolchildren (Shephard *et al.*, 1975*b*), there was a suggestion that urban children living in a medium-sized city (Trois Rivières, metropolitan population 150000) had a small advantage of working capacity relative to those living in a farming area thirty kilometres out of the city (Table 34). The authors of this experiment hypothesized that increasing mechanization of agriculture now precluded children helping their parents about the farms. City children had many opportunities for organized sport, and often had to walk several kilometres to and from school each day. On the other hand, opportunities for organized sport were few in the rural area, and the need for walking was often eliminated by a special bus that stopped at the child's house both before and after school.

Seliger (1970) made an exhaustive comparison of urban, rural, and athletic children as part of the Czechoslovakian contribution to the IBP programme. The urban sample was drawn from the suburbs and centres of large towns, while the rural sample came from both the country proper (two thirds) and from small towns (one third). Participants were selected randomly from school class lists, but certain exceptions were made, including (i) obese and asthenic children (those deviating by more than ± 2.0 SD from weight for height standards), and (ii) children judged by the school physician to be in poor health. Anthropometric observations showed that fifteen-year-old urban boys were taller and heavier than those from country areas. Since these differences were less apparent at twelve years of age, earlier maturation of the urban sample was probably responsible (Table 35). Among the twelve-year-old children, percentages of body fat and hand-grip strengths were also larger in the urban sample. A battery of performance tests was carried out, including pull-ups (boys) or

Table 34. *Indices of fitness for urban and rural samples of French Canadian children*

Sample	Predicted aerobic power (ml/kg min)	Total muscle force (ΣF, kg)	Total muscle force (ΣF, kg/kg)	Average skinfold (mm)	Heart volume (ml)	Heart volume (ml/kg)
Boys						
Urban	60.6	146	4.6	5.6	347	11.2
Rural	54.9	150	4.6	6.1	368	11.2
Girls						
Urban	49.4	143	4.0	8.4	366	10.5
Rural	42.4	139	4.5	7.6	336	10.7

Based on data of Shephard *et al.* (1974*a*, 1975*a*).

Table 35. *Indices of fitness for urban and rural samples of Czechoslovak children*

Sample	Height (cm)	Weight (kg)	Body fat (%)	Right hand-grip (kg)	Calf circum-ference (cm)	Chest circum-ference (inspiration, cm)
12 yr boys						
Urban	148.9	38.9	19.8	26.0	29.8	74.1
Rural	148.1	38.7	16.7	22.9	29.8	75.3
12 yr girls						
Urban	150.8	39.7	22.9	22.4	30.1	75.2
Rural	149.8	39.7	20.9	20.1	30.3	76.0
15 yr boys						
Urban	169.7	58.0	13.7	35.5	34.4	87.1
Rural	167.4	54.7	13.8	35.8	33.4	85.2
15 yr girls						
Urban	163.6	54.5	19.4	24.6	34.2	83.1
Rural	162.3	53.9	19.5	26.9	33.6	82.1

Based on data of Seliger (1970).

flexed arm hang (girls), sit-ups, standing long jump, 50 m sprint, endurance run–walk, 40 m shuttle run, and 2 kg ball throw. No urban/rural differences of performance were seen at twelve years, but at fifteen years the urban boys did better than the rural sample on the 50 m sprint, shuttle run and ball throw, while the urban girls also excelled on the long jump. In terms of PWC_{170} per kilogram of body weight, urban and rural boys achieved almost equally. However, urban girls had a poorer capacity than

Table 36. *Indices of working capacity for urban and rural samples of Czechoslovak children*

Sample	PWC_{170} (kg m/min)	PWC_{170}/kg	f_h (load 4) (beats/min)	Aerobic power (load 4) (ml/kg min STPD)
12 yr boys				
Urban	558	14.4	193	47.0
Rural	548	14.4	196	43.0
12 yr girls				
Urban	384	9.6	197	38.3
Rural	401	10.2	200	36.7
15 yr boys				
Urban	914	15.6	196	45.8
Rural	855	15.6	195	44.4
15 yr girls				
Urban	551	10.2	200	35.8
Rural	600	11.4	196	34.9

Based on data of Seliger (1970).

those from rural areas. The maximum aerobic power was not measured as such, but to judge from the heart rates, oxygen consumptions recorded in the fourth load of a progressive bicycle ergometer test were close to maximum values (Table 36). On this last criterion, both boys and girls from urban areas seemed slightly superior to their rural counterparts. However, in view of the discrepancy between the PWC_{170} and aerobic power data, it is possible that an emotional tachycardia developed in the rural group as maximum effort was approached. As in the Quebec example (Shephard *et al.*, 1975*b*) Seliger (1970) cautioned: 'the country population. . . is less used to similar examinations. Nor [do] repeated visits in the laboratory remove this uneasiness. . . '. In Seliger's sample, the journey to school was longer for the rural sample (average 3 km) than for the urban group (0.6 km at age twelve; 0.9–1.4 km at age fifteen). On the other hand, about a quarter of the rural children were conveyed by vehicle; this reduced their actual walking distance to an average of 1.2 km. More children practised sport in urban areas (20–52% of the samples) than in rural areas (7–39% of samples). The duration of many habitual activities was similar in twelve- and sixteen-year-old boys, with respective 24-hour caloric intakes of 2500 and 3000 kcal. However, the older boys spent less time asleep and more time on sedentary activities. Respective average pulse rates for the 24-hour period were 90 and 80 beats/min (Seliger, Bartuněk & Trefný, 1974).

In Japan, (Asahina, personal communication), a preliminary analysis of IBP studies suggested that children and young people living in rural

Table 37. *A comparison of fitness for Alpine and Milanese Italian subjects*

Sample	Aerobic power (ml/kg min STPD)	Body fat (%)	Anaerobic power (ml/kg min STPD)
20 yr men			
Milanese	51	20.0	170
Alpine	41	14.0	175
20 yr women			
Milanese	41	26.0	145
Alpine	34	26.5	145
50 yr men			
Milanese	37	24.0	115
Alpine	35	18.2	125
50 yr women			
Milanese	29	28.0	80
Alpine	31	31.3	80

Based on data of Steplock *et al.* (1971).

districts had a higher $\dot{V}_{O_2(max)}$ than those living in urban areas; however, the extent of the urban/rural gradient varied from sample to sample (see page 209).

Adults

Adult urban/rural comparisons for the more-developed nations have been relatively few. Sigurjonsson (1969) drew attention to a large gradient of ischaemic heart disease mortality between urban and rural areas of Iceland. Taking data for males between the ages of thirty-five and sixty-four, the rates per 100000 were 175.4 for Reykjavik, 151.6 for other towns, and 78.2 for rural areas. Sigurjonsson commented on the higher caloric intake and the lower body weight found in rural areas, but gave no details.

Ashcroft *et al.* (1967) compared the heights and weights of adults living in a crowded Welsh mining valley (Rhondda Fach) with a rural sample drawn from the adjacent but more prosperous Vale of Glamorgan. Men from the Rhondda were shorter and lighter than those living in the Vale. Women from the Rhondda were also shorter at all ages, but in later life became heavier than those living in the Vale.

Several other authors have failed to show advantages of fitness for those living in rural areas. Steplock, Veicsteinas & Mariani (1971) compared data from a small and isolated sub-Alpine community of Italians with results previously collected in Milan (Table 37). Fugelli (1976) examined the health of Norwegian fishermen based on the islands of Vaerøy and Røst. More than 20% of the group had disabilities that reduced their earning capacities

50% or more. However, even among the healthy men, the average maximum oxygen intake at the mean age of 40.6 years was only 40.0 ml/kg min, a value essentially comparable with that found for office workers in Oslo (Andersen, 1964). The poor working capacity was seen despite fairly heavy energy expenditures (4.5–5.5 kcal/min for different types of fishing operation). Papers from Australia (J. G. Allen, 1966) and Poland (Kozlowski *et al.*, 1969), equally, have shown that relatively heavy occupations do little to improve endurance fitness; leisure activities have more influence on physical working capacity than the steady rhythm of occupational effort.

Wolanski & Pyzuk (1972, 1973) made a detailed ecological study of adults living in several rural regions of Poland. Their groups included the population of Czorsztyn, in the low mountains of the Piening range (where tourist rafting, agriculture, and sheep breeding provide quite a high level of daily activity), and people from the forests of Kurpie (where the soil is tilled on a seasonal basis, but food is in rather short supply). The maximum oxygen intake for men of the Piening region (as predicted from a rather brief step test) ranged from 3 to 4 l/min, with no obvious age trend over the working span. In contrast, the young Kurpie men had a predicted maximum oxygen intake of only 2.9 l/min and this decreased to 2.5 l/min at the age of sixty. The Polish women had a predicted aerobic power of 2.3–2.6 l/min, irrespective of age or region of residence.

Acculturation and working capacity

The effects of acculturation can be examined longitudinally, noting the changes in the physiological characteristics of a settlement that occur with the adoption of a less primitive life-style. Alternatively, cross-sectional correlations may be made between specific indices of individual acculturation and responses to physiological testing at any give point in the process of transition.

The longitudinal approach has been exploited most fully in connection with the secular trend of standing heights (page 190). Other comparisons can be made between successive surveys, although doubts often arise about the comparability of techniques. Thus Elsner (1963) reported average skinfold readings of only 6 mm when he examined Alaskan Eskimos, but the IBP investigators at Point Barrow and Wainwright found almost double these thicknesses (Jamison, 1970).

In Igloolik, the IBP team found that the majority of the 147 adult male Eskimos still liked to regard themselves as hunters. However, an inventory of dog-teams and snow-mobiles soon established that the number of villagers who were seriously persisting with the traditional life-style was much smaller. Some twenty-five men could be regarded as relatively

Table 38. *The influence of seasons and life-style upon skinfold thicknesses and predicted aerobic power*

Life-style		Average skinfold thickness (mm)		Predicted aerobic power[a]			
				l/min STPD		ml/kg min	
		S	W	S	W	S	W
Traditional hunter	20	5.0±0.8	6.4±1.2	3.72±0.52	3.75±0.74	56.6±5.1	56.2±10.1
Transitional	22	6.1±2.7	6.7±1.8	3.64±0.76	3.63±0.71	54.9±10.9	54.9±9.2
Acculturated	18	6.7±2.8	7.9±4.4	3.43±0.64	3.38±0.49	51.2±9.6	50.1±7.8

Mean±SD of data collected for Igloolik Eskimos during summer (S, May to August 1970) and winter (W, January to March 1971). From Rode & Shephard (1973*c*) by permission of the publisher, American College of Sports Medicine.

[a] Values adjusted downwards by 8% to allow for the maximum possible overestimate of maximum oxygen intake by the prediction method.

acculturated, having accepted wage-earning employment with government or commercial agencies in the settlement. A further ninety-three were in a transitional state, living in government housing, relying mainly on government welfare payments, but eking out their allowances by a little soap-stone carving and occasional hunting trips with borrowed equipment. Only twenty-nine of the villagers could be regarded as serious hunters (Godin & Shephard, 1973). Both in summer and in winter months (Table 38), we demonstrated a clear gradient of skinfold thicknesses and of predicted aerobic power from traditional hunters, through the transitional group to acculturated salaried workers (Rode & Shephard, 1973*c*). The association between the traditional life-style and cardio-respiratory fitness became even more pronounced when hunters were classified according to the frequency of long (2–3 week) hunting trips. Those who made such excursions only three or four times per year had an aerobic power of 54.5±4.8 ml/kg min STPD, but those who reported ten or more major trips per year had an average aerobic power of 62.0±7.5 ml/kg min.

Dr Ross MacArthur of the Canadian IBP team developed a simple empirical index of acculturation, based on such items as time in school, English vocabulary, type of housing and domestic equipment, geographic mobility and wage income. His index was significantly correlated with skinfold thickness (for 132 men, $r = 0.29$, $P \sim 0.001$; for 93 women, $r = 0.34$, $P \sim 0.001$). In the men (but not in the women), the index was also negatively correlated with aerobic power (l/min STPD). Height, weight, and grip strength were unrelated to the index, but there was a suggestion of a negative correlation with leg strength in the men ($r = -0.15$, $P = 0.068$).

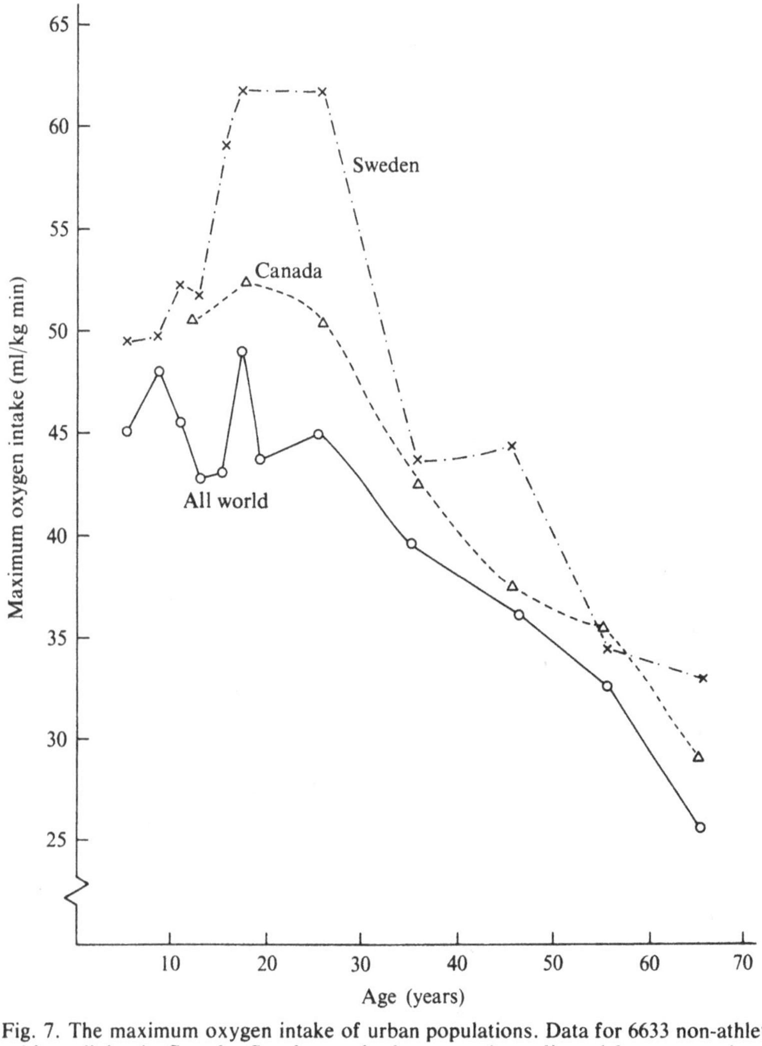

Fig. 7. The maximum oxygen intake of urban populations. Data for 6633 non-athletic male subjects living in Canada, Sweden, and other countries, adjusted for systematic errors of methodology. (For sources, see Shephard, 1966*d*. Illustration from Shephard, 1969*a*, by permission of University of Toronto Press.)

Such observations do not prove that a traditional life-style conserves a good working capacity and prevents obesity. It is equally possible that the fitter members of a community are better able to preserve their traditional culture in the difficult environment of the 1970s. Nevertheless, if working capacity data show a marked gradient with acculturation, the average results for a given community will be in error unless an appropriate proportion of the often absent hunters are tested.

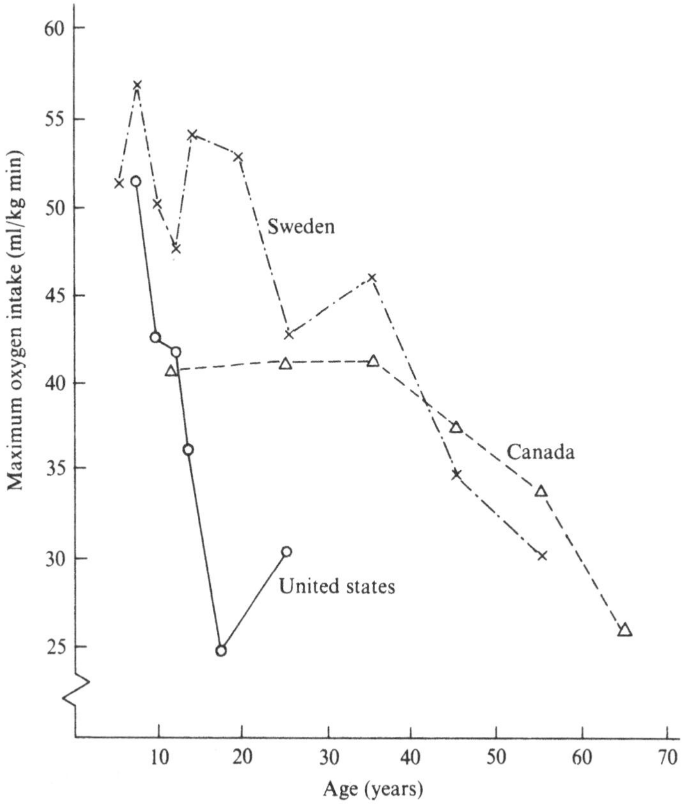

Fig. 8. The maximum oxygen intake of urban populations. Data for female subjects, including author's figures for 156 Canadians, and reported figures for 286 Scandinavians and 211 US citizens. The US sample includes 58 subjects with very low maximum oxygen intakes as reported by Rodahl *et al.* (1961). (For sources, see Shephard, 1966*d*. Illustration from Shephard, 1969*a*, by permission of University of Toronto Press.)

The ultimate in acculturation

The modern metropolitan community is presumably approaching the end-point of the acculturation process, with physical activity having little immediate survival value.

A synthesis of data on the aerobic power of men and women from the more-developed nations (Figs. 7 and 8) has been published (Shephard, 1966*d*). The major uncertainty in analysing this data was the nature of the populations tested. Often 'volunteers' included a large number of individuals with above average fitness, together with the neurotic, those fearing a coronary attack, and those with an above average interest in health (page 11). Details of the problem are well illustrated by two Canadian attempts to examine random groups of the adult population. In

Table 39. *Fitness data for a sample of adult Canadians that included representative numbers of volunteers from each of the main occupational categories*

Age (yr)	PWC_{170} (kg m/min)	PWC_{170}/kg	Right hand-grip (kg)	Skinfold thickness (mm)
Men				
18–19	1005	14.5	48	7.5
20–24	953	13.9	53	9.2
35–44	1028	13.0	53	13.5
Women				
18–19	541	10.1	28	10.8
20–24	518	9.4	31	11.0
35–44	574	9.2	33	14.5

Based on data of Metivier & Orban (1971).

the first experiment (Metivier & Orban, 1971), appropriate numbers of volunteers were selected from each of several hundred occupational categories indicated by Statistics Canada, the government census-taking organization. The national sample resynthesized in this way showed a surprisingly high level of fitness, with little tendency to ageing other than an increase of subcutaneous fat (Table 39).

An alternative approach adopted in the Prairie community of Saskatoon (population 120 000) was to select names at random from the telephone book and to invite adult respondents to the laboratory for a physical fitness test (Bailey *et al.*, 1974*a*). A panel of specially trained staff made a total of 2648 telephone contacts. Five per cent of the prospective volunteers were eliminated on the basis of a simple health questionnaire administered during the telephone conversation. A surprisingly large percentage of acceptable recruits (899 persons, 34%) agreed to visit the laboratory, the remainder being unwilling or unable to attend because of overriding commitments. All but 5% of the 899 volunteers kept their first appointment, but a further 72 were eliminated because a poor medical history was elicited by the nurse-receptionist. Thus 778 remained to report to the exercise/electrocardiograph station. Another 8% were rejected because of an abnormal resting electrocardiogram or a poor medical response to exercise (appearance of excessive fatigue, exhaustion, or e.c.g. abnormalities). A progressive step test was performed on day 1, and 713 of the original contacts returned for a progressive bicycle ergometer test on day 2. At the second visit, a few further medical disqualifications were necessary because of abnormal resting or exercise e.c.g.s. The residual population of 686 was complemented by 544 volunteers from other sources

Table 40. *Fitness of a sample of Canadians obtained in Saskatoon through randomly placed telephone contacts*

	Men		Women	
Age (yr)	Step test (ml/kg min)	Bicycle ergometer (ml/kg min)	Step test (ml/kg min)	Bicycle ergometer (ml/kg min)
15–19	47.7	43.1	39.3	34.0
20–29	40.8	36.4	38.3	30.6
30–39	38.1	32.2	35.7	27.8
40–49	34.9	26.9	32.7	24.3
50–59	32.5	25.7	32.4	21.9
60–69	30.5	22.8	29.8	19.0

Based on data of Bailey *et al.* (1974*a*)

Table 41. *Fitness data for a random sample of the Czechoslovak population*

Age (yr)	Maximum heart rate (/min)	Aerobic power (ml/kg min STPD)	Body fat (%)	Right hand-grip (kg)	Vital capacity (ml BTPS)	PWC_{170} (kg m/min)	PWC_{170}/kg
Men							
18	193	45.7	12.1	49.8	5140	1145	16.8
25	191	39.0	15.5	49.7	5493	1189	15.6
35	188	36.8	17.0	49.5	5049	1129	14.4
45	182	35.3	16.9	49.9	4870	1180	15.0
Women							
18	196	35.3	20.0	31.2	3649	614	10.8
25	192	32.0	20.1	30.0	3856	672	10.8
35	189	29.4	23.5	30.3	3609	667	10.2
45	182	27.7	25.8	30.0	3370	740	10.8

From data of Seliger (personal communication).

covering a broad socio-economic spectrum. While interest in physical activity was surprisingly high (page 11), the group tested was probably as close to a random sample of adults as can be obtained in a free society. The step test predictions of aerobic power closely mirrored values previously reported for a non-random sample of the Canadian population (Figs. 7 and 8). However, bicycle ergometer results were substantially lower, particularly in women and older men (Table 40). This presumably reflects the impact of quadriceps muscle weakness upon the bicycle ergometer predictions.

Seliger, Macek, Skranč, Pirič, Handzo, Horak, Rous & Jirka (personal

Table 42. *Twelve-minute running distances for US and Austrian soldiers*

Sample	Distance run in 12 min < 2.41 km (%)	2.41–2.82 km (%)	> 2.82 km (%)
Austrian soldiers (13763)	20.5	42.5	37.0
US soldiers (6712)	58.1	39.2	2.7

Based on data of Cooper & Zechner (1971).

communication) studied what was considered a random sample of the adult Czech population. Exercise tests were by bicycle ergometer to the point of subjective exhaustion. Results (Table 41) were much superior to the bicycle ergometer data for Saskatoon; indeed, PWC_{170} values exceeded those reported by Metivier & Orban (1971). The grip strength was not particularly remarkable by North American standards, but the percentage of body fat was substantially lower than in most Canadian and US surveys. Seliger stated that anthropometric data was consistent with a random sample; he did not give details of test disqualifications, but assuming the random nature of his group, one must accept that North American adults are less fit than some Europeans. This had been suspected previously from comparisons of smaller non-random samples of physiological data (Shephard, 1966*d*) and from differences in the twelve-minute running distances (Table 42) attained by US and Austrian soldiers (Cooper & Zechner, 1971).

There have been many other studies of working capacity in more-developed countries since 1966; however, in general the samples have been poorly defined, and the results have not added materially to existing knowledge.

Patterns of habitual activity and community survival

Passing mention of daily caloric expenditures has already been made for certain populations. At this point, practical experience of the various measuring techniques will be reviewed, activity patterns will be discussed further, and the results translated into an equation of energy balance for community survival.

Techniques of measurement

The IBP teams and other recent investigators have used a variety of techniques to study habitual activity patterns, ranging from simple observation and arbitrary grading of activity (Edwards *et al.*, 1972) through study of dietary records (Durnin, 1966; Viteri *et al.*, 1971; Edholm *et al.*, 1973) to tape-recordings of the electrocardiogram (Andersen, 1967*b*; Shephard, 1967*b*; Ekblöm & Gjessing, 1968; Sidney & Shephard, 1977), telemetry (Lammert, 1972), use of SAMI heart rate integration (Heywood & Latham, 1971; Miller *et al.*, 1972; Edholm *et al.*, 1973), and formal time and motion studies with direct measurements of oxygen consumption by Kofranyi–Michaelis respirometer (Viteri *et al.*, 1971; Edholm *et al.*, 1973; Godin & Shephard, 1973).

In many primitive communities, the technical problems of data collection have been substantial, and both the number of subjects and the range of activities studied have left much to be desired. No method has proved ideal. However, it is perhaps worth stressing that few authors have reported failures of the Kofranyi–Michaelis equipment. Although superficially simpler, the SAMI heart rate integrator has proved fallible in practice, with difficulties arising from instrument breakdowns, damage to connectors, detachment or drying of electrodes, and false counts due to poor skin contacts (Edholm *et al.*, 1973). With so many potential sources of error, the interpretation of results inevitably became somewhat subjective. Edholm *et al.* (1973) wrote: 'results obtained were accepted as satisfactory and suitable for inclusion in the analysis if at least one night heart rate within the range 40 to 90 beats/min or a daytime heart rate between 55 and 110 beats/min was obtained'. Acceptable data for day and night heart rates were obtained in 72% of the subjects in summer, and 78% in winter. SAMI results were significantly correlated with time and motion estimates of activity. Nevertheless, 'there was a very large scatter' and 'the average energy expenditure determined from heart rate measurement was substantially higher than the estimates based on timed activities'. The authors were inclined to attribute this discrepancy to the effects of posture, emotion, and body temperature on heart rate, but it seems likely that technical problems in the use of the SAMI equipment also played some part.

In the Arctic, some difficulties were encountered with icing of the valves in the Kofranyi–Michaelis respirometer (Godin & Shephard, 1973). However, such problems were circumvented by pre-heating of the equipment in the nocturnal shelter, plus enclosure of both the meter and ancillary tubing by several layers of insulation.

Activity patterns in primitive and less-developed communities

The dominant impression in primitive and less-developed communities is of the extreme range of energy expenditures, both from subject to subject, and also within the same subject on different days. Thus Miller *et al.* (1972)

noted that the energy expenditures of Jamaicans ranged from 2000 to 4700 kcal, Edholm *et al.* (1973) cited daily activities ranging from 2190 to 4560 kcal for Yemenite Jews and 2010 to 6390 kcal for Kurdish Jews, while Edwards *et al.* (1972) reported activities of 1 to 68 arbitrary units per day in the Trinidad population.

One of the earliest studies of a primitive group was made by Rennie (personal communication) some fifteen years ago. He examined a small group of traditional male Eskimos living at Anatulik Pass in Alaska; their daily energy consumption averaged 3600 kcal. The coastal Eskimos of Wainwright and Port Barrow, featured in the US IBP studies, still have some interest in fishing, but power boats now are generally used. Point Barrow has become a substantial town, and many Eskimos living there have accepted both wage-earning employment with various government agencies and a more civilized life-style. We have already noted the probable effect of this acculturation upon skinfold thicknesses (page 72).

The main activity of the Eskimos living on the west coast of Greenland is kayak fishing and hunting based upon a motor launch (Lammert, 1972). When on board the launch, pulse rates approximate normal resting levels, but when tracking seals in kayaks, pulse rates of 120–140/min may be recorded. Assuming a metabolic rather than a postural or an emotional basis for this tachycardia, the energy cost of kayak paddling would be 5–6 kcal/min. Lammert (1972) estimated that the average Greenlandic Eskimo spent 25% of the total voyage in his kayak, work at 50% of aerobic power coming in bursts of 2–3 min duration.

The Norwegian Lapps studied by Andersen *et al.* (1962) were nomadic in life-style, supplementing their reindeer herding by ptarmigan hunting, trapping, and fishing. Much of their work was done on skis or on foot, with occasional use of a reindeer sleigh. The Scandinavian IBP project studied Lapps of the Nellim region. Detailed activity studies were undertaken by Dr Ove Wilson, but the results are not yet available. The people of the region still engage in some fishing and potato growing, but many have become employed wage-earners, particularly in forest industries.

Ikai *et al.* (1971) described the Ainu people of Hokkaido, Japan, as a relatively sedentary group. Formal caloric studies were not undertaken, but the main activities were described as wood-carving and merchandizing.

Perhaps the most detailed study of circumpolar activity patterns was conducted by Godin & Shephard (1973). Their subjects were Canadian Eskimos of the Igloolik settlement. Twenty-nine of the 147 adult men in the community were serious hunters. This group made arduous field trips, varying from three days to several weeks in length, and were absent from the village for as much as 40% of the year. The pattern of hunting naturally changed with season (Fig. 1; page 20). Caloric costs for the various types of hunt were estimated from a combination of time and

Table 43. *Average daily energy cost of several forms of hunting (data for Igloolik Eskimos, Table 14, standardized to body weight of 65 kg)*

Hunt type	Energy expenditure (kcal/day)
Caribou	
Winter	3840
Summer	3900
Fishing	
Ice	4040
Summer	4440
Seal	
Boat	3440
Floe edge	2530
Ice hole	3490
Walrus	3670
Average, 8 hunts	3670

motion studies, films of principal activities, and Kofranyi–Michaelis respirometer measurements of oxygen consumption (Table 43). The averaged 24-hour caloric counts reflect neither the occasional demand for short bursts of very intensive work, nor the periodic need for several days of activity with little respite. Nevertheless, the figures are sufficient to show that hunting can demand quite high rates of energy expenditure. Between hunts, energy consumption is much lower. Time is spent in resting, conversation, and leisurely repair of equipment. Parallel observations on Eskimos employed within the community show a more steady rhythm of moderate work (Fig. 4); however, even those remaining within the settlement have some tasks that are physically quite demanding, particularly the annual clean-up of garbage (6.5–7.8 kcal/min), the unloading of the supply ship and aircraft (5.0 kcal/min), and in winter the distribution of ice to householders (5.0–5.3 kcal/min) and the delivery of goods by snowmobile (5.9 kcal/min). The average daily energy expenditure of the employed male was set at 3350 kcal. Energy expenditures of the women tended to be higher than in an industrialized society, since fewer electrical appliances were available and lack of furniture caused many tasks to be performed from a squatting or a kneeling position. Traditional tasks of the Eskimo wife included the chewing of skins (caloric cost estimated by heart rate at 5.3 kcal/min) and walking with a baby in the amauti (3.8 kcal/min). Despite their small size, the daily energy usage was 2400 kcal for married and 2300 kcal for single women.

The first detailed activity study of a tropical community was carried out by Ekblöm & Gjessing (1968). Using a twelve-hour tape-recording of

Table 44. *Daily energy expenditures of farming males living on a plateau (1500 m) in Guatemala*

Subjects	Height (cm)	Body weight (kg)	Skinfold thickness (mm)	Energy expenditure (kcal/day)
Agricultural Peons	160.5	59.6	7.8	3873
Horsemen	159.0	54.5	6.0	4271
Carpenter/mason	156.0	59.2	13.0	3530
Dairymen, herdsmen	163.0	61.5	7.2	3145
Foreman	167.0	68.3	10.3	2778

Based on data of Viteri *et al.* (1971).

electrocardiograph signals, they were able to show that Pascuans were relatively inactive. Two outdoor labourers had average working heart rates of 102 and 86 beats/min respectively, a fisherman averaged 90 beats/min, two unemployed men 87 and 97 beats/min, and three housewives 87, 91 and 101 beats/min.

Viteri *et al.* (1971) made a detailed energy balance study of men engaged in primitive agriculture on a highland plateau of Guatemala. Tasks involved use of a machete (for mowing and cutting wood), hoeing and other forms of hand cultivation. The group were considered better nourished than workers on surrounding farms, since they were able to purchase milk at a discount, and were given a daily drink of Incaparina which provided 5 g of first-class protein. With the exception of two dairymen, a herdsman and a foreman, the daily caloric expenditures were high (Table 44).

Mayer (1972) reported caloric intakes of 2600–3500 kcal in different classes of Indian worker. The lowest intakes were for those engaged in light work (clerks and mechanics) and the highest values were for ashmen, coalmen, blacksmiths, cutters, and carriers. Mayer also estimated high intakes (3300–3400 kcal) for sedentary stall-holders and supervisors. This surprising observation in part reflects the larger size of the last group; it is unlikely to be due to gross over-feeding (as Mayer suggests), for if there were a daily 800–1000 kcal surfeit of food, body weight would increase at a rate of 130 g/day!

Edholm *et al.* (1973) described the habits of the Kurdish and Yemenite Jews in some detail. Work followed a more regular six-day rhythm than in less-developed communities. On the eve of the Sabbath, the men returned from the fields an hour or two earlier than usual, but even on Saturdays there was a fair burden of essential work (feeding and milking animals), together with participation in sports such as football and basketball. Nevertheless, energy expenditures ranged quite widely about the

Table 45. *Daily energy expenditures of Kurds and Yemenites in Israel*

	Kurds		Yemenites	
	Summer (kcal/day)	Winter (kcal/day)	Summer (kcal/day)	Winter (kcal/day)
Men	3050	3110	3050	3000
Women	2250	2390	2280	2400

Based on data of Edholm *et al.* (1973).

Table 46. *Periods of the day spent lying and sitting – data for Kurdish and Yemenite men in Israel*

	Kurdish men		Yemenite men	
	Summer (min)	Winter (min)	Summer (min)	Winter (min)
Lying	505	510	445	480
Sitting	335	435	405	455

Based on data of Edholm *et al.* (1973).

mean seasonal values (Table 45). Edholm *et al.* comment: 'the farmers were also independent, could and did work according to their own plans . . . sometimes very long hours were worked, and on other days there might be only 1 to 2 hours activity'.

Some of the heavy occupational activities encountered by this group included walking in mud (8.0 kcal/min), moving irrigation pipes (7.7 kcal/min), potato picking (6.6 kcal/min), forking and scything grass (5.9–6.0 kcal/min) and manure spreading (6.3 kcal/min). The duties of the women were generally lighter, but they were occasionally required to assist with scything (11.3 kcal/min) and forking grass (4.5 kcal/min). Despite the generally active nature of farm life, substantial parts of a typical day were spent either lying or sitting (Table 46).

One other tropical group for whom accurate figures are available is the Jamaican population of Lawrence Tavern (Miller *et al.*, 1972). Here, an average energy expenditure of 3250 kcal was encountered in eight men. Detailed activity studies of Africans living under primitive conditions are restricted to a survey of !Kung bushmen (Lee, 1969); this showed a consumption of some 2250 kcal by men and 1750 kcal by women. It is also known that after migration to mining camps, the Bantu is given (and presumably consumes) a diet with an average caloric value of 4000 kcal/

day (Wyndham, 1973). Some duties in the mines involve hard physical work, demanding up to 50% of aerobic power throughout the working shift.

A caloric deficiency may cause poor performance not only by restricting immediate activity, but also by limiting the learning experiences of the developing child. Heywood & Latham (1971) found some evidence of the latter effect in a group of Colombian children living near Bogota. Apfelbaum, Bostsarron & Lacatis (1971) have shown a similar parallel between starvation and restriction of activity in the laboratory. On the other hand, Gandra & Bradfield (1971) found little restriction of activity from the quite severe anaemia (average haemoglobin level 8.8 g/dl) encountered in a Brazilian fishing village near Sao Paulo.

Activity patterns in more-developed countries

Activity studies on the populations of more-developed countries have often used nutritional methodology. Thus Durnin & Passmore (1967) reported the daily energy expenditures of Scottish men to range from 2520 kcal in office workers through 3000 kcal in construction workers and 3280 kcal in steel workers to 3550 kcal in farmers, 3660 kcal in miners, and 3670 kcal in forestry workers. Often, even among 'heavy' industrial workers, the most significant activity occurs during leisure hours (J. G. Allen, 1966; Montoye, 1971; Buskirk *et al.*, 1971) – the maintenance of physical capacity now depends not on daily work but upon voluntary activity.

Activity questionnaires were distributed to a random sample of 3875 US men and women by the President's Council on Physical Fitness and Sports (1973). This survey indicated that some 45% of all adult Americans, whether men or women, took no deliberate exercise. Of those electing voluntary activity, the reasons cited (in order of decreasing importance) were maintenance of good health, elevation of mood ('it makes me feel good'), loss of weight, pleasure, and medical advice. Reasons cited for inactivity were lack of time, adequacy of occupational effort, medical advice and age. The two commonest forms of voluntary activity were walking and bicycle riding, both being slightly more prevalent in metropolitan than in non-metropolitan areas. In general, voluntary physical activity was most prevalent in the young, better educated and more affluent segment of the US population. Some 22% of the sample claimed to be walking almost every day, often for twenty minutes or longer. A substantial proportion of the sample had engaged in competitive sports at school, but the only competitive sport attracting more than 10% of the adult population was ten-pin bowling.

An analogous line of questioning was pursued in a study of Canadians

Table 47. *Hours spent by Canadians in sport and other forms of vigorous activity*

	Participation (% of sample)			
	1–3 hr/week		> 3 hr/week	
Age	Sport (%)	Other activity (%)	Sport (%)	Other activity (%)
14	23.3	26.3	30.5	24.2
20–24	19.5	17.7	14.1	11.0
35–44	12.1	10.9	8.2	5.6
65–69	2.9	3.7	3.1	3.4

Based on data collected by Statistics Canada.

living in Saskatoon (Bailey *et al.*, 1974*a*). There may have been some bias of this sample towards an interest in physical activity because of the associated step and bicycle ergometer tests. Nevertheless, results were not greatly dissimilar to the US survey. Some 14% of the men and 9% of the women claimed to be taking vigorous exercise (walking a mile or more, jogging, tennis, squash, or swimming four or more lengths) four or five times a week; a further 26% of the men and 27% of the women were taking equivalent exercise two or three times a week.

Details of a third survey, by Statistics Canada, are not yet available, but preliminary information shows that there was a progressive drop in both sports participation and other vigorous activities with ageing (Table 47).

By way of comparison, a survey of 2000 fourteen-year-old Swedish children (Engström, 1972) showed that boys devoted an *average* of 5 hours per week to active leisure pursuits such as soccer, ice-hockey, and swimming, while girls devoted 3.5 hours to swimming, cycling, and walking. Only 5% of the boys and 10% of the girls in Sweden had no active leisure pursuits.

Andersen (1967*b*) reported daily heart rates for arbitrarily selected groups of Norwegians. Among sedentary office workers, pulse rates of 80–90 beats/min were usual; peaks greater than 100 beats/min were rare, typically coinciding with pauses for a smoke or a visit to the washroom. Moderately heavy industrial workers had average pulse rates of about 110 beats/min, with peaks to 130 beats/min, while lumberjacks showed average rates of 130 beats/min, peaks greater than 160 beats/min, and 5–10% of the day with pulse rates greater than 140–150 beats/min.

Glagov *et al.* (1970) used watch-type counters to monitor the heart rates

Table 48. *Periods of day at specified heart rate – tape-recorded data on eight relatively sedentary men living in Toronto*

	Heart rate (/min)			
	< 100	100–119	120–139	140–170
Mean period (min)	1329.4	91.0	16.4[a]	2.0[b]
Range (min)	1155–1404	4–269	0–54	0–10

Based on data of Shephard (1967*b*).
[a] Two of the eight had no readings greater than 120/min.
[b] Six of the eight had no reading greater than 140/min.

of one hundred US men, about a half of whom were faculty and students. Night-time pulse rates averaged 48 to 80 beats/min, and daytime values ranged from 71 to 106 beats/min. The success of the watch-type counters relative to the SAMI apparatus perhaps deserves emphasis. When repeat counts were made on subsequent days, seventy-eight of the hundred subjects showed a difference of less than 10 beats/min while awake, and eighty-three of the hundred less than 10 beats/min difference while asleep. The instruments were said to have performed satisfactorily in 90% of trials.

Combined diary and e.c.g. tape-recordings on a group of relatively sedentary Canadian city-dwellers (Shephard, 1967*b*) showed similar results to the Norwegian and US studies. Pulse rates greater than 120 beats/min were a rarity, and were generally caused by emotional rather than physical stress (Table 48).

Habitual activity and community survival

The principles involved in drawing up a balance sheet of energy income and expenditures for a given geographic region are discussed by R. B. Lee (1969) and Kemp (1971). From a short-term viewpoint, it matters little to a more-developed country if energy expenditures far exceed the energy-generating potential of the region. However, in a primitive community, survival depends on the delicate balance of income and expenditure. It was the intention of the IBP programme that detailed surveys of energy resources should be matched against the demands of existence for a number of habitats, taking account not only of human nutrition and energy output, but also other sources of energy including animals, vegetables and stores of fossil fuel. Such studies have been completed, at least partially, for !Kung Bushmen (R. B. Lee, 1969), and Arctic (Foote & Greer-Wootten, 1966) and Canadian (Godin & Shephard, 1973) Eskimos.

R. B. Lee (1969) stressed the elementary nature of the economy of some 248 !Kung Bushmen living in the Dobe area of Botswana. Production and consumption of food was immediate in terms of space and time, all available food being consumed by the local group within a period of forty-eight hours. There was a minimum surplus accumulation, a minimum production of capital goods and an absence of agriculture and domestic animals (except for the occasional dog). All able-bodied persons had to seek food each week of the year, and individual groups of tribesmen were largely self-sustaining; camps ranged in size from nine to twenty-nine, and the percentage of dependents (children and the aged) from 9.1 to 66.7% (average 38.7%). Living space was defined largely by the availability of water. For the seven dry months of the year, water was available only at eight permanent water holes, and the Bushmen did not travel more than a day's walk from these locations (an area with a radius of about ten kilometres). However, in the summer months, seasonal pools of water could be found elsewhere, and the total available territory expanded to about 2600 km^2. The normal strategy was to collect and eat the desirable foods that were the least distance from standing water, and the cost of gathering such foods as the mongono nut rose progressively as the area immediately around the water hole was harvested; there was a particularly steep rise in the cost of foraging as the round-trip journey increased from nineteen to twenty-six kilometres, since now overnight journeys were required. Lee suggested that one useful measure of the current energy balance was the ratio man-days of work to man-days of consumption. The !Kung Bushman could, on average, support himself and his dependents by 2.2 days of foraging per week, leaving 4.8 days free for other activities. The average caloric need of the community was 2000 kcal per day, and Lee estimated that this demand was well covered by 2.2 days of activity per week (Table 49).

The Alaskan and Canadian Eskimos form much less independent segments of society. Foote & Greer-Wootten (1966) analysed the energy balance of Eskimos living on the upper reaches of the Kobuk river in western Alaska. Human energy expenditures were incurred for fishing trips, and for subsequent processing of the fish, but the consumption of fossil fuels by the fishing boats was often as much as two orders greater. Early in the summer fishing season (July), the main energy gain was from whitefish, but later in July and during August salmon gave a greater caloric yield. In August, additional calories reached the settlement through the stock-piling of spruce firewood. During the year of study, the total energy expenditure incurred by fishing parties from 9 July to 26 August was approximately 12.8×10^5 kcal, and the energy value of the fish and firewood that they obtained was 16.3×10^5 kcal. Considering simply the human component of the energy expenditures, a very favourable balance

Table 49. *Daily energy resources of the !Kung Bushman*

Food	Contribution to diet (%)	Per capita consumption: Weight (g)	Protein (g)	Energy yield (kcal)
Meat	37	230	34.5	690
Mongono nuts	33	210	56.7	1260
Other vegetables	30	190	1.9	190
Total	100	630	93.1	2140

Based on data of R. B. Lee (1969).

might be inferred. However, taking account also of the consumption of fossil fuels, balance was only attained under optimal fishing conditions – for example, from 9 July to 31 July the fishermen spent 7.4×10^5 kcal for a yield of only 3.8×10^5 kcal. Further, stability of the system now depends on the ability of the Eskimos to convert a portion of their salmon catch into money to replenish their stocks of gasoline.

Igloolik is far north of the tree-line, and the Eskimos of this region are currently heavily dependent on fossil fuels, particularly the heating oil delivered by the supply ship early each September. Diesel-powered generators also provide energy for cooking, lighting, washing machines and other power appliances used within the settlement. On hunting trips, heating and lighting of igloos and much of the cooking still depends on the combustion of animal oils, but substantial quantities of fossil fuels are used to power snow-mobiles, outboard motors, and a larger Peterhead fishing vessel. In the depths of winter, dog-teams still provide the most reliable source of motive power for hunting. The energy balance of the Igloolik community is particularly interesting from the nutritional standpoint. The human energy expenditures of the village are summarized in Table 50 (Godin & Shephard, 1973). For the purpose of this calculation, it has been assumed that hunters are on active hunting trips for an average of three days per week, and are engaged in sedentary pursuits within the settlement for the remaining four days. Likewise, because of the somewhat irregular pattern of wage-earning work, labourers are assumed active for four rather than five days per week. No measurements of energy expenditures have been made on children or the elderly; values for this segment of the community have been assumed to equal those for white subjects of the same weight, a minimal statement of actual expenditures. The potential energy input from the principal forms of hunting is summarized in Table 51. The total, 1.44×10^8 kcal, must be reduced by 20% to

Table 50. *Annual energy expenditure of the Igloolik community*

Group	Group size	Energy expenditure (Gcal/yr)
Males		
Hunters	29	17.1[a]
		14.8[b]
Labourers	25	17.1[c]
		10.1[b]
Sedentary	86	78.5
Elderly	7	5.9
Children		
0–5 yr	51	11.8
5–10 yr	47	23.8
10–15 yr	34	28.5
Females		
Married	81	71.0
Single	37	31.1
Elderly	3	2.2
Children		
0–5 yr	62	14.4
5–10 yr	45	22.8
10–15 yr	27	18.1
Total		367.2

From Godin & Shephard (1973), by permission of William Heinemann Medical Books Ltd.

[a] Value for 161 days spent hunting.

[b] Value for 204 days spent resting.

[c] Value for 161 days spent working.

Table 51. *Potential income of food energy for Igloolik region, 1969, from hunting and fishing*

Game	Number	Weight per animal (kg)	Total weight (Mg)	Percentage edible	Metabolic value (kcal/kg)	Total metabolic value (Gcal)
Caribou	800	70	56.0	40	960	22
Polar bear	16	700	11.2	40	1050	5
Fish	5000	5	25.0	50	1760	22
Bearded seal	55	250	13.8	27	1280	5
Ringed seal	3648	50	182.4	25	1280	58
Walrus	150	700	105.0	26	1170	32
Total						144

From Godin & Shephard (1973), by permission of William Heinemann Medical Books Ltd.

allow for wastage of edible carcasses through skinning and the pillaging of caches by other Arctic animals. Some meat is undoubtedly eaten by the 500–700 village dogs, although these feed mainly on blubber, blood and various internal organs not regarded as suitable for human consumption. Our estimates thus suggest that some 31% of the community's dietary needs come from the hunt. The remaining caloric requirements are met by purchases from the village store, particularly tinned milk, sugar, 'cookies', flour and soft drinks. The present balance of traditional and cash foods still provides an adequate quantity of first-class protein, although increased use of cooked meals has reduced the precarious vitamin intake from such sources as raw fish. Game yields of the region cannot readily be increased, and if there is a further expansion of the settlement population, it seems likely that the nutritional status of the villagers will deteriorate.

6 Working capacity and constitution

Some physiologists have held that the unusual performance of the top athlete is based largely on inherited traits – as P. O. Åstrand (1967) put it: 'I am convinced that anyone interested in winning Olympic gold medals must select his parents very carefully.' Garn *et al.* (1960) have marshalled evidence for a packaged transfer of the genes that contribute to success in competitive sports; in particular, they note a mesomorph is much more likely to be the offspring of large-framed parents than of either large- and small-framed parents or two small-framed parents. This chapter will bring together data on the inheritance of working capacity, including formal genetic studies and more empirical comparisons of performance where constitutionally different subjects have lived in the same environment or subjects of similar initial constitution have lived for many generations in contrasting environments.

Genetic studies

(a) Variance of data

Working capacity data from isolated communities might be expected to show a small coefficient of variation. Inbreeding in itself is unlikely to be sufficient to reduce the variance of a complex trait such as working capacity, and P. T. Baker (1974) has argued that if selection for an adverse environment is still proceeding, only a proportion of the population will show favourable adaptive traits, and the variance of data for the community may be increased. However, the life-style of an isolated population tends to be more homogenous than that of a large industrial city with respect to such factors as physical activity and environmental stress. Thus, Cruz-Coke, Donoso & Barrera (1973) found significant correlations between the systemic blood pressures of marriage partners in Aymara Indian communities. On the other hand, representative groups from circumpolar, tropical, and high-altitude regions show coefficients of variation in aerobic power that are remarkably uniform and similar to those encountered in the city community (Table 52).

(b) Genetic markers

In medical conditions such as hypertension and ischaemic heart disease, a preponderance of specific blood groupings can be demonstrated (Cruz-Coke *et al.*, 1973). This implies that the genes responsible for the inherited

Table 52. *Coefficient of variation for measurements of maximum oxygen intake (ml/kg min STPD) expressed as a percentage for a given age and sex. Date for Toronto whites compared with data for selected populations from circumpolar, high altitude and tropical regions*

	Coefficient of variation (%)				
Population	20–29 yr	30–39 yr	40–49 yr	50–59 yr	Author
Canada					
Toronto whites					
Men	16	17	24	19	Shephard (1969*a*)
Women	10	12	30	28	Shephard (1969*a*)
Igloolik Eskimos					
Men	18	14	10	31	Shephard (1974*c*)
Women	13	15	14	5	Shephard (1974*c*)
Chile					
Aymara Indian men	14	12	12	—	Donoso *et al.* (1974*a*)
Ethiopia					
Highland men	23	15	11	30	Andersen (1971)
Lowland men	13	16	26	16	Andersen (1971)
Japan (Hokkaido)					
Ainu men	13	18	16	—	Ikai *et al.* (1971)
Zaïre					
Hoto men	10	—	—	—	Ghesquière (1971)
Twa men	16	—	—	—	Ghesquière (1971)

component of atherosclerosis and those responsible for blood groupings are transferred in unison. It is less certain that a packaged genetic transfer applies to the complex of physiological variables determining physical working capacity. Nevertheless, a wide range of blood groupings and other genetic markers were recorded by the IBP team. We thus thought it worthwhile to explore whether the blood groupings were related to physiological variables. Three-way classifications of the Igloolik population were made with respect to (i) measurements of standing height and (ii) aerobic power. We then compared the frequency of the various blood groupings in short, average and tall Eskimos, and in those with low, average, and high levels of physical working capacity (Shephard & Rode, 1973). Application of χ^2 statistics to our tables revealed no significant associations between blood groupings and either height or aerobic power. With the exception of the blood pressure studies mentioned above (Cruz-Coke *et al.*, 1973), there seem to have been no other attempts to relate genetic markers and physiological data.

(c) Twin studies

A standard method of examining the inheritance of biological variables is to compare variances in monozygous and dizygous twins. In the simplest of such analyses, the estimated contribution of heredity (H_{Est}) to the total variance is calculated by the equation:

$$H_{Est} = \frac{(\sigma^2_{DZ} - \sigma^2_{DZm}) - (\sigma^2_{MZ} - \sigma^2_{MZm})}{(\sigma^2_{DZ} - \sigma^2_{DZm})} \times 100$$

where σ^2_{DZ} is the dizygous variance, σ^2_{MZ} is the monozygous variance, and σ^2_{DZm}, σ^2_{MZm} are the methodological components of the variances. The model assumes that the variance found in monozygous twins reflects a simple summation of environmental and methodological contributions, while that of the dizygous twins is augmented by genetic effects. Other fundamental assumptions are that monozygous and dizygous twins encounter the same range of environmental influences as the general population, with no interaction of effects between genes and environment.

Most of the primitive groups examined by IBP investigators had too small a total population to permit twin studies of this type. However, Klissouras (1971) has used an analysis of variance to estimate the inherited component of physical working capacity in Canadian children, his subjects being fifteen monozygous and ten dizygous male twins living in the Montreal area. Criteria of monozygosity included morphological traits (hair colour and texture, ear-lobe shape, eye colour, and iris pattern), dermatoglyphic analysis, and serological examination. The methodological variance was assessed from replicated examinations of seven subjects; one weakness of the study was that some of the replicated subjects 'did not participate in the study but were well accustomed to the experimental procedures'. His results apparently showed inheritance as the major determinant of variance in maximum oxygen intake, maximum blood lactate and maximum heart rate (Table 53). However, the validity of this conclusion rests on the adequacy of the allowance made for methodological error; the precision suggested for both aerobic power and maximum heart rate seems excessive for young children who have only made one previous visit to the laboratory. Further, the basic model tacitly assumes that both groups of twins have been exposed to a full possible range of environments. In fact, individual pairs of twins are often exposed to rather similar environments, particularly if they are identical.

In the Klissouras study, the possible environmental contribution was further reduced by examining a group of children with similar 'upbringing, living standards, and leisure time activity'. It is thus hardly surprising that the overall variance of data was low even for the dizygous twins, and that a large proportion of the variability was attributed to genetic factors.

Table 53. *Analysis of variance for monozygous and dizygous twins*

Variable	σ^2_M	σ^2_{MZ}	σ^2_{DZ}	H_{Est} (%)
$\dot{V}_{O_2(max)}$/kg	1.30	15.37	214.05	93.4
Maximum blood lactate (mg/dl)	31.70	156.57	701.88	81.4
Maximum heart rate (/min)	2.63	40.75	273.81	85.9

Based on data of Klissouras (1971).
σ^2_M = sum of squares due to methodology; σ^2_{MZ} = variance in monozygote; σ^2_{DZ} = variance in dizygote; H_{Est} = estimated contribution of heredity to total variance.

The same method of analysis was applied to fifteen pairs of male twins and fourteen pairs of female twins, children of Jyvaskyla, Finland, aged ten to fourteen years (Komi, Klissouras & Karvinen, 1973). Many attributes of physical performance were measured, but for the majority of tests variance was similar in monozygous and dizygous twins. Among the boys, three notable exceptions were the Margaria test of anaerobic power (H_{Est} = 99.2%), the patellar reflex time (H_{Est} = 97.5%) and reaction time (H_{Est} = 85.7%); however, among the girls equal variances were noted even for these measurements. The authors thus speculated that performance was more susceptible to environmental factors in girls than in boys.

A useful corrective to such extravagant genetic claims was published by the same laboratory (Klissouras, 1972). This report concerned a single set of monozygotic twins. Both had experienced a very active childhood, engaging in athletic training from the age of eight to fifteen. At the time of examination (age twenty-one), one twin had become a rather sedentary salesman, driving 1600 km per week and participating only occasionally in unorganized sports such as swimming and golfing. His brother, on the other hand, continued year-round strenuous training for football and ice-hockey. At this stage, the active twin had a 47.3% advantage of maximum oxygen intake (respective values 2.37 and 3.49 l/min STPD), and a 61.0% advantage of maximum oxygen debt (respective values 4.13 and 6.65 l STPD); however, the two boys still had very comparable resting (54, 54) and maximum (184, 182) heart rates.

A further experiment used four pairs of thirteen-year-old monozygous twins and four pairs of sixteen-year-old monozygotes (Klissouras & Weber, 1973). Ten weeks of deliberate training was administered to one twin of each pair while the other was left to 'his usual kinetic routine'. In the thirteen-year-olds, there was no difference in the end-results for 'trained' and 'untrained' twins. However, this was largely because both made big gains of aerobic power over the ten-week period (12.5, 16.0%

respectively). Klissouras & Weber (1973) called this a 'growth spurt', but it is hard to envisage such dramatic growth over ten weeks; it seems more likely that the unregulated twins shared in the additional activity of their trained partners on an unofficial basis. Among the sixteen-year-old twins, the 'trained' group made a 20.8% gain of aerobic power, while the 'untrained' group gained only 3.2%.

In reviewing the work of Klissouras, Cotes (1974) draws attention to the fallacy of assuming that environmental factors are comparable for mono and dizygous twins. His own studies of vital capacity in British children show a small variance in monozygous boys and girls and in dizygous girls, but a larger variance in dizygous boys. He suggests this may reflect more diversity of environment for the dizygous boys than for the other classes of twin.

(d) Comparisons of siblings

Variance can also be partitioned on the basis of parent–parent, parent–offspring, and fullsib–sib data correlations (Cruz-Coke *et al.*, 1973). According to current principles of quantitative genetics (Li, 1961), a full partition of variance should distinguish four components – environmental and the additive, dominant and interactive parts of genetic variance. The additive part reflects differences between homozygotes, summed over genes. The dominant component is due to deviation of the heterozygote from the value intermediate between two contrasting homozygotes, while the epistatic fraction is a measure of interaction between loci. A possible mode of analysis (Cavalli-Sforza & Bodmer, 1971) can be summarized as follows:

Component of variance	Equation
Additive (V_A)	$V_A/V_p = 2r_{p-o}$
Dominant (V_D)	$V_D/V_P = 4(r_{s-s} - r_{p-o})$
Genetic (V_G)	$V_G/V_P = V_A/V_P + V_D/V_P$
Environmental (V_E)	$V_E/V_P = 1 - (V_A/V_P + V_D/V_P)$
Phenotypic (V_P)	$V_P = 1$

In these equations, r_{p-o} is the parent–offspring correlation, and r_{s-s} is the sib–sib correlation. To date, this approach has been applied to the inheritance of blood pressure (Table 54), but does not seem to have been exploited in the analysis of other physiological variables. The Aymara Indians (Cruz-Coke *et al.*, 1973) are of particular interest in the context of IBP. The highlanders are primitive agriculturalists and semi-nomadic shepherds. The lowlanders are of similar genetic stock, 78% having

Table 54. *Components of phenotypic variation of diastolic blood pressure (subjects of both sexes, as collected by Cruz-Coke et al., 1973)*

	Components of phenotypic variance			
		Heredity		
Population	Environment (%)	Additive (%)	Dominant (%)	Author
Chile				
Arica	44.2	44.2	11.6	Cruz-Coke *et al.* (1973)
Aymara highlanders	72.2	16.4	10.4	Cruz-Coke *et al.* (1973)
Jamaica	36.6	37.8	25.6	Miall *et al.* (1972)
United States				
Boston	0	34.0	72.0	Zinner, Levy & Kass (1970)
Tecumseh	66.0	14.0	20.0	B. C. Johnson, Epstein & Kjelsberg (1965)
Wales	34.6	36.6	28.8	Miall & Oldham (1963)

migrated from the highlands. However, they use more advanced agricultural methods and are subject to urban influences from the nearby port of Arica. In both regions, environment accounts for a substantial part of the variance in diastolic blood pressure. However, the highlanders show no significant increase of pressures with age, whereas in the lowlanders pressures are age-related and in some instances progress to frank hypertension. Cruz-Coke *et al.* (1973) thus speculate that the hypertensive phenotype is related directly to the acculturation process.

All authors to date have demonstrated that the genetic component of the variance in diastolic pressures has a 'dominant' portion (Table 54). This is a strong argument in favour of the view that blood pressure is transmitted through only a few major genes; a consistent dominant effect through many loci would be most unlikely. Among the Aymara lowlanders, the genes associated with MN and rhesus blood groupings are specifically implicated, higher levels of blood pressure being recorded for R_1R_1 and NN homozygotes.

Wolanski (1969) has drawn attention to one potential problem with this type of analysis. He found significantly different correlations for maternal–sibling and paternal–sibling relationships, and suggested that such discrepancies arose from non-genetic inheritance – the influence of the respective parents as mediated by differences in foetal, neonatal and childhood environments.

(e) The superb athlete

Another possible approach to the partitioning of variance between genes and environment is to consider how far the characteristics of the superb athlete can be attributed to selection of a member of the general population with unusual genetic potential. The highest figure yet reported for aerobic power is 92 ml/kg min, seen in a Finnish cross-country skier (Åstrand, personal communication). In contrast, the aerobic power of an average young male worker in Toronto is 48±8 ml/kg min (Shephard, 1969*a*). Assuming the variance within the general population to be normally distributed, the best that could be anticipated from an exhaustive search of Canadians would be an individual with an aerobic power four standard deviations greater than the population mean, or 80 ml/kg min. The difference between 80 and 92 ml/kg min must represent a response to the athlete's regimen of long and arduous training. At a minimum, 12/44 (27%) of the difference between himself and his compatriots is environmental rather than genetic. However, much of the variance of aerobic power in the general population is also environmental. A very simple activity questionnaire was able to describe 36% of the variance in exercise responses among a population of British soldiers (Shephard & Callaway, 1966). Equally, the recent survey of Saskatoon citizens (Bailey *at al.*, 1974*a*) showed a striking difference of aerobic power between subjects who said they exercised very frequently (four or five times a week) and those who exercised infrequently (once a week or less), the average discrepancy amounting to 31.4% in the men, and 19.5% in the women. We may thus reassess factors producing the superb athlete as follows:

Athletic training 27%
Population factors 73%

Habitual activity $\frac{36 \times 73}{100} = 26\%$

Genetic factors unrelated to habitual activity $\frac{64 \times 73}{100} = 47\%$

Ardent geneticists could attempt to reassert their position by arguing that differences in habitual activity and willingness to undergo athletic training are determined by genetic factors; however, the training responses required by the present analysis (15% gain of aerobic power in athletes, 31.4% gain in sedentary men, 19.5% gain in sedentary women) are of the order anticipated from compulsory training programmes, and as such have been demonstrated not only in the general population but also in pairs of monozygous twins (Klissouras, 1972; Klissouras & Weber, 1973).

Environmental studies

(a) Similar environment, differing constitution

Early reports from IBP investigators were quite optimistic that genetically determined differences of fitness would be revealed by comparisons of distinctive ethnic groups living in the same environment. Thus Andersen (1969) drew attention to data subsequently published by Ghesquière (1971) on the Bantu Negroes (Ntomba) and the pygmoid people (Twa) living side by side in the same jungle community. Health conditions, nutrition, and way of living were said to be 'closely the same', although the pygmoids were hunters, while the Bantu Negroes made a living from primitive agriculture in addition to fishing and hunting. When aerobic powers were compared on a per kilogram basis, the pygmoid (47.4 ml/kg min) had a 15% advantage over the Bantu (42.6% ml/kg min), and Andersen wrote 'this difference is believed to demonstrate a genetically determined difference in fitness for work'. However, even if the basic premise of equal environment is accepted, there remain doubts regarding an appropriate method of data standardization, since the two groups differed widely in standing height. If aerobic power was expressed per cm of stature, differences between the two populations became quite small (14.7 ml/cm min for the Bantu and 15.1 ml/cm min for the Twa).

Mining supervisors long cherished the belief that Bantus from the Basuto tribe had a better tolerance of hard physical labour than other Bantu tribesmen; accordingly, Basutos were selected for shaft-sinking tasks that involved the rapid shovelling of broken rocks. However, a formal study of 338 Bantu mining recruits from Xhosa, Basuto, Pondo, Hlubi, Shangaan, Baca, M'Pedi, Bechuana, Swazi and Zulu origin disclosed no significant inter-tribe differences of either body weight or maximum oxygen intake (Wyndham *et al.*, 1966*c*). It was thus concluded that if indeed the Basuto men were better employees, the difference resided in their motivation rather than in their physiological characteristics. A later comparison of Pedi and Venda tribesmen (Wyndham, 1973) showed some differences of aerobic power while the two populations were living in their traditional life-styles on the reserves; however, maximum oxygen intakes became closely comparable after they had migrated to the mining camps (Table 31).

Recent immigrations to Israel have provided convenient material for genetic studies. Populations of very different origin now live under similar environmental and ecological conditions; particularly on the kibbutzim, levels of physical activity, nutrition, and socio-economic variables are unusually well controlled. C. T. M. Davies *et al.* (1972) compared Kurdish and Yemenite Jews who had settled in the Negev desert; there were small differences of predicted maximum oxygen intake between the two

Table 55. *Differences of directly measured oxygen intake between four segments of the Israeli population*

Origin of population	Height (cm)	Weight (kg)	Body fat (%)	Maximum heart rate (/min)	Maximum oxygen intake l/min STPD	Maximum oxygen intake ml/kg min STPD	Maximum oxygen intake ml/kg min FFW
Europeans	173	70.7	14.6	188	2.80	40.0	46.4
Iraqis	169	62.6	10.6	187	2.77	43.8	48.6
North Africans	168	59.4	9.9	190	2.49	41.3	46.4
Yemenites	165	52.5	7.7	186	2.60	49.2	53.6

Based on data of Glick & Schvartz (1974).

populations (respective values 44.4 and 46.9 ml/kg min for the men, and 29.0 and 35.4 ml/kg min for the women); however, sample size was less than optimum (page 9) and these differences were said to be insignificant. The authors' conclusion appears to be correct when oxygen intake is reported relative to gross body weight, but as Glick & Shvartz (1974) have pointed out the difference becomes significant if the divisor is lean body mass. Glick & Schvartz compared data on four groups of Israelis whose parents had migrated respectively from Europe (Germany, Austria, Poland, Russia, Rumania, Hungary, and Czechoslovakia), North Africa (Morocco, Tunisia and Algeria), Iraq (the Kurds also originated from the mountains of Iraq) and the Yemen. Both in terms of predicted aerobic power (179 subjects) and direct measurement of $\dot{V}_{O_2(max)}$ (35 subjects), the Yemenites were significantly superior to the other three groups (Table 55). The authors ruled out differences of activity and diet between subjects, and concluded 'ethnic group differences in PWC in this study were, therefore, largely a result of genetic differences between these groups'. However, interpretation of the data is complicated by substantial differences of body build; 'more work is needed to determine relationships between ethnic origin, anthropometric characteristics, and PWC'.

Substantial differences of aerobic power can also be demonstrated between the Ainu, the several genetically distinct Eskimo settlements, and white residents of the circumpolar regions (Fig. 5; Table 21; Shephard, 1974*c*); however, in these communities levels of activity and standards of nutrition vary so much from one population to another that it seems impossible to distinguish the respective roles of heredity and environment.

(b) Differing environment, similar initial constitution

Another early postulate of IBP investigators was that nature might select strong and hardy humans for life in the physically most stressful climates, while weaker people would survive in warm and pleasant regions. Exposure over successive generations might produce true ethnic group differences, based on genetic selection, while exposure within a given generation would manifest itself as an environmental variance within a genetically homogenous population (Andersen, 1967*c*).

We have noted above (page 94) that one of the first requirements of such a hypothesis – a substantial evolutionary pressure from disease, food shortages and a hazardous physical environment – has existed and to some extent persists in the circumpolar regions. Further, although the Eskimos of Alaska and the Canadian north coast had a common origin, serological studies now reveal substantial inter-group differences in genetic markers (Dossetor *et al.*, 1971; Simpson & McAlpine, 1976); at least in Greenlandic Eskimos a number of disorders with probable relations to known polymorphic systems (diabetes, duodenal ulcer, and multiple sclerosis) are currently very rare (Harvald, 1976). In terms of such measures of working capacity as aerobic power, there is a suggestion that populations from cold and/or mountainous regions perform better than those from pleasant, warm climates (compare Tables 15, 17 and 21). However, the difference is by no means remarkable, and could easily be explained on the basis of greater current physical activity rather than a more permanent change in the genetic constitution of the circumpolar and mountain peoples.

(c) The paradox of weak natural selection

How may we explain the paradox that formal genetic studies suggest inheritance of a large part of individual differences in working capacity, while there is apparently little resultant natural selection for specific habitats?

The first weakness of the original postulate lies in a failure to quantitate evolutionary pressures in the various geographic regions. Because a particular tropical island seems like paradise to a scientist who is making a brief visit, this does not prove that the evolutionary forces of disease, malnutrition and inter-tribal warfare have operated less vigorously than in a bleak Arctic settlement. Secondly, the inheritance of favourable characteristics, readily demonstrated for an athlete, tends to be event-specific. A person well-qualified as a wrestler is at a substantial disadvantage in distance running, and vice versa. Unfortunately, the challenges presented by a primitive environment are characterized by diversity rather than specificity, and there is no guarantee that a genetic variant giving an advantage in one type of situation may not have a negative adaptive value with respect to some other problem of environment. Further, as in many athletic contests, there is no guarantee that brawn will triumph over skill and motivation. In our studies of the Eskimo hunter (Godin & Shephard, 1973), we were much impressed with the disparity between energy expended and the rewards accruing to the individual. One Eskimo set forty-eight traps, and by an 11 kilometre journey was able to capture twenty-one foxes in the space of twenty-four hours. Another set twenty-four traps on each of two occasions, journeying 24 and 136 kilometres, yet managed to capture nothing! Clearly, the first man was a much more skillful hunter, and was using his resources of physical working capacity more effectively.

Even in the most primitive societies, it is debatable how often it has been necessary to use working capacity to the full. According to our estimates, traditional Canadian Eskimos now engage in vigorous activity on only three days of an average week; however, this pattern is dictated more by the desired standard of living than by specific survival pressures. Among the more primitive !Kung Bushmen, an average of 2.2 days of activity per week is sufficient to ensure survival (R. B. Lee, 1969), and a man who is poorly endowed from the viewpoint of either physique or intelligence can still survive by working a somewhat longer week.

Wolanski (1970) has questioned the usefulness of trying to distinguish environmental and genetic influences, pointing to the importance of non-genetic inheritance through what he terms the 'maternal regulator' – foetal nutrition, lactation, and the subsequent environment that a mother chooses for her developing offspring. Certainly, the problem is more complex than seems to have been realized at the outset of the IBP investigations, and is unlikely to be resolved through any simplistic comparisons of contrasting peoples or contrasting habitats.

7. The physical working capacity of the athlete

One interesting feature of man's physical working capacity is that his potential for performance can be improved through appropriately chosen programmes of voluntary activity ('physical training'). Gains realized in this way are largely independent of genetic and environmental considerations. A central objective of the IBP study of working capacity was to examine the full possible range of physical fitness in various societies. It thus became relevant to collect data on athletes, the extreme in the processes of selection and conditioning, contrasting such observations with results for the general metropolitan population – 'Homo sedentarius'.

The original IBP plans called for the testing of athletes in conjunction with the 1968 Olympic games. A fair number of measurements were made in Mexico City, but interpretation of the results was complicated by differences in adaptation to the test environment (both an altitude of 2240 m and a population of unfamiliar gastro-enteric micro-organisms). Many athletes and their coaches had irrational fears of altitude, training schedules were interrupted, and some participants were unwilling to give maximum endurance effort in close proximity to a major international contest.

Problems also arose from the specificity of athletic fitness. Unfortunately, many physiologists failed to distinguish either the level of competition or the class of event for which a competitor was prepared – for example, 'swimmers' included an amorphous amalgam of men participating in 100 m and 5000 m events, 'soccer players' were a mixture of wing forwards and goalkeepers, and 'yachtsmen' handled everything from a fourteen-foot dinghy to a seventy-foot schooner. Equally, too little thought was given to the test mode. A cyclist performs best on a bicycle ergometer, while a runner is likely to achieve his highest maximum oxygen intake during uphill treadmill running. The present chapter will consider data sport by sport, and where possible will distinguish the class of event, the level of competition and the method of testing. The relative importance of selection and training to the shaping of the successful athlete has been discussed in the previous chapter.

Table 56. *Physiological characteristics of cyclists, classified with respect to event*

Event	Height (cm)	Weight (kg)	Skinfold thickness (mm)	Lean muscle mass (kg)	PWC_{170} (kg m/min)	PWC_{170}/kg	$\dot{V}_{O_2(190)}$ (l/min STPD)	$\dot{V}_{O_2(190)}$/kg (ml/kg min)
Sprint	173.2	74.0	5.7	66.7	1502	20.4	4.03	54.7
Chase	175.6	73.1	5.5	66.7	1600	21.9	3.88	53.1
Motor-paced	173.6	73.6	5.9	66.7	1721	23.4	4.71	64.2
Track-race	174.4	73.5	5.6	66.7	—	—	—	—
Road-race	175.1	70.5	4.8	65.2	1687	23.9	4.35	61.7

Based on data of Vank (1973) and Placheta *et al.* (1973).

Table 57. *Physiological characteristics of cyclists classified with respect to ethnic grouping*

Ethnic grouping	Height (cm)	Weight (kg)	Skinfold thickness (mm)	Lean muscle mass (kg)	PWC_{170} (kg m/min)	PWC_{170}/kg	$\dot{V}_{O_2(190)}$ (l/min)	$\dot{V}_{O_2(190)}$/kg (ml/kg min)
Africa	172.9	66.7	4.31	62.7	1396	21.8	3.79	59.2
Canada	171.8	66.6	4.21	62.8	1538	23.0	3.93	59.0
Europe	177.0	74.8	5.35	68.5	1716	23.0	4.39	58.9
Japan	170.2	66.3	5.22	60.8	1468	22.1	3.81	57.7
Mexico	168.3	64.0	5.37	58.5	1622	25.4	4.34	67.9

Based on data of Vank (1973) and Placheta *et al.* (1973).
Of the seventy competitors, forty-three were involved in road-racing (distance events).

Cyclists

Brno data

The amateur cyclists' world championship in Brno (1969) provided opportunity to apply comparable bicycle ergometry and other measurement techniques to cyclists drawn from many parts of the world (Rouš, 1973). Five different categories of competitor were distinguished (sprint, chase, motor-paced, track-race, and road-race); of these, only the last is a long-distance event. Participants were relatively light and extremely thin, this being particularly true of the road-racers (Table 56). Both in absolute terms and relative to body weight, the aerobic power of the sprint and chase cyclists was unremarkable. The motor-paced and road-race competitors had fairly high levels of aerobic power, but were inferior to those commonly described in international-class endurance cyclists. This may reflect partly the level of competition, and partly the reporting of data at a pulse rate of 190 beats/min rather than at a plateau of oxygen consumption. European competitors were taller and heavier than those from other parts of the world, but this seems due to a high proportion of sprinters in the European sample (Table 57). The relative aerobic power of the Mexicans is large; however, it has yet to be determined whether this indicates good training, selection, or a specific adaptation to the altitudes of Mexico City.

Physical characteristics

Saltin & Åstrand (1967) reported that the third competitor in the 1964 Olympics (a Swede) was tall (189 cm) with a relatively light body weight (75 kg). Ishiko (1967) examined fifteen potential Japanese competitors in the Tokyo Olympic games; an average weight of only 63.4 kg was recorded for a height of 167.9 cm. Archer *et al.* (1965) found somewhat higher weights in twenty-four British potential international competitors (mainly road-racers); in the men, the average weight was 160 pounds (72.7 kg) for a height of 70½ inches (179.1 cm), while in the women the weight was 131 pounds (59.4 kg) for a height of 64½ inches (163.8 cm). Gedda, Milani-Comparetti & Brenci (1968) made many anthropometric measurements on contestants in the Rome Olympics of 1960. They commented specifically on the large bicristal diameter and the well-developed vital capacity of the cyclists. Monte, Severini & Angella (personal communication to author) found that the large vital capacity (5.79 litres compared with an age and height predicted value of 4.60 litres) could be developed at least equally well while mounted on a bicycle with dropped handlebars – indeed, trained cyclists could develop a larger maximum voluntary ventilation mounted than unmounted. Di Prampero, Limas & Sassi (1970) examined six Cubans and one Guatemalan cyclist at the Mexico City Olympics; they estimated body fat as 12%. Berg (1972) used a tritium dilution technique on a group of adolescent boys who belonged to a Swedish bicycle racing club. The preferred distance was not specified, but daily caloric intake was 4000 kcal. Body fat was 8% of body weight, while height, weight and other aspects of body composition were within limits proposed for average Swedish children.

Exercise response

Brooke & Davies (1973) compared the observed oxygen uptake of cyclists during a race over macadmized roads with that predicted from air and rolling resistance (Whitt, 1971). Approximately 75% of aerobic power was used for 200 min, at the high mechanical efficiency of 24.6%. Hollman *et*

Table 58. *Working capacity of national and international cyclists*

Competition level	Max. oxygen intake (ml/kg min)	Author
German national	80	Hollmann (1972)
Olympic games	50[a]	Di Prampero *et al.* (1970)
Olympic games	80	Saltin & Åstrand (1967)
Pan American games	70.9±5.8	G. R. Cumming (1970)
Swedish national	74	Saltin & Åstrand (1967)

[a] Step test prediction.

al. (1971) examined a team of fourteen juvenile racing cyclists in West Germany, and found that unlike normal subjects they developed a higher pulse rate and respiratory minute volume during treadmill running than during an equivalent submaximal exercise on the bicycle ergometer. During cycling, respiratory rate is related to pedalling rate, but is not necessarily synchronized with it (Hamley, 1963; Kay, Petersen & Vejby-Christensen, 1974).

There is no great evidence that the maximum heart rate of cyclists is lower than that of sedentary subjects. Saltin & Åstrand (1967) reported a value of 187/min in their 24-year-old Olympic competitor, and in the Brno championships it was found quite practical to report oxygen consumptions at a pulse rate of 190/min.

Some of the earlier studies of working capacity are reported as Harvard step-test scores. Perhaps because of differences in musculature for stepping and cycling, the values are not outstanding – among British potential international competitors, reported scores are 107 for men and 90 for women (Archer *et al.*, 1965), while among potential Japanese male entrants to the Tokyo Olympiad scores averaged 125 (Ishiko, 1967). On the other hand, maximum oxygen intake data for international competitors (Table 58) is generally high, the only exceptions being the figures reported by Scano & Venerando (1968) – where maximum heart rates do not seem to have been reached – and the results of Di Prampero *et al.* (1970) for one Guatemalan and six Cuban cyclists. The latter observations were made in Mexico City, and performance may have been adversely affected both by incomplete adaptation to altitude and by use of a step test procedure. The importance of selection as opposed to training is illustrated by the data of Berg (1972); in adolescent boys, the average maximum oxygen intake was only 55 ml/kg min despite arduous training (4000 kcal/day) for bicycle racing.

The only observation of anaerobic capacity seems the terminal lactate

Table 59. *Physical characteristics of track competitors*

Sample	Height (cm)	Weight (kg)	Leg length (cm)	Sitting height (cm)	Body fat (%)	Skin-fold (mm)	Author
			Sprint (100–200 m)				
India 8M	—	62.8	—	—	7.8	—	Malhotra *et al.* (1972*a*)
Japan[a] 17M	173.5	65.1	—	—	—	—	Ishiko (1967)
Negro 3M	178.7	74.3	86.2	92.5	—	5.1	Tanner (1964)
White 12M	176.6	71.8	83.1	93.5	—	4.6	Tanner (1964)
			Middle distance (400 m)				
Asian 1M	174.2	63.8	81.0	93.2	—	4.1	Tanner (1964)
India 3M	—	58.8	—	—	8.1	—	Malhotra *et al.* (1972*a*)
Japan[b] 17M	173.5	65.1	—	—	—	—	Ishiko (1967)
Negro 5M	179.0	70.0	88.1	91.0	—	4.9	Tanner (1964)
White 11M	185.4	75.6	88.8	96.6	—	4.6	Tanner (1964)
			Long distance (800–1500 m)				
New Zealand 1M	179.8	76.0	—	—	—	6.0	Carter *et al.* (1967)
S Africa							
3M Bantu	169.5	63.3	—	—	—	—	Leary & Wyndham (1965)
1M White	176.5	72.3	—	—	—	—	Leary & Wyndham (1965)
2M	178.9	63.7	—	—	—	—	Wyndham (1969)
Sweden							
5M	177.2	66.1	—	—	—	—	P. O. Åstrand (1964)
2M	189.0	72.7	—	—	—	—	Saltin & Åstrand (1967)
USA 3M	185.3[c]	71.3	—	—	—	—	Daniels & Oldridge (1970)
White 16M	180.5	68.9	87.7	92.8	—	4.4	Tanner (1964)
			Very long distance (3000–10000 m)				
India[d] 7M	—	62.1	—	—	8.4	—	Malhotra *et al.* (1972*a*)
Japan[e] 9M	165.3	54.8	—	—	—	—	Ishiko (1967)
Negro 2M	174.5	62.9	84.9	89.6	—	4.7	Tanner (1964)
S Africa							
3M Bantu	165.0	53.6	—	—	—	—	Leary & Wyndham (1965)
2M White	181.6	64.8	—	—	—	—	Leary & Wyndham (1965)
Sweden 3M	175.3	60.0	—	—	—	—	Saltin & Åstrand (1967)
USA							
9M best	177.5	66.3	—	—	7.1	—	Costill (1967)
8M poorer	177.4	68.5	—	—	9.5	—	Costill (1967)
3M	180.0	65.7	—	—	—	—	Daniels & Oldridge (1970)
1M	188.0	73.1	—	—	—	—	Costill *et al.* (1971)
9M	173.3	63.4	—	—	—	—	Fox *et al.* (1972)
11F	169.4	57.2	—	—	15.2	—	Wilmore & Brown (1974)
White 19M	174.4	60.8	83.2	91.2	—	3.9	Tanner (1964)
			Marathon				
Asian 1M	170.5	61.1	77.0	93.5	—	4.0	Tanner (1964)
Japan[f] 9M	165.3	54.8	—	—	—	—	Ishiko (1967)
S Africa 6M	176.7	63.6	—	—	—	—	Wyndham (1969)
White 9M	171.1	59.9	81.6	89.5	—	3.9	Tanner (1964)

Table 59. (*cont.*)

Sample	Height (cm)	Weight (kg)	Leg length (cm)	Sitting height (cm)	Body fat (%)	Skin-fold (mm)	Author
				Walkers			
India[f] 7M	—	62.1	—	—	8.4	—	Malhotra *et al.* (1972*a*)
Japan 6M	166.8	57.8	—	—	—	—	Ishiko (1967)
S Africa 1M	177.8	65.7	—	—	—	—	Wyndham & Strydom (1971)
UK 4M	184.2	72.3	—	—	—	—	Menier & Pugh (1968)
White 6M	177.0	66.6	83.1	93.9	—	4.1	Tanner (1964)
				Orienteering			
Sweden							
6M	183.8	74.4	—	—	—	—	Saltin & Åstrand (1967)
3F	162.7	58.0	—	—	—	—	Saltin & Åstrand (1967)

[a] Includes middle distance. [b] Includes sprint.
[c] Sample includes Jim Ryun (Height 189 cm).
[d] Includes walkers. [e] Includes marathon.
[f] Includes long distance.

value of 14.2 mmol/l reported for the Swedish Olympic champion (Saltin & Åstrand, 1967).

Runners and walkers

Unfortunately, a number of authors (Hukuda & Ishiko, 1966; Novak, Hyatt & Alexander, 1968; Gedda *et al.*, 1968; G. R. Cumming, 1970; Šprynarova & Pařízková, 1971) have grouped data for 'runners' or 'track and field' competitors without reference to event. Although this serves to increase sample size, it is an unsatisfactory method of data treatment, since there are substantial differences in physiological characteristics for the different classes of competition.

Physical characteristics

No one body build seems mandatory, and some very successful runners (for example, Jim Ryun of the USA) have departed widely from the average configuration for their distance. Nevertheless, the majority of runners are fairly short and relatively light (Table 59). The heaviest are the middle-distance competitors. Tanner (1964) reported an average weight of 75.6 kg in eleven white 400 m runners, and one well-known 800–1500 m runner from New Zealand (Peter Snell) weighed 76.0 kg (Carter *et al.*, 1967). Tanner (1964) commented that Snell had the typical build of a 400 m man, but presumably gained an advantage from his powerful muscles during the final sprint of the longer race. The lightest men (Table 59) choose ultra-long distance and marathon events. Although the walkers sometimes cover as great a total distance (20–50 km) as the marathon runners (42 km), the former are generally heavier, as are orienteers. The Negroes examined by Tanner (1964) did not differ greatly in weight from his white athletes, but in all classes of event the Asiatic subjects were both shorter and lighter. Athletes from India (Malhotra *et al.*, 1972*a*) and South African Bantu (Leary & Wyndham, 1965) are also short and light.

With regard to standing height, the tallest runners elect to participate in middle-distance events. Sprint competitors are short, with short legs; this feature may help their performance by raising the natural frequency of oscillation of the legs, thereby increasing the economy of forced movements (Shephard, 1972*b*). Measurements of limb circumferences and soft-tissue radiography (Tanner, 1964) show that the sprint and middle-distance men are well muscled, with particular development of the thigh region. The shortest men of all are the marathon runners, but in this group shortness is attributable to a lack of sitting height rather than leg length. Certainly, the work load of the heart is lightened by a small body, and Tanner (1964) has argued that slender muscles are also more easily perfused. While characteristics of height must be attributed to a selection of event by suitably endowed competitors, muscular development may be in part a response to subsequent training. It is also arguable that the large pelvic diameters of distance walkers are a mechanical response to the unusual walking action rather than an inherited anatomical peculiarity.

The secular trend among atheletes can be examined by comparing statistics for the Amsterdam Olympiad of 1928 (Kohlrausch, 1929) and the Rome games of 1960 (Tanner, 1964). As with the general population, runners have become taller, but the optimum body proportions for most events seem to have remained much the same. In Rome, the sprinters were 4 cm taller and 7 kg heavier than in Amsterdam. Corresponding trends for the 400 m competitors were 8 cm and 10 kg, for middle-distance runners 4 cm and 10 kg, and for long-distance runners 4 cm and 1 kg.

Negro runners in the sprint and middle-distance events are characterized by longer limbs, narrower hips, wider bones and narrower calf muscles than their white counterparts. The lighter, shorter and slimmer body gives the Negro an obvious advantage when running, and it has been argued that the lighter calf has a lower moment of inertia, thus permitting more rapid recovery movements (Tanner, 1964). Athletics traditionally has offered social mobility to the North American Negro, and it may be that selective pressures have operated more forcibly for black than for white competitors. However, most of the 'advantages' of body form ascribed to the Negro athlete can also be discerned in the non-athletic negroid population (Metheny, 1939).

Asians seem characterized by particularly short legs. However, this potential advantage in sprint events is offset by weak musculature, and to date competitive performances generally have been poor. The Bantu is also short and light relative to white South Africans, but here the two groups achieve very comparable track results (Leary & Wyndham, 1965).

All runners are extremely thin. Novak *et al.* (1968) quote an average of 3.7% body fat for 'track athletes', and Šprynarova & Pařízková (1971) in presenting a figure of 6.3% fat for runners comment that this is a lower average percentage than for all other sportsmen. Costill (1967) notes an appreciable difference between good (7.1% fat) and poorer (9.5% fat) cross-country runners. The one disparate report is a prediction of body fat by the formula of Durnin & Ramahan (1967); this yields figures of 11% for distance competitors, 12% for middle-distance runners and 13% for sprinters, all South American contestants at the Mexico City Olympiad (Di Prampero *at al.*, 1970).

There have been rather few measurements of lung volumes in track competitors. Results seem within the normal range for more sedentary adults. Novak *et al.* (1968) reported a vital capacity of 6.31 l in their track team, as opposed to 6.98 l in their swimmers. Costill (1967) found little difference of vital capacity between good (5.79 l) and poorer (5.63 l) cross-country runners; however, he remarked that the good runners had a larger maximum voluntary ventilation than the poorer performers (respective values 171 and 156 l/min BTPS). Carter *et al.* (1967) comment that while Snell had a vital capacity only 13% above the Baldwin standards, his maximum voluntary ventilation (249.2 l/min BTPS) was 54% higher than predicted, thus allowing him to develop a large ventilation (147 l/min BTPS) while running. Static lung volumes also seem relatively independent of the class of event. Ishiko (1967) found very similar vital capacities for distance and 'non-distance' runners (4.14, 4.20 l BTPS). Similarly, Malhotra *et al.* (1972*b*) observed no relationship between vital capacity or one-second forced expiratory volume and running distance in a group of Indian athletes:

	Vital capacity (l BTPS)	$FEV_{1.0}$ (l BTPS)
Sprint	4.40	3.55
Middle distance	4.36	3.63
Long distance	4.38	3.54

Table 60. *Working capacity of track competitors*

Sample	Max oxygen intake (l/min STPD)	Max oxygen intake (ml/kg min)	Max. heart rate (/min)	Max. blood lactate (mg/dl)	Author
		Sprint (100–200 m)			
Canada					
2M	—	53.2	—	—	G. R. Cumming (1970)
2F	—	44.6	—	—	G. R. Cumming (1970)
Germany 5M	—	56	—	—	Hollmann (1972)
India 8M	3.19	51.1	181	84.0	Malhotra *et al.* (1972*a*)
Japan 9M	2.03	34.4	—	—	Ishiko (1967)
S America	—	47[a]	—	—	Di Prampero *et al.* (1970)
Various 7M	4.12	55.5	—	—	Cerretelli & Radovani (1960)
		Middle distance (400 m)			
Germany 5M	—	63	—	—	Hollmann (1972)
India 3M	3.29	55.2	183	117	Malhotra *et al.* (1971*a*)
S America 6M	—	52[a]	—	—	Di Prampero *et al.* (1970)
Sweden					
4M	4.9	67	—	—	Saltin & Åstrand (1967)
3F	3.1	56	—	—	Saltin & Åstrand (1967)
Various 5M	4.49	64.3	—	—	Cerretelli & Radovani (1960)
		Long distance (800–1500 m)			
Australia 1M	5.04	76.6	194	—	P. O. Åstrand (1952)
Germany 5M	—	70	—	—	Hollman (1972)
India 3M	3.29	55.2	183	104.7	Malhotra *et al.* (1972*b*)
Japan 8M	2.51	45.3	—	—	Ishiko (1967)
New Zealand 1M	5.50	72.3	187	—	Carter *et al.* (1967)
S Africa					
3M Bantu	3.73[b]	60.7[b]	196[b]	—	Leary & Wyndham (1965)
1M White	4.31[b]	59.6[b]	168[b]	—	Leary & Wyndham (1965)
2M	4.25[c]	66.4[c]	—	—	Wyndham *et al.* (1969)
S America 5M	—	65[a]	—	—	Di Prampero *et al.* (1970)
Sweden					
5M	4.81	72.8	194	109.8	P. O. Åstrand (1964)
5M	5.4	75	—	—	Saltin & Åstrand (1967)
2M	5.45	75	177	124.2	Saltin & Åstrand (1967)
USA					
2M	—	{60.2 / 74.4}	—	—	Robinson (1938)
3M	—	75.2 (79.0)[d]	—	—	Daniels & Oldridge (1970)
Various 12M	4.25	67.8	—	—	Cerretelli & Radovani (1960)
		Very long distance (3000–10000 m)			
Canada 2M	—	66.1	—	—	G. R. Cumming
India					
7M	3.27	54.2	185	116	Malhotra *et al.* (1972*a*)
1M	3.66	63.1	173	—	Malhotra *et al* (1972*b*)
S Africa					
3M Bantu	3.52[b]	65.8[b]	199[b]	—	Leary & Wyndham (1965)
2M White	4.07[b]	62.8[b]	179[b]	—	Leary & Wyndham (1965)

Table 60. (*cont.*)

Sample	Max oxygen intake (l/min STPD)	Max oxygen intake (ml/kg min)	Max. heart rate (/min)	Max. blood lactate (mg/dl)	Author
Sweden 3M	4.75	79.2	198	119.7	Saltin & Åstrand (1967)
USA					
1M	5.35	81.5	—	—	Robinson (1937)
5M	4.92	72.6	—	—	Kollias, Moody & Buskirk (1967)
3M	—	72.5 (76.2)[d]	—	—	Daniels & Oldridge (1970)
11F	3.35	59.1	180.4	—	Wilmore & Brown (1974)
			Marathon		
India 3M	3.28	55.1	183	77.7	Malhotra *et al.* (1972*b*)
S Africa 6M	3.95[b]	62.4[b]	—	—	Wyndham (1969)
UK 1M	4.15[c]	70.0[c]	—	—	Menier & Pugh (1968)
USA					
1M	4.7	76	—	—	Dill (1965)
2M	3.78	64.4	185	—	Costill & Winrow (1970)
1M	5.09	69.7	188	88.8	Costill *et al.* (1971)
9M	—	70.0	185	—	Fox *et al.* (1972)
			Walkers		
India 3M	3.18	50.2	182	82.0	Malhotra *et al.* (1972*b*)
S Africa 1M					
Running	4.05	61.6	—	—	Wyndham & Strydom (1971)
Walking	3.70	56.3	—	—	Wyndham & Strydom (1971)
Sweden 4M	4.7	71	—	—	Saltin & Åstrand (1967)
UK 4M					
Running	4.15[e]	57.4[e]	—	—	Menier & Pugh (1968)
Walking	4.33[e]	60.0[e]	—	—	Menier & Pugh (1968)
			Orienteering		
Sweden					
9M	5.4	77	—	—	Saltin & Åstrand (1967)
5M	5.87	77	189	123.3	Saltin & Åstrand (1967)
5F	3.4	59	—	—	Saltin & Åstrand (1967)
3F	3.47	60.1	192	112.5	Saltin & Åstrand (1967)

[a] Data obtained by step test in Mexico City. Anaerobic power = 180 ml/kg min.
[b] Data obtained in Johannesburg (altitude 1830 m).
[c] Data obtained in Johannesburg. One subject attained 5.1 l/min, 78.0 ml/kg min at sea level.
[d] After altitude training.
[e] Data obtained in the Pyrenees (altitude 1800 m).

Working capacity

Ishiko (1967) noted a marked gradient of Harvard step test scores from sprinters and middle-distance runners (111.0), through walkers (126.9) to the extremely fit long-distance runners (160.2). On the other hand, Costill (1967) found relatively low scores in collegiate cross-country runners, with little difference between good (117) and poorer (110) competitors.

The maximum oxygen intake of runners has generally been determined on the treadmill, although two authors who obtained rather low results used a step test (Di Prampero *et al.*, 1970) and a bicycle ergometer (Malhotra *et al.*, 1972*b*) respectively. The hard surface, confined running area and steep slope of the average laboratory treadmill differ somewhat from the usual athletic track, but nevertheless problems of the specificity of training create less problems of data interpretation for the runner than for many classes of sportsman. Treadmill running even seems a satisfactory mode of exercise for the distance walker; one laboratory found a slightly higher $\dot{V}_{O_2(max)}$ during walking than during running (Menier & Pugh, 1968), but another laboratory found larger values running than walking (Wyndham & Strydom, 1971). The disagreement may be related to the speed at which the champions walk more efficiently than ordinary people; Menier & Pugh (1968) found an advantage at 8 km/hr, but Wyndham & Strydom (1971) reported a normal efficiency at speeds of less than 9.7 km/hr.

The maximum heart rates of runners are quite variable. P. O. Åstrand (1964) found normal values in five Swedish 1 mile runners (average 194) as in Landy (P. O. Åstrand, 1952), but a later paper from the same laboratory showed two 1500 m champions with an average maximum heart rate of only 177 (Saltin & Åstrand, 1967). The mean for forty-four male runners competing over all distances from 100 metres to a marathon course is 189.6, slightly less than the figure anticipated in a sedentary young man.

The prime physiological determinant of track performance varies with distance. A sprint demands anaerobic power, middle distances anaerobic capacity, and long distance aerobic power. It is thus not surprising that laboratory measurements show a progressive augmentation of maximum oxygen intake from the sprinters through 400 m runners to competitors over 800–1500 m (Table 60). Particularly high maximum oxygen intakes have been reported for Lash (81.5 ml/kg min) and Landy (76.6 ml/kg min). Competitors over very long distances (> 1500 m) tend to have lower absolute $\dot{V}_{O_2(max)}$ readings, but values per kilogram of body weight remain high (for instance, 82 ml/kg min in Keino, and 79.2 ml/kg min in three Swedish champions). Marathon runners use only about 75% of their aerobic power (E. L. Fox & Costill, 1972), and success depends not only on $\dot{V}_{O_2(max)}$ but also on running technique and maintenance of heat and fluid balances. It is thus not surprising that even on a relative basis the $\dot{V}_{O_2(max)}$ of marathon runners is lower than that for competitors over shorter distances. A typical average seems 70 ml/kg min, although Dill (1965) estimated that Clarence De Mar had a $\dot{V}_{O_2(max)}$ of 76 ml/kg min at the peak of his career (age 36). Long-distance walkers generally have a moderate $\dot{V}_{O_2(max)}$ (60–62 ml/kg min), although Saltin & Åstrand (1967) reported an average of 71 ml/kg min for four Swedish champions. Orienteering involves 11–16 km of cross-country running between check-points, using a map and compass. The top competitors have a well-developed aerobic power, both in absolute units (5.87 l/min) and in relative terms (77 ml/kg min).

There have been striking improvements in athletic records over the past few decades (Craig, 1968; Jokl & Jokl, 1968). It is thus remarkable that the distance runners of today have a maximum oxygen intake that is no larger than that recorded for Lash (Robinson, Edwards & Dill, 1937). Presumably, much of the gain in performance must be attributed to improvements in track, equipment, and technique of running, including a greater emphasis on full utilization of the potential oxygen debt.

The data are rather limited with respect to racial considerations. Indian athletes apparently have a poor maximum oxygen intake that matches their poor track records. However, the Indian data were collected on a bicycle ergometer, with rather low maximum heart rate and blood lactate readings, and it must be questioned whether a true maximum oxygen intake was realized. Ishiko (1967) also obtained extremely low readings on Japanese athletes. On the other hand, Leary & Wyndham (1965) found a very similar relative $\dot{V}_{O_2(max)}$ in Bantu and white athletes.

There have been suggestions that runners develop an exceptional tolerance of anaerobic metabolism. Di Prampero *et al.* (1970) reported that sprint, middle- and long-distance runners had a large anaerobic power relative to other classes of athlete. However, it could be questioned whether running experience gave the track group some advantage in performing the required stair-climbing sprint. Maximum blood lactate concentrations (Table 60) are generally in the high normal range; such data probably reflect high intramuscular concentrations of lactate in the light long-distance competitors.

With regard to muscle strength, G. R. Cumming (1970) found the runners somewhat superior to swimmers, both on grip strength (55.9 versus 53.2 kg) and on summated muscle strength (3.49 versus 3.36 kg/kg). However, Costill (1967) observed a lower grip strength in good cross-country runners than in poorer competitors (respective values 49 and 52 kg), with no difference of explosive force (vertical jump 47.5 cm) between the two groups. Malhotra *et al.* (1972*a*) reported little difference of vertical jump between sprinters and distance competitors, although his sprinters were superior in terms of broad jump and chin-up scores:

	Vertical jump (cm)	Broad jump (cm)	Chin ups – total work (kg m)
Sprint	36	267	206
Middle distance	42	214	146
Long distance	37	214	107

Many long-distance runners fare poorly when subjected to detailed performance-test evaluations. Thus S. R. Brown (1966) applied an eighteen-item 'motor efficiency' classification test to Peter Snell. He passed only eight of the eighteen items, namely one of three balance tests, one of three flexibility tests, none of the three agility tests, two of three strength tests, three of five endurance tests and the sole power test. His overall score was poorer than that of 56% of sedentary university students! Roger Bannister reputedly had a similar dismal score when he tried this test battery in 1952.

Other track and field events

Physical characteristics

The physical characteristics of competitors in other track and field events are summarized in Table 61. The biathlon and pentathlon subjects seem of a similar build to middle-distance runners, the Japanese being smaller than their European counterparts. Hurdlers are heavier, and are distinguished by their long legs. The North American Negro seems at an advantage and the small Asian at a disadvantage in this class of contest. Steeplechasers have the lightweight build previously remarked for cross-country runners. The jumpers are tall, with leg length being a particular advantage in the high jump. Body weights are lower for the long jump and the triple jump than for the high jump and the pole vault. Again, Asian competitors are at a disadvantage in terms of leg length; they also have a low body weight, and thus probably lack muscular power. The throwers are tall, powerfully built individuals, with an extremely high body weight. Lack of endurance fitness is suggested by relatively high skinfold readings.

Malhotra *et al.* (1972*a*) measured the lung volumes of their jumpers and throwers. Values for the jumpers (VC = 4.46, $FEV_{1.0}$ = 4.07 l BTPS) were superior to those for runners, and the results for the throwers (VC = 5.61, $FEV_{1.0}$ = 4.79 l BTPS) were extremely high for Indian subjects.

Working capacity

Ishiko (1967) found a very high Harvard step test score (152.6) in four Japanese pentathlon competitors, aspirants to the Tokyo Olympiad; in contrast, six Japanese throwers achieved a very poor score (94.0).

Formal measurements of aerobic power show quite high readings for biathlon and pentathlon competitors (Table 62), but poorer results in a German decathlon team. Scores for the throwers and jumpers, on the other hand, are poorer than in the general population.

Di Prampero *et al.* (1970) found the highest anaerobic powers of their

Table 61. *Physical characteristics of track and field competitors other than runners*

Sample	Height (cm)	Weight (kg)	Leg length (cm)	Sitting height (cm)	Body fat (%)	Skin-fold (mm)	Author
			Biathlon				
Sweden 1M	185	73	—	—	—	—	Saltin & Åstrand (1967)
			Pentathlon				
Italy 4M	—	—	—	—	10	—	Di Prampero *et al.* (1970)
Japan 4M	171.6	68.2	—	—	—	—	Ishiko (1967)
Sweden 1M	188	77	—	—	—	—	Saltin & Åstrand (1967)
			110 m hurdles				
Negro 1M	191.0	90.5	91.4	99.6	—	5.2	Tanner (1964)
White 3M	182.8	78.4	87.7	95.1	—	4.4	Tanner (1964)
			400 m hurdles				
Asian 1M	172.7	62.0	78.3	94.4	—	4.4	Tanner (1964)
White 5M	180.6	71.0	86.6	94.0	—	4.5	Tanner (1964)
			Steeplechase				
Negro 1M	176.0	64.6	85.1	90.9	—	4.0	Tanner (1964)
White 4M	179.2	64.8	86.2	93.0	—	3.8	Tanner (1964)
			High Jump				
Negro 2M	191.5	85.1	96.4	95.1	—	4.9	Tanner (1964)
White 8M	188.7	76.7	91.0	97.7	—	5.0	Tanner (1964)
			Long jump				
Asian 2M	172.3	58.8	78.7	93.6	—	4.0	Tanner (1964)
Negro 1M	187.0	72.9	89.5	97.5	—	3.7	Tanner (1964)
White 2M	181.5	71.5	87.8	93.7	—	4.1	Tanner (1964)
			Pole vault				
White 2M	186.0	78.4	87.3	98.7	—	5.2	Tanner (1964)
			Triple jump				
Asian 2M	175.6	61.1	80.5	95.1	—	3.8	Tanner (1964)
White 3M	183.1	71.6	86.7	96.4	—	4.7	Tanner (1964)
			Unspecified jumping events				
Indian 3M	172.6	64.7	—	—	8.9	—	Malhotra *et al.* (1972*a*)
			Discus throwing				
White 2M	192.4	105.7	90.8	101.6	—	7.9	Tanner (1964)
			Hammer throwing				
Asian 2M	175.7	80.9	78.6	97.2	—	6.6	Tanner (1964)
White 2M	188.8	101.3	91.9	96.9	—	7.7	Tanner (1964)
			Javelin throwing				
White 2M	186.5	92.9	88.4	98.1	—	7.4	Tanner (1964)
			Shot putting				
White 6M	190.8	105.0	89.8	100.9	—	8.0	Tanner (1964)
			Unspecified throwing events				
Indian 4M	191.3	100.1	—	—	15.0	—	Malhotra *et al.* (1971*a*)

Table 62. *Working capacity of track and field competitors other than runners*

Sample	Max oxygen intake (1/min STPD)	Max oxygen intake (ml/kg min STPD)	Max. heart rate (/min)	Max. blood lactate (mg/dl)	Author
			Biathlon		
Sweden 1M	5.71	78.2	—	115.2	Saltin & Åstrand (1967)
Sweden 5M	5.4	73	—	—	Saltin & Åstrand (1967)
			Decathlon		
Germany 5M	—	58	—	—	Hollmann (1972)
			Pentathlon		
Italy 4M	—	57[a]	—	—	Di Prampero *et al.* (1970)
Sweden 1M	5.73	74.4	—	146.7	Saltin & Åstrand (1967)
			Jumpers		
India 2M	2.90	44.8	187	109.1	Malhotra *et al.* (1972*a*)
Japan 2M	3.30[b]	—	—	—	Nakanashi *et al.* (1966)
Japan –M	—	59.4	—	—	Ishiko & Acki (1973)
			Throwers		
Canada 1M	—	38.2	—	—	G. R. Cumming (1970)
Canada 3F	—	38.4	—	—	G. R. Cumming (1970)
India 4M	3.90	38.9	184	99.6	Malhotra *et al.* (1972[a])
Japan –M	—	54.9	—	—	Ishiko & Acki (1973)

[a] Measured in Mexico City using a step test. Anaerobic power = 220 ml/kg min STPD.
[b] Oxygen debt = 9.5 l.

series in four pentathlon contenders. Equally, Nakanashi *et al.* (1966) noted a large oxygen debt in two Japanese jumpers.

With regard to muscle strength, G. R. Cumming (1970) found the hand-grip of the throwers to be superior to that of all other athletes, with readings of 68.6 kg in the men and 49.1 kg in a woman javelin thrower. On the other hand, the summated index of muscle strength (3.68 kg/kg) was ranked fifth of fourteen sports.

Swimmers

Physical characteristics

The basic characteristics of competitive swimmers are summarized in Table 63. As in many other sports, racial differences reflect general population statistics; thus, the Japanese athletes are much shorter and lighter than those from Europe or North America. Canadian data show little relationship between body weight and contest duration, although the distance competitors (400 to 1500 m) are somewhat thinner than those participating in sprint and medium-distance events (50 to 200 m). An

Table 63. *Physical characteristics of competitive swimmers*

Sample	Height (cm)	Weight (kg)	Body fat (%)	Skinfold (mm)	Author
Argentine, Guatemala					
F	—	—	12.5	—	Di Prampero *et al.* (1970)
Canada					
M (sprint)	181.1	75.0	6.9	10.4	Shephard *et al.* (1973*a*)
M (medium distance)	178.0	74.6		9.9	Shephard *et al.* (1973*a*)
M (long distance)	179.0	74.9		7.5	Shephard *et al.* (1973*a*)
Czechoslovakia					
M	182.3	79.1	8.5	—	Šprynarova & Pařízková (1971)
F	166.2	63.9	19.2	—	Šprynarova & Pařízková (1969)
Europe					
M	182.5	79.6	—	—	Hukuda & Ishiko (1966)
F	170.8	65.2	—	—	Hukuda & Ishiko (1966)
Japan					
M	170.8	68.7	—	—	Hukuda & Ishiko (1966)
M	171.9	70.1	—	—	Ishiko (1967)
M	170.1	70.0	—	—	Miyashita, Hayashi & Furuhashi (1970)
F	159.0	55.6	—	—	Hukuda & Ishiko (1966)
F	164.1	59.3	—	—	Miyashita *et al.* (1970)
Sweden					
F	167.0	64.0 53.7	—	—	Saltin & Åstrand (1967)
USA					
M	182.9	78.9	5.0	—	Novak *et al.* (1968)
M	181.2	73.7	—	—	Dixon & Faulkner (1971)

increase of body fat is an advantage in marathon swimming (Pugh *et al.*, 1960), but over the usual competitive distances there seems little tendency to obesity, perhaps because most swimming pools are heated to 25 °C.

A full range of anthropometric measurements was made on Canadian university-class swimmers (Shephard *et al.*, 1973*a*). However, there were surprisingly few differences from control subjects. In agreement with the previous observations of Gedda *et al.* (1968), we found a large biacromial diameter and sitting height. The broad chest presumably reflects partly a large vital capacity and partly good development of the shoulder musculature. On the other hand, the antero-posterior diameter of the chest was less than in control subjects, thus facilitating hydroplaning. The ratio of wrist breadth to femoral bicondylar diameter was somewhat increased (respective values 0.595 in swimmers, 0.571 in controls) due to both muscularity of the arms and the competitive advantage of light legs that can be held in the horizontal position during swimming (Rennie *et al.*, 1973).

Over normal competitive distances, swimmers rely on a large vital capacity to maintain their buoyancy. Many authors have reported lung volumes 10–15% above age and height predictions in successful contestants (Cureton, 1951; P. O. Åstrand *et al.*, 1963; Bloomfield & Sigerseth, 1965; Saltin & Åstrand, 1967; Novak *et al.*, 1968; Andrew, Becklake & Guleria, 1972; Shephard *et al.*, 1973*a*). In confirmation of Bloomfield & Sigerseth (1965), we found (Shephard *et al.*, 1973*a*) that the average vital capacity was somewhat larger in distance swimmers (6.46±0.61 l) than in the sprinters (5.71±0.97 l) despite the greater standing height of the latter group. In the distance competitors, vital capacity was quite closely correlated with both overall performance and endurance. Sustained endurance-type activity in the growing child can lead to at least a temporary acceleration in the development of vital capacity (Ekblöm, 1969; Engström *et al.*, 1971). Nevertheless, the clear separation of data for sprint and distance competitors, with the absence of an unusual vital capacity in the marathon swimmers (Pugh *et al.*, 1960) suggests that the observed lung volumes are more a consequence of selection than of any biological response to repeated swimming.

The one-second forced expiratory volume was at least as large in the distance men as in the sprinters, but the distance competitors expelled a smaller percentage of their vital capacity in one second (81.3±7.1% as against 86.9±5.0%) because of their substantial advantage of vital capacity. Residual gas volumes (1.52 l in the sprinters, 1.51 l in the distance men) were normal values.

Exercise response

When running on the treadmill or operating a bicycle ergometer, the maximum heart rate of the competitive swimmer is close to the figure anticipated for a sedentary person of comparable age (Table 64). However, lower maximum heart rates are seen in the water, whether competing in a pool (Magel, McArdle & Glaser, 1969), swimming against tethering weights (Magel, 1971) or swimming in a flume (Holmér, 1972). This seems partly a consequence of face immersion and breath-holding, and partly a response to greater body cooling when swimming (Åstrand, personal communication). On the other hand, the maximum cardiac output is greater during tethered swimming than when running (Dixon & Faulkner, 1971).

Well-trained swimmers hypoventilate during exercise, with a high oxygen extraction, a high expired CO_2 concentration, and a low respiratory gas exchange ratio (P. O. Åstrand *et al.*, 1963; Dixon & Faulkner, 1971;

Table 64. *Maximum heart rate of competitive swimmers*

Sample	Exercise mode	Max. heart rate (/min)	Author
1 F	BI	204	Saltin & Åstrand (1967)
1 F	TM	196	Saltin & Åstrand (1967)
13 M	TM	192	Šprynarova & Pařízková (1971)
6M; 3F	FL	183	Holmér (1972)
6M; 3F	BI	191	Holmér (1972)
6M; 3F	TM	193	Holmér (1972)
18M	TM	195	Shephard *et al.* (1973*a*)

BI = bicycle ergometer; FL = flume; TM = treadmill.

Table 65. *Working capacity of competitive swimmers*

Sample	Exercise mode	Max. oxygen intake (l/min STPD)	Max. oxygen intake (ml/kg min STPD)	Author
Collegiate				
17M	TM	4.20	55.5	Magel & Faulkner (1967)
7M	TM	4.88	65.4	Novak *et al.* (1968)
6M	TM	4.26	57.8	Dixon & Faulkner (1971)
5M	TM	3.83	49.4	McArdle *et al.* (1971)
International				
10M	TM	3.99	54.0	Cerretelli & Radovani (1960)
1M	TM	4.07	59.6	Leary & Wyndham (1965)
20M	TM	4.36	62.4	Miyashita *et al.* (1970)
3M	ST	—	51.0[a]	Di Prampero *et al.* (1970)
5M	BI	—	57.3	G. R. Cumming (1970)
5M	?BI	—	61.0	Hollman (1972)
1M	FL	5.59	69.9	Holmér (1972)
4F	TM	2.75	54.0	Miyashita *et al.* (1970)
National				
6M	BI	5.00	67.0	Saltin & Åstrand (1967)
5F	BI	3.20	57.0	Saltin & Åstrand (1967)
10F	TM	2.92	46.0	Šprynarova & Pařízková (1969)
Top				
13M	TM	4.50	56.9	Šprynarova & Pařízková (1971)
Trained				
3M	FL	4.48	62.2	Holmér (1972)
12F	FL	3.17	58.3	Holmér (1972)

BI = bicycle ergometer; FL = flume; ST = step test; TM = treadmill.
[a] Measurements taken in Mexico City.

Šprynarova & Pařízková, 1971; Holmér, 1972). Limitation of respiration is particularly marked during the front crawl, butterfly, and breast stroke. Breathing becomes dependent on the speed of arm movement, and can only occur when coordinated with head and arm movements.

Ishiko (1967) reported a Harvard step test score of 127 in twenty-five potential Japanese contestants in the Tokyo Olympiad. Recent research has led to the accumulation of a substantial volume of data on the aerobic power of swimmers at both the international and collegiate levels of competition (Table 65). Most authors have used standard laboratory modes of exercise (bicycle, treadmill, or step), and the results could be unrealistic when effort is undertaken mainly by the arms, with the body immersed horizontally in the water. Certainly, moderately trained university-level swimmers achieve lower values of maximum oxygen intake during swimming than during treadmill exercise (Goodwin & Cumming, 1966; Dixon & Faulkner, 1971; McArdle *et al.*, 1971). However, the difference becomes smaller (Holmér, 1972) or even disappears (Dixon & Faulkner, 1971) when dealing with elite competitors. Åstrand (personal communication) has provided an interesting example of this phenomenon in a pair of monozygotic twins, one of whom was a well-trained swimmer, and the other a recreational swimmer. Treadmill figures for maximum oxygen intake were closely comparable (3.61, 3.56 l/min in the two subjects), but there were substantial inter- and intra-individual differences of aerobic power when swimming in the flume.

	Trained twin (l/min)	Untrained twin (l/min)
Breast stroke	3.63	2.82
Front crawl	3.36	2.71
Arm stroke	2.74	1.84
Leg kick	3.37	2.72

Since the body weight of a swimmer is largely supported by water displacement, the absolute aerobic power has more significance than figures relative to body weight. Group means for international and collegiate levels of competition can be summarized as follows:

National/international	M	4.53 l/min	(N = 40)	60.5 ml/kg min	(N = 53)
	F	2.96	(N = 19)	50.6	(N = 19)
Collegiate	M	4.31	(N = 38)	57.4	(N = 38)

As would be anticipated from the differences of body weight, the Japanese competitors are at some disadvantage of absolute aerobic power relative

Table 66. *Working capacity of collegiate swimmers, classified by event*

Event	Max. oxygen intake (l/min STPD)	Max. oxygen intake (ml/kg min STPD)	Oxygen debt (l STPD)	Max. blood lactate (mg/dl)	Arm ext. force (kg)	Knee ext. force (kg)
Sprint ($N = 8$)	4.37	58.3	7.67	152	43.4	66.0
Middle distance ($N = 4$)	4.34	55.4	7.95	146	48.5	76.2
Long distance ($N = 6$)	4.89	65.4	7.72	138	49.1	81.3
Control ($N = 4$)	3.73	50.7	6.64	140	39.9	79.5

Data of Shephard *et al.* (1973*a*).

to those from Europe and the United States. Classifying data according to the type of event (Shephard *et al.*, 1973*a*), the aerobic power is higher (Table 66) in long-distance competitors (400–1500 yards) than in the short- and middle-distance swimmers (50–200 yards).

Surprisingly little of the advantage in aerobic power is retained after retirement from competition. Eriksson, Lundin & Saltin (1973*b*) examined female ex-swimmers, and demonstrated a maximum oxygen intake of only 2.13 l/min, 'almost below the normal value for Swedish females today'. On the other hand, the large heart and lung volumes persisted unchanged. A three-month training programme produced a 14% gain of aerobic power, an increase 'of the same magnitude as that obtained in persons who had never trained'. Thus, it seems that once the effects of training have been lost through disuse, the ex-athlete finds it as difficult as anyone else to restore his or her physical condition.

In the Toronto study (Table 66), the long-distance swimmers had powerful arm and leg muscles. However, among the sprinters, arm strength was not outstanding, and knee extension strength was actually less than in the controls drawn from laboratory staff. As with the wrist width/leg width ratios, this probably reflects the importance of light legs that will float to a horizontal position during fast swimming (Rennie *et al.*, 1973). Hukuda & Ishiko (1966) pointed out the low muscular strength of the Asiatic competitors compared with their European rivals. Respective hand-grip forces ('both hands') were 94.9 and 109.8 kg in the men and 62.0 and 69.8 kg in the women. Even the European figures are not particularly impressive, although they coincide closely with measurements reported by G. R. Cumming (1970) for Canadian contestants in the Pan American games (53.2 and 36.5 kg for a single hand-grip in men and women respectively). S. R. Brown, Pomfret & Parsons (1966) suggested that while the hand-grip of Canadian swimmers was poor, back and leg strengths were above average. On the other hand, Cumming's data shows a poor total strength (arm, hand, hip and knee) relative to many other sportsmen, despite an 'international status the highest any athletic team Canada has ever had'. Hukuda & Ishiko (1966) reported back strengths of 148.9 and 185.0 kg for Japanese and European men, and 103.1 and 119.3 kg for Japanese and European women contestants.

Skill is presumably more vital to the sprint swimmer than explosive force, and Di Prampero *et al.* (1970) found a relatively low anaerobic power (140 ml/kg min) in swimmers tested at the Mexico City Olympiad. On the other hand, the anaerobic capacity, as measured by oxygen debt and blood lactates following maximal exercise is well up to average levels (Table 65; P. O. Åstrand *et al.*, 1963; Saltin & Åstrand, 1967; Holmér, 1972). Van Huss & Cureton (1955) found a low coefficient of correlation between the magnitude of the oxygen debt and swimming performance, equally in sprint and in distance competitors. One possible explanation of this anomaly is that whereas the sprinter produces more lactic acid, the distance man (by virtue of his large vital capacity) can sustain a larger alactate oxygen debt.

Diving

Japanese divers (Ishiko, 1967), potential contestants at the Tokyo Olympiad, were substantially lighter (height 165.1 cm, weight 59.1 kg) than their swimming colleagues. They were also less fit, with a Harvard step test score of only 105.6.

Canadian data from the Pan American games (G. R. Cumming, 1970) confirms this picture. In the men, the aerobic power was only 54.2 ml/kg min, with a PWC_{170} of 17.9 kg m/kg min, and in the women the corresponding figures were 46.8 ml/kg min and 11.6 kg m/kg min. The average hand-grip of male divers was only 45.3 kg. On the other hand, when expressed per unit of body weight the summated muscle strength of the male contestants (3.31 kg/kg) was comparable with that of the swimmers (3.36 kg/kg).

Water polo

Gedda *et al.* (1968) noted a number of anthropometric features of water polo contestants in the Rome Olympiad of 1960. Both height and weight were above average, as were arm span, bicristal diameter, and vital capacity. On the other hand, the biacromial diameter was less than anticipated.

The aerobic power seems much as in swimmers. At the Pan American

Table 67. *Physical characteristics of participants in water sports*

Sample	Height (cm)	Weight (kg)	Body fat (%)	Skin-fold (mm)	Vital capacity (1 BTPS)	Author
Canoe						
Czechoslovakia						
16M	177.9	75.1	—	—	—	Seliger *et al.* (1969)
Hawaii						
2M	—	74.5	—	—	—	Horvath & Finney (1969)
Japan						
4M	172.3	71.3	—	—	—	Ishiko (1967)
Kayak/white-water						
Canada						
7M	173	64.0	11.6	5.9	4.77	Sidney & Shephard (1973)
3M (older)	180	79.2	15.7	9.4	5.16	Sidney & Shephard (1973)
2F	166	57.3	23.8	9.2	3.70	Sidney & Shephard (1973)
Czechoslovakia						
13M	178.0	76.2	—	—	—	Seliger *et al.* (1969)
13F	165.5	62.2	—	—	—	Seliger *et al.* (1969)
Japan						
6M	172.1	67.7	—	—	—	Ishiko (1967)
Rowers						
Belgium						
19M	182.2	79.8	16.8	—	6.04	De Pauw & Vrijens (1971)
Brazil						
11M	182.1	80.7	—	—	—	Kiss *et al.* (1973)
Canada						
13M	182.4	85.3	11.8	5.5	5.96	Wright *et al.* (1976)
Germany						
4M	191.0	95.6	—	—	—	Mellerowicz & Hansen (1965)
8M	—	92.4	—	—	—	Nowacki *et al.* (1967)
21M	187.0	87.0	—	—	—	Weiderman *et al.* (1968)
Italy						
5M	188.6	84.9	—	—	—	Di Prampero *et al.* (1971)
Japan						
23M	186.0	82.2	—	—	—	Hirata (1966)
23M	178.8	75.6	—	—	—	Ishiko (1967)
New Zealand						
26M	186.1	82.6	—	—	—	Hagerman & Howie (1971)
S Africa						
10M	178.5	76.4	—	—	—	Strydom, Wyndham & Greyson (1967)
USA						
26M	190.1	88.2	—	—	—	Hagerman *et al.* (1972)
Yugoslavia						
–M	181.2	80.0	—	—	—	Medved (1966)
Sailing						
Canada						
10M	181.5	84.1	19.9	—	—	Niinimaa *et al.* (1974)
Japan						
12M	168.6	62.4	—	—	—	Ishiko (1967)

games in Winnipeg (G. R. Cumming, 1970), the Canadian water polo team had a slightly higher maximum oxygen intake than the swimmers (58.3 and 57.3 ml/kg min respectively), but a slightly lower PWC_{170} (18.5 and 20.5 kg m/kg min respectively). The Manitoba provincial team had a lower maximum oxygen intake (53.3 ml/kg min, 4.07 l/min, according to a report by Goodwin & Cumming, 1966). Muscle strength (G. R. Cumming, 1970) was high both in absolute terms (hand-grip force 62.5 kg) and relative to body weight (summated strength 3.73 kg/kg).

Rowing

Physical characteristics

Hay (1967) has suggested that there is a strong relationship between anthropometric measurements and rowing skills. A substantial standing height (Bloomfield, Blanksby & Elliott, 1973) with a long trunk and arms are helpful both in applying leverage to the oars and in developing a good stroke (Table 67). Such anatomical features are particularly well demonstrated by German and US oarsmen (Mellerowicz & Hansen, 1965; Hagerman, Addington & Gaensler, 1972); however, Japanese rowers also are much taller than their compatriots in many other sports (Hirata, 1966; Ishiko, 1967).

There is little displacement of the centre of gravity of the body during rowing, and in consequence some very successful contestants – including the German teams (Mellerowicz & Hansen, 1965; Nowacki, Krause & Adam, 1967) – are extremely heavy and powerfully muscled. The Japanese (Hirata, 1966) are lighter than German rowers of similar height, but nevertheless they are very heavy relative to most Asian athletes. Belgian rowers (De Pauw & Vrijens, 1971) also carry a substantial burden of body fat. On the other hand, Canadian oarsmen (who follow a vigorous endurance training programme when the weather does not permit rowing) are quite thin, their average skinfold readings being only 52% of those for a sedentary young man of 'ideal' body weight (Wright, Bompa & Shephard, 1976). There is some evidence that the regional distribution of fat is altered in the rowers, with decreases concentrated in regions of vigorous body movement (supra-iliac, suprapubic and knee folds).

Vital capacities are about 10% above normal values predicted from age and standing height (De Pauw & Vrijens, 1971; Wright *et al.*, 1976).

Working capacity

There has been considerable discussion of the specificity of methods proposed for estimating the working capacity of the rower. Harvard fitness test scores (125.5; Ishiko, 1967) are only moderate by athletic standards.

Table 68. *Working capacity of participants in water sports*

Sample	Exercise mode	Max. oxygen intake (l/min STPD)	Max. oxygen intake (ml/kg min STPD)	Anaerobic capacity (l O_2 debt)	Max. heart rate (/min)	Max. blood lactate (mg/dl)	Author
Canoe							
Canada							
2M	BI	—	51.8	—	—	—	G. R. Cumming (1970)
Hawaii							
2M	BO	3.79	50.8	5.55	181	73	Horvath & Finney (1969)
Sweden							
4M	BI	5.1	70.0	—	—	—	Saltin & Åstrand (1967)
Kayak/white-water							
Canada							
7M	TM	3.83	60.0	6.60	198	146	Sidney & Shephard (1973)
3M (older)	TM	4.45	55.1	7.24	184	127	Sidney & Shephard (1974)
2F	TM	2.80	49.2	4.61	193	121	Sidney & Shephard (1973)
Rowers							
Argentina, Italy							
7M, 10M	SM ST	—	52[a]	—	—	—	Di Prampero *et al.* (1970)
Belgium							
19M	BI	4.63	58.0	—	—	—	De Pauw & Vrijens (1971)
Brazil							
11M	SM BI	4.06	50	—	—	—	Kiss *et al.* (1973)
Canada							
13M	TM	4.85	56.9	8.32	—	127.5	Wright *et al.* (1976)
Germany							
4M	BI	5.90	61.8	—	—	—	Mellerowicz & Hansen (1965)
8M	BI	6.16	66.6	—	—	—	Nowacki *et al.* (1967)
5M	—	—	70	—	—	—	Hollmann (1972)
Italy							
6M	TM	5.00	58.7	—	—	—	Cerretelli & Radovani (1960)
4M 1M	TM BA	5.01	59.0	—	—	—	Di Prampero *et al.* (1971)
S. Africa							
10M	TM	3.63	47.5	—	—	—	Strydom *et al.* (1967)
Sweden							
5M	BI	5.1	62.0	—	—	—	Saltin & Åstrand (1967)
USA							
26M	SM RM	5.52	62.6	—	—	—	Hagerman *et al.* (1972)
24M	TM	—	66.1	—	—	—	Schwartz (1973)
Yugoslavia							
–M	—	4.73	—	—	—	—	Medved (1966)
Sailing							
Canada							
10M	TM	4.10	49.7	8.8	196	106.4	Niinimaa *et al.* (1975)

BA = basin; BI = bicycle ergometer; BO = boat; RM = rowing machine; SM = submaximum; ST = step test; TM = treadmill.

[a] Data is taken in Mexico City. Anaerobic power = 155 ml/kg min.

However, it is hardly fair to use a stair-climbing test when the event itself calls for weight-supported use of the arm, back and leg muscles. Some authors have developed rowing ergometers (for example, Niu *et al.*, 1968). These may have a role in the winter training of oarsmen, but as Di Prampero *et al.* (1971) point out: 'rowing in a basin with practically still water is an entirely different exercise from actual rowing'. Indeed, because the task is abnormal, the athlete may show a peripheral limitation of effort that he does not experience in normal practice of his sport (Schwartz, 1973; Wright *et al.*, 1976). An alternative possibility is to install gas collection apparatus in the boat itself (R. Jackson, unpublished data). This is at best cumbersome, and current indications are that the well-trained rower can develop at least 95% of his treadmill maximum oxygen intake while in his boat (J. Daniels, personal communication).

Because the sport is weight-supported, more regard must be paid to the absolute than to the relative aerobic power (Table 68). In terms of the former criterion, the German oarsmen generally have a substantial competitive advantage although one Canadian rower has developed a $\dot{V}_{O_2(max)}$ of 6.9 l/min (Klavora, 1973). Radiographic estimates of heart volume (Mellerowicz & Hansen, 1965; Wiederman *et al.*, 1968; De Pauw & Vrijens, 1971) equally show the Germans to have a larger heart than their Belgian competitors, although in relative units the Belgians have a small advantage.

The anaerobic power of rowers is not outstanding (Di Prampero *et al.*, 1970), but probably as a consequence of a large lean body mass, the total oxygen debt following exhausting exercise is greater than in many classes of athlete (Wright *et al.*, 1976). Canadian oarsmen, who are taught to immobilize their wrists and pull the oars with hooked fingers, have only a moderate hand-grip force (54.5 kg; Wright *et al.*, 1976). However, perhaps because of a different rowing technique, much higher values (average 75 kg) are found in Japanese contestants (Yamakawa & Ishiko, 1966). The knee extension strength is good in Canadian (Wright *et al.*, 1976) and Belgian (De Pauw & Vrijens, 1971) contestants, with an extremely high figure (208 kg) reported on US rowers (Hagerman *et al.*, 1972).

Canoeing and kayak paddling

Information regarding participants in canoeing, kayak and white-water events is rather sketchy. An early German paper (Wohlfeil, 1928) set the energy cost of recreational canoeing at only 3–7 kcal/min. More recent work by Seliger *et al.* (1969) has shown caloric expenditures of 9.7 kcal/min in canoeing and 33.8 kcal/min in kayak paddling, while Horvath & Finney (1969) quote figures of 6–9 kcal/min for the Polynesian outrigger canoes found in Hawaii.

The physical characteristics of Japanese (Ishiko, 1967) and Canadian (Sidney & Shephard, 1973) competitors can be compared with rowers of the same nationality (Table 67). In both countries, the canoe and kayak paddlers are shorter and lighter than the rowers. The Canadian white-water paddlers show progressive selection with age, competitors persisting in the sport if they have appropriate anthropometric and physiological characteristics. The older men are thus taller and have a greater lean body mass than young paddlers. In the young white-water competitors, the body fat content is much as in rowers, but the older participants have a larger percentage of fat, apparently without loss of competitive advantage.

Vital capacities of the white-water paddler, whether young or old, are close to the height predicted standards for sedentary subjects of the same age (Sidney & Shephard, 1973).

With regard to working capacity (Table 68), the Swedish data suggest that canoeists have an absolute aerobic power that is the equal of rowers (Saltin & Åstrand, 1967); indeed, in relative terms, the canoeists have the higher maximum oxygen intake. Statistics for the Canadian white-water paddlers illustrate the opposing effects of ageing and progressive selection of participants (Sidney & Shephard, 1973). Men that persist with the sport into early middle age have a higher absolute aerobic power but lower relative values than younger competitors.

Maximum heart rates as measured on the treadmill are no lower than in sedentary subjects of the same age (Sidney & Shephard, 1973). However, the data on two Polynesian outrigger canoeists (Horvath & Finney, 1969) suggests that somewhat lower heart rates may be realized in a boat than on the treadmill.

Because the lean body mass is quite small, the total oxygen debt after exhausting exercise is lower for canoeists and kayak paddlers than for rowers (Sidney & Shephard, 1973).

Sailing

Ishiko (1967) has reported very low Harvard fitness scores in Japanese yachtsmen (average 74.8), and many authors have suspected that dinghy-sailing makes few physiological demands upon a competitor.

In calm weather, success in the handling of a fourteen-foot dinghy depends very much upon experience and skill, but when sailing in a stiff breeze, specific anthropometric and physiological characteristics are significantly correlated with peer-ratings of sailing ability (Niinimaa *et al.*, 1974). In particular, a high centre of gravity is helpful in counterbalancing the craft during 'hiking', while the thigh muscles must be strong, well-developed and have an unusual tolerance of sustained contractions.

Table 69. *Physical characteristics of participants in team sports*

Sample	Height (cm)	Weight (kg)	Body fat (%)	Skin-fold (mm)	Vital capacity (1 BTPS)	Author
American football						
USA						
25M	183	91.2	10.0	—	—	Behnke *et al.* (1942)
72M	179.3[a]	92.6[b]	—	—	—	Costill *et al.* (1968)
Tackles, centres	184.2	107.4	—	—	—	Costill *et al.* (1968)
Ends	183.6	84.0	—	—	—	Costill *et al.* (1968)
Guards	180.6	95.1	—	—	—	Costill *et al.* (1968)
Backs	179.6	83.8	—	—	—	Costill *et al.* (1968)
16M	185	96.4	13.8	—	6.82	Novak *et al.* (1968)
Linemen	185	108.9	—	—	—	Novak *et al.* (1968)
Ends, linebackers	185	99.2	—	—	—	Novak *et al.* (1968)
Backs	185	87.4	—	—	—	Novak *et al.* (1968)
27M						
Linemen, linebackers	187	96.0	17.6	—	—	Kollias *et al.* (1972)
Backs, ends	183	83.3	13.7	—	—	Kollias *et al.* (1972)
51M	189	102.8	12.5	—	—	Balke (1972)[c]
Baseball						
USA						
10M	182.7	83.3	14.2	—	6.58	Novak *et al.* (1968)
Basketball						
Brazil						
16M	192.2	86.7	—	—	—	Kiss *et al.* (1973)
16F	169.1	59.2	—	—	—	Kiss *et al.* (1973)
9F	167.9	61.0	—	—	—	Kiss *et al.* (1973)
Czechoslovakia						
15M	182.3	74.7	—	—	—	Seliger (1966)
Japan						
12M	183.4	75.5	—	—	—	Ishiko (1967)
Handball						
Brazil						
12M	179.8	72.4	—	—	—	Kiss *et al.* (1973)
Japan						
–M	168.0	64.4	—	—	4.47	Hukuda & Ishiko (1966)
Rumania						
–M	179.0	77.3	—	—	5.68	Hukuda & Ishiko (1966)
Ice-hockey						
Argentina						
16M	—	—	15	—	—	Di Prampero *et al.* (1970)
Canada						
16M (forwards)	—	75.9	—	9.1	5.30	H. J. Green & Houston (1975)
8M (defence)	—	81.6	—	9.3	5.87	H. J. Green & Houston (1975)
2M (goal)	—	68.6	—	6.7	5.52	H. J. Green & Houston (1975)
Czechoslovakia						
15M	174.4	65.1	—	—	—	Seliger (1966)
Japan						
18M	167.6	60.6	—	—	—	Ishiko (1967)
Rugby football						
UK						
11M	179.0	77.6	12.7	—	—	Williams *et al.* (1973)
62M						
Forwards	—	—	—	11	—	Bell (1973)
Backs	—	—	—	8	—	Bell (1973)

Table 69. (*cont.*)

Sample	Height (cm)	Weight (kg)	Body fat (%)	Skin-fold (mm)	Vital capacity (1 BTPS)	Author
Soccer						
Belgium						
23M	173.7	69.1	—	—	5.28	Damoiseau *et al.* (1966)
4M						
Attack	170.5	71.5	—	—	—	Deroanne *et al.* (1971)
Defence	180.0	76.5	—	—	—	Deroanne *et al.* (1971)
Yugoslavia						
1M (goalkeeper)	176.5	74.6	—	—	4.74	Smodlaka (1947)
1M (back)	176.4	75.1	—	—	4.98	Smodlaka (1947)
1M (centre half)	178.6	75.4	—	—	4.87	Smodlaka (1947)
1M (outside half)	167.6	66.9	—	—	4.15	Smodlaka (1947)
1M (centre forward)	175.7	75.4	—	—	4.89	Smodlaka (1947)
1M (inside forward)	174.3	68.3	—	—	4.54	Smodlaka (1947)
1M (wing forward)	168.3	65.4	—	—	4.24	Smodlaka (1947)
Zaïre						
22M	172	66	—	—	4.5	Ghesquière (1971)
Volleyball						
Brazil						
19F (14.5 yr)	165.4	58.5	—	—	—	Kiss *et al.* (1973)
Czechoslovakia						
16M	179.6	76.2	—	—	—	Seliger (1966)
-F	171.6	70.6	—	—	—	Hukuda & Ishiko (1966)
Japan						
16M	183.7	74.6	—	—	—	Ishiko (1967)
-M	181.6	74.8	—	—	—	Hukuda & Ishiko (1966)
-F	168.1	63.3	—	—	—	Hukuda & Ishiko (1966)
Brazil						
35M	176.0	74.4	10.2	—	—	de Rose (1973)
Czechoslovakia						
12M	178.8	72.5	11.6	—	—	Zelenka, Seliger & Ondrey (1967)
16M	175.1	73.2	13.3	—	5.07	Seliger *et al.* (1970)
Germany						
14M	178.5	77.6	—	—	4.6	Hollmann *et al.* (1962)
11M	176.9	75.8	—	—	4.6	Hollmann *et al.* (1962)
Japan						
19M	171.2	66.4	—	—	—	Ishiko (1967)
-M	166.8	66.2	—	—	4.32	Hukuda & Ishiko (1966)
UK						
9M	174.6	69.4	12.4	—	—	Williams *et al.* (1973)
USA						
17M	176	70.6	9.9	—	—	Serfass (1971)
USSR						
-M	172.0	75.1	—	—	5.60	Hukuda & Ishiko (1966)
Poland						
-M	183.3	80.9	—	—	—	Hukuda & Ishiko (1966)

[a] Variation 182.9 to 173.5 cm with ability group.

[b] Variation 98.5 to 86.3 kg with ability group.

[c] Cited by Kollias *et al.* (1972).

Table 70. *Working capacity of participants in team sports*

Sample	Exercise mode	Max. oxygen intake (l/min STPD)	Max. oxygen intake (ml/kg min STPD)	Anaerobic power (ml/kg min)	Max. heart rate (/min)	Author
American football						
USA						
72M						
Tackles, centres	—	—	—	156	—	Costill *et al.* (1968)
Ends	—	—	—	158	—	Costill *et al.* (1968)
Guards	—	—	—	167	—	Costill *et al.* (1968)
Backs	—	—	—	171	—	Costill *et al.* (1968)
16M	TM	5.65	59.7	—	185+	Novak *et al.* (1968)
27M						
Linemen, linebackers	TM	4.76	49.8	—	186	Kollias *et al.* (1972)
Backs, ends	TM	4.14	49.9	—	186	Kollias *et al.* (1972)
51M	TM	5.20	50.6	—	—	Balke (1972)[a]
Baseball						
USA						
10M	TM	4.47	52.3	—	188+	Novak *et al.* (1968)
Basketball						
Brazil						
16M	SM BI	4.34	50	—	—	Kiss *et al.* (1973)
16F	SM BI	2.32	38	—	—	Kiss *et al.* (1973)
9F	BI	2.64	44.4	—	—	Kiss *et al.* (1973)
Canada						
2M	BI	—	53.0	—	—	G. R. Cumming (1970)
29F	—	—	38.7	—	—	Higgs (1973)
Germany						
5M	BI	—	59	—	—	Hollmann (1972)
Field-hockey						
Canada						
1M	BI	—	51.6	—	—	G. R. Cumming (1970)
Germany						
5M	BI	—	63	—	—	Hollmann (1972)
Handball						
Brazil						
12M	SM BI	3.34	46	—	—	Kiss *et al.* (1973)
Germany						
5M	BI	—	62	—	—	Hollmann (1972)
Ice-hockey						
Argentina						
16M	SM ST	—	46[b]	140[b]	—	Di Prampero *et al.* (1970)
Canada						
17M	SK	4.04	54.7	—	193	Ferguson *et al.* (1969)
12M (forwards)	TM	4.44	57.9	202	189	H. J. Green & Houston (1975)
4M (defence)	TM	4.67	55.7	195	191	H. J. Green & Houston (1975)
2M (goal)	TM	3.83	54.8	188	205	H. J. Green & Houston (1975)
Czechoslovakia						
13M[c]	BI	4.32	54.6	—	185	Seliger *et al.* (1972)

Table 70. (*cont.*)

Sample	Exercise mode	Max. oxygen intake (l/min STPD)	Max. oxygen intake (ml/kg min STPD)	Anaerobic power (ml.kg min)	Max. heart rate (/min)	Author
Germany						
5M	BI	—	58	—	—	Hollmann (1972)
USA						
6M	TM	4.05	55	—	—	Faulkner (1969)[d]
Rugby football						
UK						
11M	BI	—	50.3	—	—	Williams *et al.* (1973)
Soccer						
Belgium						
23M	TM	4.23	59.1	—	—	Damoiseau *et al.* (1966)
4M						
Attack	TM	3.75	52.4	—	197	Deroanne *et al.* (1971)
Defence	TM	4.18	54.6	—	202	Deroanne *et al.* (1971)
Brazil						
35M	SM BI	4.54	61	—	—	de Rose (1973)
Canada						
2M	BI	—	51.2	—	—	G. R. Cumming (1970)
Ethiopia						
5M	SM ST	—	42[b]	170[b]	—	Di Prampero *et al.* (1970)
Germany						
14M	BI	5.1	65.7	—	—	Hollmann *et al.* (1962)
11M	BI	4.6	61.5	—	—	Hollmann *et al.* (1962)
Italy						
10M (goalkeepers)	SM ST	—	46.9	188	—	Caru *et al.* (1970)
19M (backs)	SM ST	—	51.0	184	—	Caru *et al.* (1970)
19M (half-backs)	SM ST	—	52.1	181	—	Caru *et al.* (1970)
23M forwards)	SM ST	—	51.0	180	—	Caru *et al.* (1970)
15M (wing forwards)	SM ST	—	52.3	174	—	Caru *et al.* (1970)
9M (centre forwards)	SM ST	—	49.6	186	—	Caru *et al.* (1970)
UK						
9M	BI	—	57.8	—	—	Williams *et al.* (1973)
USA						
-M	—	4.57	58.9	—	183	Serfass (1971)
Zaïre						
22M	BI	3.46	53	—	—	Ghesquière (1971)
Volleyball						
Brazil						
19F (14.5 yr)	SM BI	1.93	33	—	—	Kiss *et al.* (1973)
Canada						
6M	BI	—	51.9	—	—	G. R. Cumming (1970)

BI = bicycle ergometer; SK = skating; SM = submaximum; ST = steptest; TM = treadmill.
[a] Cited by Kollias *et al.* (1972).
[b] Measured in Mexico City.
[c] Max. blood lactate = 132 mg/dl.
[d] Cited by Ferguson *et al.* (1969).

Team sports

The physiological demands imposed by the various team sports have certain similarities – in particular, a demand for repeated short bursts of vigorous energy expenditure throughout a game of 60 to 90 minutes duration (Stanescu, 1970; Seliger, Navara & Pachlopniková, 1970; Seliger *et al.*, 1972). It is therefore convenient to consider jointly the research findings for American football, baseball, basketball, field-hockey, handball, ice-hockey, rugby and association football, and volleyball (Tables 69 and 70).

As with other sports, there is a problem of specificity. Where possible, discussion will thus be made not only of the sport in general, but also of differences in body characteristics that fit a player for a particular team position.

American football

The element of body contact is particularly emphasized in American football. It is thus not surprising that those electing to play this sport are very heavy (Table 69), the characteristic being most strongly developed in the professional team (the 'Green Bay Packers') studied by Balke (cited in Kollias *et al.*, 1972). Partly on account of the secular trend, and partly through more stringent selection of players, the current generation of professionals are taller, heavier, and fatter than those examined by Behnke, Feen & Weltham (1942). It is difficult to achieve currently required standards of body weight without at the same time becoming fat, and many clubs have found it necessary to impose penalties upon players who report to the training camp with an excess of adipose tissue.

Di Giovanna (1943) noted that linemen had broader chests and hips than the backfield. More recent studies show other differences with playing position. Costill *et al.* (1968) found the heaviest players were the offensive linemen (tackles and centres); the next largest were the defensive linemen (guards) and the lightest were the ends and backs. Novak *et al.* (1968) found no difference of standing height between playing positions, but commented on the gradation of body weight from backfield through ends and linebackers to four massive linemen with an average body weight of 108.9 kg. The backs were also the leanest, with about 5% body fat, while the two heaviest linemen had 26% fat. Kollias *et al.* (1972) again reported that the interior linemen and linebackers were physically larger and possessed more body fat than the backs and ends.

In keeping with their powerful body build, Novak *et al.* (1968) also demonstrated a very large vital capacity (some 119% of the height and age predicted value) and a large maximum voluntary ventilation (106% of predicted) in their team of college-level football players.

The absolute levels of aerobic power compare favourably with those for many other classes of sportsman; indeed, it is not uncommon for football players to have maximum oxygen intakes in excess of 6 l/min (Table 70). On the other hand, because of an enormous weight, the relative aerobic power of the footballer is low. Kollias *et al.* (1972) found no difference between playing positions, but Novak *et al.* (1968) commented that whereas the backs had levels greater than 50 ml/kg min, the nationally rated 'best' lineman achieved only 39.1 ml/kg min. Such poor scores are hardly surprising in view of selection for a game that demands an energy expenditure of only 6.5 kcal/min (Yamaoka, 1965) with a training programme that emphasizes 'wind sprints' and weight-lifting. However, it may account for the occasional exhaustion of supposed star players in the fourth quarter of a strenuous game with long plays.

H. H. Clarke (1973) found football players were about 7% stronger than basketball and baseball players in terms of a summated strength index (grip, back and leg dynamometry, chinning, push-ups, and vital capacity). Costill *et al.* (1968) reported a somewhat larger anaerobic power for guards and backs than for the ends and offensive linemen; however, when expressed per unit of body weight, none of the values for American footballers were outstanding.

Baseball

Novak *et al.* (1968) commented on the wide range of body characteristics of baseball players. They are often shorter than the general population (Di Giovanna, 1943). Body weights can range from 65.2 kg for an in-fielder to 96.7 kg for the pitcher, with a corresponding range in percentages of body fat, from 1% for the in-fielder to 20% for pitchers and outfielders (Novak *et al.*, 1968). The aerobic power of the baseball player is only a little higher than that of the average working man, but the summated strength index is 15% higher than for the general population (H. H. Clarke, 1973). Arm girth is particularly well developed (Di Giovanna, 1943).

Basketball

Basketball players gain a competitive advantage from a standing height and arm span that enables them to guide the ball close to the basket (Gedda *et al.*, 1968). This is reflected in statistics from both Japan and Brazil (Ishiko, 1967; Kiss *et al.*, 1973), basketball players being some 12 cm taller than soccer and handball players from the same nations. Basketball depends heavily on anaerobic work, with an average energy expenditure of 16.2 kcal/min during play (Seliger, 1966). The weight of the average team member is not outstanding in view of his unusual height, and the aerobic power is typically low, both in absolute and relative terms.

H. H. Clarke (1973) puts the summated strength of basketball players some 15% above that of the general population. G. R. Cumming (1970) ranks the grip strength fourth of fourteen sports for male players (average 61.5 kg) and second only to javelin throwers (average 42.6 kg) in female participants; however, he places the total strength (kg/kg) low in the athletic listing, with scores of 3.21 for men and 2.87 for women.

Field-hockey

Height, biacromial diameter and bicristal diameter are less than average in field-hockey players (Gedda *et al.*, 1968), while the arms are longer than anticipated. A single measurement of aerobic power on a field-hockey player (G. R. Cumming, 1970) disclosed the undistinguished value of 51.6 ml/kg min. However, five German hockey players averaged 61 ml/kg min (Hollmann, 1972).

Six of Cumming's players had hand-grips averaging 58.0 kg, with a summated muscle strength of 3.37 kg/kg.

Handball

Handball players are fairly short and light, particularly the Japanese participants. The Brazilian players have low levels of aerobic power (Table 70), but Hollmann's sample averages 62 ml/kg min.

Ice-hockey

Czechoslovak hockey players are relatively short and light (Seliger, 1966). Canadian players are accustomed to a rougher game, and perhaps for this reason all except the goal-tenders are heavier than their Czech counterparts, with a fair amount of subcutaneous fat.

Seliger (1966) has found an energy expenditure of 26.8 kcal/min during periods of play – the game is thus dependent largely upon the anaerobic release of energy. Ferguson, Marcotte & Montpetit (1969) reported that ice-hockey players developed closely comparable aerobic powers during skating and treadmill running; however, more recent studies (Green, Houston & Thomson, 1974) have suggested that only 84% of the treadmill maximum oxygen intake can be realized during skating. The Czech data and several reports from Canada and the USA set the aerobic power around 55 ml/kg min. In relative units, values are similar for forwards, defence, and goal-tenders (Ferguson *et al.*, 1969; Green & Houston, 1975), but because they are lighter, the goal-tenders have a lower absolute maximum oxygen intake. The anaerobic power, as measured by the Margaria staircase sprint, is very high in the muscular Canadian players

(Green & Houston, 1975) but quite poor in the Argentinians tested by Di Prampero *et al.* (1970).

Rugby football

Bell (1973) and Williams, Reid & Coutts (1973) have examined certain anthropometric features of the British rugby football player. As in American football, there is selection for contact, with forwards being heavier and fatter than the backs. First-class forwards are also taller than more indifferent players, while first-class backs are shorter than their second-class colleagues. The smallest players are the outside and scrum halves, while the tallest are the locks and the number eights. Bell (1973) points out the adaptive advantage of these characteristics: 'The scrum half needs to be near the ground because much of his work involves collection and distribution from ground level, while the outside half needs to be evasive and to accelerate quickly.'

The centre forward has a leg length about 10 cm greater than that of the scrum half, when adjusted for differences of trunk length. This gives him a good stride. The scrum half, on the other hand, needs stability, and this is provided by wide hips.

The energy expenditure during play (11.6 kcal/min) is substantially higher than in American football (Yamaoka, 1965). Nevertheless, the university-level team studied by Williams *et al.* (1973) had a poorer aerobic power than their sample of British professional soccer players and the university-level American footballers of Novak *et al.* (1968). The average grip strength (62.1 kg) is quite high, with forwards showing a greater explosive strength, back strength and grip strength than backs (Evans, 1973).

Soccer

The character of soccer has changed somewhat over the last thirty years. However, a substantial proportion of the energy demands are still met by aerobic means, with an average caloric expenditure of 13.1 kcal/min during play (Seliger, 1966; Seliger *et al.*, 1970). The emphasis remains on agility, balance and speed, and players are still shorter and lighter than in many team sports (Table 69), these characteristics being particularly marked in the outside halves and the wing forwards (Smodlaka, 1947). Gedda *et al.* (1968) has commented on the high leg/trunk length ratio of the soccer players. The percentage of body fat and the vital capacity are both lower than in men playing American football.

The aerobic power apparently varies little with playing position (Caru *et al.*, 1970). The highest figures are found for German and Brazilian teams

Table 71. *Physical characteristics of participants in other sports*

Sample	Height (cm)	Weight (kg)	Body fat (%)	Skinfold (mm)	Vital capacity (1 BTPS)	Author
Boxing						
Czechoslovakia						
15M	171.7	63.2	—	—	—	Šprynarova & Pařízková (1971)
Japan						
2M (fly/bantam)	161.1	53.3	—	—	—	Ishiko (1967)
3M (feather/light)	168.9	60.7	—	—	—	Ishiko (1967)
3M (welter/middle)	176.3	72.2	—	—	—	Ishiko (1967)
1M (light heavy)	179.5	82.0	—	—	—	Ishiko (1967)
Fencing						
Japan						
13M	169.2	58.9	—	—	—	Ishiko (1967)
Gymnastics						
Czechoslovakia						
10F	162.3	56.5	16.8	—	—	Šprynarova & Pařízková (1969)
Japan						
10M	160.4	57.7	—	—	—	Ishiko (1967)
Norway						
12F	159	48.9	—	9.0	—	Oseid & Hermansen (1973)
USA						
7M	178.5	69.2	4.6	—	5.8	Novak *et al.* (1968)
Judo						
Japan						
2M (light)	163.0	67.0	—	—	—	Ishiko (1967)
2M (middle)	173.1	80.5	—	—	—	Ishiko (1967)
5M (heavy)	181.0	97.1	—	—	—	Ishiko (1967)
Shooting						
Japan						
6M (clay)	166.3	56.8	—	—	—	Ishiko (1967)
6M (pistol)	166.9	60.7	—	—	—	Ishiko (1967)
3M (rifle)	169.8	67.0	—	—	—	Ishiko (1967)
Speed skating						
Sweden						
2M	181.5	74.5	—	—	—	Saltin & Åstrand (1967)
Alpine skiing						
Czechoslovakia						
9M	176.6	74.8	7.4	—	—	Šprynarova & Pařízková (1971)
Sweden						
5F	164.2	58.7	—	—	—	Saltin & Åstrand (1967)
Nordic skiing						
Sweden						
5M	174.0	67.4	—	—	—	Saltin & Åstrand (1967)

Table 71. (*cont.*)

Sample	Height (cm)	Weight (kg)	Body fat (%)	Skinfold (mm)	Vital capacity (1 BTPS)	Author
Table tennis						
Czechoslovakia						
15M	170.8	68.7	—	—	—	Seliger (1966)
Weight-lifting						
Czechoslovakia						
14M	166.4	77.2	9.8	—	—	Šprynarova & Pařízková (1971)
Wrestling						
Japan						
4M (fly/bantam)	161.2	58.9	—	—	—	Ishiko (1967)
–M (fly/bantam)	158.9	56.3	—	—	4.30	Hukuda & Ishiko (1968)
4M (feather/light)	159.0	69.8	—	—	—	Ishiko (1967)
4M (welter/middle)	176.6	81.6	—	—	—	Ishiko (1967)
4M (light heavy)	176.3	98.8	—	—	—	Ishiko (1967)
–M (light heavy)	175.3	88.9	—	—	4.97	Hukuda & Ishiko (1967)
USA						
8M	176.5	74.8	—	—	—	Ribisl & Herbert (1970)
USSR						
–M (fly/bantam)	155.0	58.8	—	—	3.95	Hukuda & Ishiko (1968)
–M (light heavy)	181.8	103.0	—	—	5.59	Hukuda & Ishiko (1968)

(Hollmann *et al.*, 1962; de Rose, 1973); the latter emphasize endurance training, can run an average of 3200 m in twelve minutes and have fared very well in international competitions. The anaerobic power is relatively good irrespective of playing position (Caru *et al.*, 1970; Di Prampero *et al.*, 1970), but grip strength is only moderate in both Japanese (47.7 kg) and Russian (55.1 kg) players (Hukuda & Ishiko, 1968).

Volleyball

Volleyball players are a little taller and heavier than soccer players of the same nationality (Table 69). Both the Harvard fitness index (115; Ishiko, 1967) and the maximum oxygen intake (51.9 ml/kg min; G. R. Cumming, 1970) indicate only a moderate level of fitness in those playing this sport. Seliger (1966) sets the caloric demand at no more than 7.4 kcal/min. The back strength and grip strength seem moderate athletic values (Hukuda & Ishiko, 1968), grip scores being 51.5 kg for Japanese men, 32.0 kg for Japanese women, 59.2 kg for Polish men and 43.7 kg for Czech women.

Table 72. *Working capacity of participants in other sports*

Sample	Max. oxygen intake (l/min STPD)	Max. oxygen intake (ml/kg min STPD)	Anaerobic power (ml/kg min STPD)	Max. heart rate (/min)	Max. blood lactate (mg/dl)	Author
Archery						
Sweden						
3F	2.3	40	—	—	—	Saltin & Åstrand (1967)
Badminton						
Germany						
5M	—	55	—	—	—	Hollmann (1972)
Boxing						
Canada						
3M	—	54.5	—	—	—	G. R. Cumming (1970)
Fencing						
Sweden						
5M	4.2	59	—	—	—	Saltin & Åstrand (1967)
3F	2.4	43	—	—	—	Saltin & Åstrand (1967)
Golf						
Germany						
5M	—	54	—	—	—	Hollmann (1972)
Gymnastics						
Canada						
1M	—	42.1	—	—	—	G. R. Cumming (1970)
3F	—	43.1	—	—	—	G. R. Cumming (1970)
Czechoslovakia						
10F	2.40	42.5	—	—	—	Šprynarova & Pařízková (1969)
Germany						
5M	—	36	—	—	—	Hollmann (1972)
Sweden						
6M	3.9	60	—	—	—	Saltin & Åstrand (1967)
USA						
7M	3.83	55.5	—	178+	—	Novak *et al.* (1968)
Judo						
Canada						
3M	—	49.1	—	—	—	G. R. Cumming (1970)
Germany						
5M	—	45	—	—	—	Hollmann (1972)
Speed skating						
Sweden						
2M	5.85	78.5	—	189	118.8	Saltin & Åstrand (1967)
3M	5.8	78	—	—	—	Saltin & Åstrand (1967)
6F	3.1	53	—	—	—	Saltin & Åstrand (1967)

Table 72. (*cont.*)

Sample	Max. oxygen intake (l/min STPD)	(ml/kg min STPD)	Anaerobic power (ml/kg min STPD)	Max. heart rate (/min)	Max. blood lactate (mg/dl)	Author
Alpine skiing						
Czechoslovakia						
6M	4.66	62.4	—	—	—	Šprynarova & Pařízková (1971)
Sweden						
6M	4.2	68	—	—	—	Saltin & Åstrand (1967)
3F	3.1	51	—	194	108.0	Saltin & Åstrand (1967)
Nordic skiing						
Sweden						
5M	5.6	82	—	179	124.6	Saltin & Åstrand (1967)
5F	3.8	63	—	—	—	Saltin & Åstrand (1967)
Table tennis						
Sweden						
3M	3.8	59	—	—	—	Saltin & Åstrand (1967)
3F	2.4	43	—	—	—	Saltin & Åstrand (1967)
Tennis						
Canada						
1M	—	55	—	—	—	G. R. Cumming (1970)
Germany						
5M	—	61	—	—	—	Hollmann (1972)
Weight-lifting						
Canada						
2M	—	55.9	—	—	—	G. R. Cumming (1970)
Czechoslovakia						
14M	3.29	43.6	—	193	—	Šprynarova & Pařízková (1971)
Sweden						
3M	4.5	56	—	—	—	Saltin & Åstrand (1967)
Wrestling						
Canada						
3M	—	53.8	—	—	—	G. R. Cumming (1970)
Germany						
5M	—	52	—	—	—	Hollmann (1972)
Sweden						
10M	4.5	56	—	—	—	Saltin & Åstrand (1967)
Various						
10M	—	53	190	—	—	Di Prampero *et al.* (1970)

Other miscellaneous sports

Archery

Archery has traditionally been the resort of girls who wish to escape the rigours of a required physical education programme. Durnin & Passmore (1967) set the energy demand at only 3.2–5.7 kcal/min, and Saltin & Åstrand (1967) showed that the aerobic power of Swedish national champions was no higher than in average university students (Table 72).

Badminton

Badminton has a somewhat higher energy cost than archery (6.3 kcal/min; Durnin & Passmore, 1967), and the aerobic power of German champions (55 ml/kg min STPD) is a little higher than that of the average worker (Hollmann, 1972).

Boxing

Boxing is a surprisingly vigorous sport. Estimates of caloric expenditures are 14.8 kcal/min (Seliger, 1966) and 9.0–14.4 kcal/min (Durnin & Passmore, 1967). Body build varies (Table 71) from an extremely light frame in the fly and bantam weight categories to quite tall and heavy individuals in the heavyweight categories (Ishiko, 1967).

In keeping with the vigour of this activity, the Harvard fitness scores of boxers are quite high (159.1 for fly and bantam weights, 131.9 for feather and light weights, 128.3 for welter and middle weights, and 132.7 for light heavyweights; Ishiko, 1967). However, G. R. Cumming (1970) found an aerobic power of only 54.5 ml/kg min in three Canadian boxers (category of competition unspecified). The grip strength of the Canadian boxers (50.5 kg) was low, ranking thirteenth out of fourteen sports, but the summated strength index (3.63 kg/kg) ranked sixth of fourteen sports.

Fencing

Ishiko (1967) reported the body build of thirteen Japanese fencers, all aspirants to the Tokyo Olympic games; their characteristics were much as in feather weight boxers (Table 71). Gedda (1968) commented on positive deviations in the length measurements of fencers attending the Rome Olympic games.

The Harvard fitness score of the Japanese team (105.0; Ishiko, 1967) was relatively low. Saltin & Åstrand (1967) quote a maximum oxygen intake of 59 ml/kg min for Swedish male champions, and 43 ml/kg min for female champions. Again, these figures are very low compared to those obtained by the same authors on other classes of sportsmen.

Golf

The energy cost of playing golf (3.7 kcal/min, Getchell, 1968; 5.0 kcal/min, Durnin & Passmore, 1967) is insufficient to make any contribution to the maintenance of endurance fitness. The aerobic power of 54 ml/kg min (Hollmann, 1972) quoted for top German players seems too large to be attributed to several kilometres of walking per day and it must reflect either initial selection or endurance training undertaken independently of golf practice.

Gymnastics

Gedda *et al.* (1968) remark on negative deviations of length, body weight and vital capacity in gymnasts attending the Rome Olypmpic games. However, a wide range of body builds and varying degrees of obesity seem compatible with success in gymnastics (Table 71).

The Harvard fitness score of Japanese gymnasts is extremely low (92.9, Ishiko, 1967). Equally, Oseid & Hermansen (1973) comment that female gymnasts in Norway have no advantage of aerobic power over their sedentary compatriots, and this view is supported by data from Canada, Czechoslovakia and Germany. However, Swedish and US gymnasts have rather higher levels of endurance fitness.

The grip strength of fifteen Candian male gymnasts was above average at 57.9 kg; on the other hand, the grip strength of four female gymnasts (37.0 kg) was relatively poor (G. R. Cumming, 1970). The summated strength indices (4.18 kg/kg for men; 3.43 kg/kg for women) rated second of fourteen sports for the men, and first of five sports for the women.

Harvard fitness scores are moderate (126.9 for lightweight, 121.0 for middle weight and 107.0 for heavyweight competitors, Ishiko, 1967). However, measurements from Canada and Germany (weights unspecified) quote poor relative levels of maximum oxygen intake (49.1 and 45 ml/kg min respectively).

Shooting

Data on the various shooting events might be thought useful, since it provides the reference point of a class of activities that make minimal physical demands and rely almost entirely upon the skill of the competitor. Gedda *et al.* (1968) note a broad biacromial diameter with negative deviations of arm span and vital capacity in shooting contestants at the Rome Olympics. Ishiko's (1967) figures for height and weight are rather typical of the average Japanese citizen, with average Harvard fitness scores of only 87.4, 86.9 and 33.3 in the three categories of contestant. However, G. R. Cumming (1970) has noted a well-developed grip strength (average 63.3 ml/kg min) and an above average summated strength (3.57 kg/kg).

Speed skating

The Swedish speed skaters seen by Saltin & Åstrand (1967) were relatively tall, and had high levels of both absolute and relative aerobic power (Tables 71 and 72).

Alpine skiing

Alpine skiing makes moderate demands on the oxygen transport system, with reported energy expenditures of 11.1 kcal/min (Seliger, 1966) and up to 10 kcal/min (Durnin & Passmore, 1967). The main physical demands are for balance and vigorous isometric activity of the leg muscles when cornering. Nevertheless, both Swedish and Czech champions have moderately high levels of maximum oxygen intake (Table 72).

Nordic skiing

Nordic skiing provides the most vigorous of all forms of sustained muscular activity. The body build of the champions is not remarkable, but the relative maximum oxygen intake is higher than for any other sport (Tables 71 and 72). One Norwegian cross-country skier currently has an aerobic power of 92 ml/kg min (Åstrand, personal communication).

Table tennis

Seliger rates table tennis as a relatively light activity (5.4 kcal/min). The maximum oxygen intakes of Swedish champions (59 ml/kg min in men, and 43 ml/kg min in women) are consistent with this view.

Tennis

Durnin & Passmore (1967) quote an energy expenditure of 5.7–8.5 kcal/min for tennis players, the upper end of this range being higher than the estimated caloric cost of badminton. Hollmann's data (1972) shows a somewhat higher aerobic power in tennis than in badminton players (Table 72).

Weight lifting

Weight lifters have a shorter lifting distance if they are not too tall. Gedda *et al.* (1968) note the negative deviation of lengths in weight lifters attending the Rome Olympiad. On the other hand, chest circumferences and body weights are above average. The selection of a favourable weight/height ratio is well illustrated by Czech statistics (Šprynarová & Pařízková, 1971). The maximum oxygen intake is at the low level encountered in boxers and wrestlers; this is hardly surprising, since the total duration of lifting is commonly no more than three seconds.

Wrestling

Wrestlers show the wide range of body builds previously remarked for boxers. Stability is an obvious advantage to a wrestler and Gedda *et al.* (1968) comment on a positive deviation of biacromial diameter, sitting height and chest circumference, with a negative deviation of standing height and arm span. G. R. Cumming (1970) also remarks on the short legs of the wrestler.

The Harvard fitness scores are mediocre for the three lighter classes of wrestler (119.3, 117.9 and 123.8; Ishiko, 1967), becoming very poor for the light heavyweight category (98.8). Measurements of PWC_{170} (12.8 kg/kg min; Ribisl & Herbert, 1970) and aerobic power (Table 72) confirm this view. On the other hand, the anaerobic power (190 ml/kg min; Di Prampero *et al.*, 1970) is quite well developed. Grip strength varies with category; while figures of 44.1 and 45.2 kg are obtained for Japanese and Russian contestants in the fly and bantam weight classes, readings rise to 58.5 and 58.8 kg for light heavyweights from the same nations (Hukada & Ishiko, 1968). G. R. Cumming (1970) quotes a grip strength of 54.2 kg in eight Canadian wrestlers, with a summated strength index of 3.70 kg/kg; unfortunately, the class of wrestling is unspecified in this report.

Lessons from the athlete

Perhaps the main lesson to be drawn from this brief survey of the working capacity of the athlete is the tremendous specialization of contestants. Diligent searching of national populations reveals individuals with anthropometric and physiological characteristics that depart from population averages by two and even three standard deviations. Early physiological studies that merely distinguished 'athletes' and 'non-athletes' or 'technique', 'strength' and 'endurance' sports were far too simplistic. There is a specific profile that contributes to success in almost every type of sport, and for many activities the optimum physical and physiological endowment varies with both playing position and style of play.

If the game demands such a body build, most nations can find players who are taller than 190 cm and weigh more than 100 kg. However, it is difficult to achieve such great weights without an increase of body fat and a diminution of relative aerobic power. The lowest percentages of body fat, around 5%, are seen in distance runners. Participants in sports that rely upon buoyancy (swimming) or use of the chest muscles have vital capacities 20% above age and height standards for the sedentary population. Massive athletes in weight-supported sports can develop an absolute aerobic power of over 6 l/min, while lighter endurance competitors can attain a relative aerobic power of over 90 ml/kg min. Contests that require explosive muscular force attract competitors with an anaerobic power of up to 220 ml/kg min, some 50% above values for the average young man. Data on muscle strength is best documented for hand-grip force – in some classes of athlete, thi is 30% higher than in the sedentary population.

How important is it for the would-be athlete to possess the peculiarities of body build and physiological characteristics shown in the above tables? G. R. Cumming (1970) comments that an ideal body helps performance, but is by no means essential. In many types of competition, it is possible to catalogue atypical individuals who have achieved striking success. A crucial factor is skill – the use that is made of natural endowment. A further important variable is the persistence of skill in the tense atmosphere of an international arena (Sidney & Shephard, 1973). Nevertheless, if the sport is physically demanding, correlations of 0.5–0.9 can be demonstrated between competitive success and relevant physiological measurements (Shephard *et al.*, 1973*a*; Sidney & Shephard, 1973; Niinimaa *et al.*, 1974), while profiles that combine the major anthropometric and physiological variables can account for 95% or more of the variance in performance of quite closely matched competitors (Shephard, 1975*c*).

8. The growth of working capacity

In this chapter, we shall examine the influences of environment and heredity upon the growth and development of working capacity, considering also problems of methodology, activity patterns and facets of athletic performance peculiar to the child.

Considerations of methodology

Types of survey

There are three potential approaches to the study of growth and development (Shephard & Rode, 1975). The traditional method is well exemplified by the IBP study in Saskatoon (Bailey *et al.*, 1974*b*) and an analogous study of French Canadian schoolchildren (Shephard *et al.*, 1974*b*). Cohorts of children are followed in a longitudinal fashion, with observations being repeated at intervals of six to twelve months, for a total period of ten to fifteen years. This technique allows a rather precise definition of growth spurts, but it is expensive and time-consuming, particularly if the tests to be used are more than simple measurements of body dimensions. Further, in primitive societies affected by rapid acculturation, there is a danger that normal processes of growth will be confounded with changes brought about by alterations in habitual physical activity, diet and other environmental variables.

The cross-sectional survey is a much cheaper alternative, used for example in many of the Czechoslovakian and Japanese contributions to the IBP (Seliger, 1970; Hasegawa *et al.*, 1966; Ikai, Shindo & Miyamura, 1970; Matsui *et al.*, 1971). In a relatively stable industrialized society, a useful overall picture of development is obtained. Thus Hermansen & Rodahl (1976) found little difference in the apparent growth of aerobic power in northern Norway when this was assessed by cross-sectional rather than longitudinal studies. However, growth spurts are necessarily flattened and sometimes displaced by the averaging of cross-sectional data (Harrison *et al.*, 1964). Furthermore, if the community is undergoing rapid acculturation, observed differences of physiological variables between younger and older children reflect not only the course of growth but also the response to cumulative differences in the environmental experiences of the various age groups. This point was well demonstrated in a study of children in Stuttgart (Tanner, 1962). Food shortages experienced during World War II led to a reversal of the previous secular trend to an increase of height and weight, the stunting of growth being more marked in adolescents than in younger children. Many of the population developed a 'catch-up' phase of accelerated growth after the war, as food again

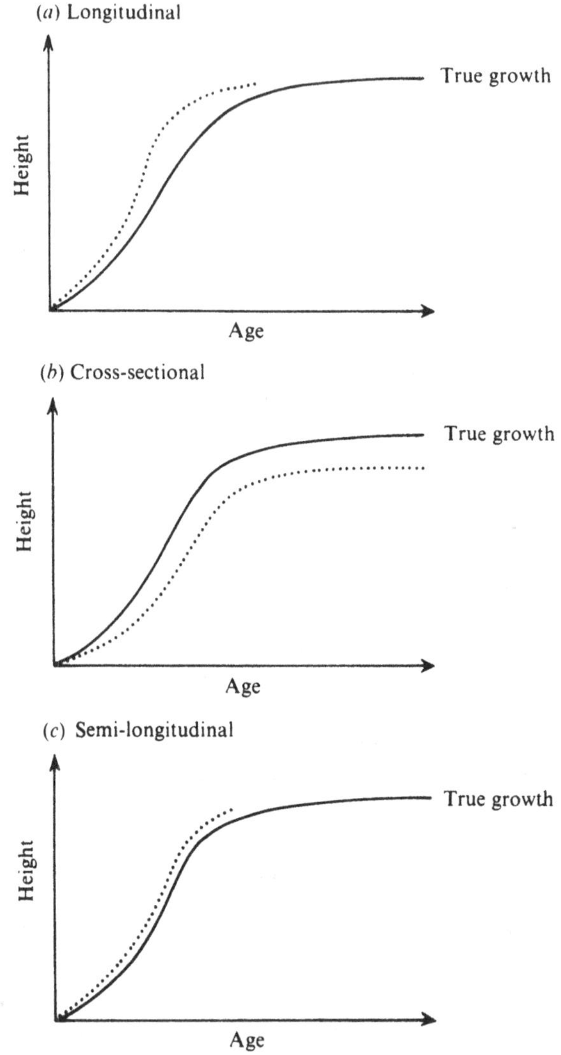

Fig. 9. Systematic errors in the analysis of growth and development by (*a*) longitudinal, (*b*) cross-sectional, and (*c*) semi-longitudinal methods (Shephard & Rode, 1975). Broken lines indicate the apparent form of the growth curve in a situation of rapid acculturation, whilst solid lines indicate the true form.

became plentiful. Throughout both the period of privation and the subsequent recuperation, a cross-sectional survey would necessarily have given a misleading impression of the course of normal growth. Naturally, many primitive societies face comparable episodic shortages of food due to such hazards as a succession of hard winters or sustained droughts.

The types of systematic error that develop in growth curves during acculturation are illustrated in Fig. 9. Improved health and nutrition exaggerate the apparent growth rate if a child is followed longitudinally. On the other hand, the older children in a cross-sectional survey have battled a more hostile environment, and their statistics lead to underestimation of the growth rate for the community. Anthropological data are not subject to learning artefacts if the observer is initially well trained. However, the more sophisticated measures of working capacity require cooperation from the subject, and there is then a danger that improved test scores may reflect not the process of growth, but rather habituation and a learning of test procedures by the child (Shephard, 1969*b*).

The compromise solution, adopted for example in IBP studies of the growth and development of working capacity among the Igloolik Eskimos (Rode & Shephard, 1973*d*), is to combine a cross-sectional study with a brief longitudinal study, lasting but one year (the so-called 'semi-longitudinal' approach). However, a full longitudinal study remains necessary to examine the effects of vigorous training upon growth (page 228).

Measurement of working capacity in children

The IBP human adaptability project established a specific working party to consider the problems of exercise testing in children (Shephard *et al.*, 1968*c*). Certain difficulties have been noted in previous chapters. It is hard to persuade children to sustain maximal effort, and perhaps for this reason there are often difficulties in defining a plateau of maximum oxygen intake (page 26). Subsidiary criteria of maximum effort are equally unsatisfactory. Maximal blood lactates (80 mg/dl) are less than in young adults and an anxious child may approach his maximal heart rate when exercising at no more than 80% of maximum oxygen intake (Shephard *et al.*, 1974*a*). There are wide differences in reported heart rate maxima. The classical material of P. O. Åstrand (1952) noted values of 210–215/min in young children, but most other authors have terminated their tests at pulse rates of 195–200/min (Shephard *et al.*, 1968*c*; Shephard, 1971*c*).

Was Åstrand particularly successful in motivating his children to an all-out effort, with other investigators obtaining 'near-maximal' rather than 'maximal' data? There is some experimental evidence against such a conclusion. The average terminal lactate readings in the Toronto schoolchildren studied by the IBP working group (Shephard *et al.*, 1968*c*) were 78.7 mg/dl in the boys and 77.1 mg/dl in the girls, almost identical with the figures reached in the Stockholm series. When the Toronto sample was subdivided to distinguish three categories of maximal effort ('good', 'moderate' and 'poor') on the basis of terminal blood lactates and observer ratings of performance (Table 12), even the children making a good effort (average blood lactate readings 105 mg/dl in the boys and 109 mg/dl in the girls) had final heart rates (195±7, 199±7 respectively) much lower than the figures reported by Åstrand (1952). The reason for the high values in the Swedish children thus remains in doubt. From the practical point of view, the directly measured maximum oxygen intake exceeded the figure predicted by the Åstrand nomogram in children making a good or a moderate effort, and only in those making a poor effort (seven of twenty-three boys, five of twenty-four girls) was there a small underestimate by the direct measurement (9% in the poor boys, 1% in the poor girls). It is thus likely that in populations where heart rates averaged 195–200/min, about three-quarters of the group reached their maximum oxygen intake and that downward biassing by the remaining quarter of the subjects diminished the average $\dot{V}_{O_2(max)}$ by only 1–2%.

If submaximum tests are used to predict aerobic power, uncertainty regarding the maximum heart rate is an equal source of embarassment with young children. The Åstrand nomogram applies an age correction equivalent to assuming a maximum heart rate of 212/min in the boys and 207/min in the girls. Other procedures assume lower maxima and apparently lead to substantial underestimation of the true maximum (see, for example, Hermansen & Oseid, 1971).

In view of the difficulties encompassing both maximum and submaximum tests, G. R. Cumming (1974) has proposed that three- to six-year-old children be examined by timing a progressive treadmill walk. The child starts on a 10% slope, and the gradient is increased by 1% per minute to a final value of 18 to 28% (depending on fitness and motivation) with final heart rates in the range 160 to 224/min. Klimt, Pannier & Paufler (1974) and Mrzena & Máček (1974) have each developed rather similar tests for pre-school children.

When dealing with older children, most authors formally associated with the IBP have measured the oxygen cost of work rather than the corresponding work load. Nevertheless, there are many population studies of PWC_{170}. Early studies of mechanical efficiency suggested that when children rode the bicycle ergometer, they attained a rather lower figure than the supposed adult average of 23% (Shephard, 1971*c*). One important variable is the design of bicycle; not all authors have taken the care to modify their ergometer to permit efficient exercise by children. Klimt & Voigt (1971) have suggested the optimum is a work load of 12 kg m/min per kg (close to the PWC_{170}), with a pedal rate of 50 revolutions per minute, a knee joint angle of 150 degrees, and a crank length increasing from 13 cm at six years to 15 cm at eight and ten years. A further variable is the rate of increase in work load. Substantially higher PWC_{170} figures are obtained from a continuously increasing task (as used by some German and Dutch laboratories) than from the step-wise increase of load proposed

Table 73. *Net efficiency of bicycle ergometry in children*

Population	Age (yr)	Work load (kg m/min)	Net mechanical efficiency (%)	Author
Canada				
Eskimo boys	11–13	470	18.4	Shephard & Rode (unpublished data)
	14–16	658	18.8	Shephard & Rode (unpublished data)
	17–19	788	18.9	Shephard & Rode (unpublished data)
Eskimo girls	11–12	400	19.6	Shephard & Rode (unpublished data)
	13–15	446	18.8	Shephard & Rode (unpublished data)
	16–19	493	18.6	Shephard & Rode (unpublished data)
White boys	10–12	240	20.1	Shephard *et al.* (1968*c*)
	10–12	360	25.6	Shephard *et al.* (1968*c*)
	10–12	420–450	22.2	Shephard *et al.* (1968*c*)
	10–12	540–600	22.8	Shephard *et al.* (1968*c*)
White girls	10–12	240	19.4	Shephard *et al.* (1968*c*)
	10–12	360	22.7	Shephard *et al.* (1968*c*)
	10–12	420–450	22.8	Shephard *et al.* (1968*c*)
	10–12	540–600	26.6	Shephard *et al.* (1968*c*)
Czechoslovakia				
Rural boys	12	235	22.5	Seliger (1970)
	12	352	25.6	Seliger (1970)
	12	467	26.8	Seliger (1970)
	12	846	34.4	Seliger (1970)
	15	326	26.2	Seliger (1970)
	15	487	28.3	Seliger (1970)
	15	649	28.5	Seliger (1970)
	15	1121	33.8	Seliger (1970)
Urban boys	12	235	19.4	Seliger (1970)
	12	352	22.7	Seliger (1970)
	12	467	24.1	Seliger (1970)
	12	846	30.2	Seliger (1970)
	15	326	22.5	Seliger (1970)
	15	487	25.0	Seliger (1970)
	15	649	25.9	Seliger (1970)
	15	1121	30.6	Seliger (1970)
Norway				
Boys	10	357	22.6	Andersen & Ghesquière (1972)
	14	567	22.3	Andersen & Ghesquière (1972)
Girls	10	276	20.8	Andersen & Ghesquière (1972)
	14	484	22.0	Andersen & Ghesquière (1972)
USA				
White girls	7–9	300	24.7	Wilmore & Sigerseth (1967)
	7–9	600	25.5	Wilmore & Sigerseth (1967)
	7–9	900	27.0	Wilmore & Sigerseth (1967)
	10–11	300	28.7	Wilmore & Sigerseth (1967)
	10–11	600	29.9	Wilmore & Sigerseth (1967)
	10–11	900	26.2	Wilmore & Sigerseth (1967)
	12–13	300	23.8	Wilmore & Sigerseth (1967)
	12–13	600	26.5	Wilmore & Sigerseth (1967)
	12–13	900	24.9	Wilmore & Sigerseth (1967)

for IBP testing (Mocellin *et al.*, 1971*a*). Data from the IBP working party (Shephard *et al.*, 1968*c*), from the Czech IBP studies (Seliger, 1970) and from Norway (Andersen & Ghesquière, 1972) but not from the US (Wilmore & Sigerseth, 1967) support the view that at light loads mechanical efficiency is less than in the adult. However, there seems a progressive increase of efficiency with loading (Table 73), perhaps partly because of the onset of anaerobic work, and at the higher loadings the apparent net efficiency is at least as great as in older subjects. Because of this trend, a high efficiency (at least 25%) can be calculated from the slope of the oxygen consumption/work performance line (Rutenfranz & Mocellin, 1967; Andersen *et al.*, 1971). The Czechoslovakian studies suggest a slightly higher mechanical efficiency for rural communities. However, Eskimo children (like their adult counterparts) have a low mechanical efficiency.

Sample definition

In rural areas and primitive communities, sampling problems can arise because substantial numbers of children do not attend school regularly. The ages of the population may also be uncertain. Ashcroft & Lovell (1966) found that in Jamaica, only 65–75% of children on the school roll were available for measurement of height and weight at any one visit. However, those who were absent did not seem to bias growth curves unduly relative to samples known to be representative. A more serious source of difficulty was that many children had been assigned approximate birth dates coinciding with the commencement of the school term. Thus, 21.5% of children were recorded as born in January and 13.6% in June. Abbie (1967) suggested that estimates of the age of Australian aboriginal children were accurate to within one month in the first year of life, to three months up to two years of age, to within six months from two to five years, to within one year from ages five to twelve, and to within two years from ages twelve to eighteen. Sinnett & Whyte (1973*a*, *b*) and C. T. M. Davies, Mbelwa & Doré (1974) have also commented on difficulties in establishing the ages of children in New Guinea and in Tanzania respectively. In other areas such as St Kitts–Nevis (Ashcroft, Buchanan & Lovell, 1965*b*) and the Canadian Arctic (Rode & Shephard, 1973*d*), there has been a compulsory registration of births for quite a number of years.

A further age problem can arise in comparing results between investigators. One author may record data over the age span 10½–11½ years, another 11–12 years, and a third 10½–11 and 11–11½ years. Detailed examination of apparently discordant results should take account of this factor although its importance can be reduced if data are standardized, for example per unit of body height or per unit of body weight.

Standardization of data

There are various possible bases of data standardization, and the choice of method becomes quite critical when assessing the effects of training upon growth, when comparing boys and girls, and when contrasting populations that differ in maturity. Thus, Andersen & Ghesquière (1972) commented that when expressed per unit of body weight, the maximal oxygen intake of Norwegian girls dropped from age ten to fourteen, whereas in boys it increased. However, when data were expressed relative to standing height, there was an increase in both sexes, since much of the disadvantage of the older girls was due to an increase in their percentage of body fat. Again, it might be concluded that a specific training regimen or a particular environment had increased the working capacity of a group of children, when in fact it had done no more than alter the timing of the adolescent growth spurt, with its associated gains of strength in boys and of body fat in girls. Two possible approaches to this dilemma are (i) to seek independent assessments of maturation such as skeletal age, and (ii) to express the various measurements of working capacity as theoretically determined power functions of standing height.

Radiographic assessment of maturity

If one is dealing with the fairly homogenous adolescent population of an industrialized nation, radiographic measurements of bone age can make a useful contribution to the description of variance in both performance tests and laboratory measurements of working capacity (G. R. Cumming, Garand & Borysyk, 1972; G. R. Cumming, 1973*a*). However, in younger children it is difficult to establish bone age more closely than to the nearest six months, and the reported values have a marked correlation with calendar age ($r = 0.94$); thus, radiographic observations do not add significantly to the description of working capacity provided by calendar age, height and weight (Shephard *et al.*, 1975*b*).

The effects of physical activity upon skeletal maturation have been examined in a large population of French Canadian schoolchildren drawn equally from an urban and a rural milieu (Shephard *et al.*, 1975*b*). Multiple regression equations for the maturation of the wrist (Greulich & Pyle, 1959) and the mandible (Garn, Rohmann & Silvermann, 1967) were calculated by scaling females, a rural environment and an hour of additional required physical activity per day as one, leaving males, an urban environment and control activity patterns scaled as zero. The resultant equations were:

$$\text{Age (wrist)} = 4.56\ (\text{sex}) + 0.39\ (\text{milieu}) - 3.59\ (\text{activity}) - 13.0$$
$$\text{Age (mandible)} = 1.33\ (\text{sex}) + 4.53\ (\text{milieu}) + 1.47\ (\text{activity}) - 0.28$$

Significant effects on wrist maturation were demonstrated for sex ($F_{1,614} = 23.6$) and added physical activity ($F_{1,614} = 13.5$). Suzuki (1970) also described a negative effect of physical activity upon maturation of the wrist. In the case of the mandible, the main effect was an advancement of maturation in the rural milieu ($F_{1,614} = 33.9$); the speeding of maturation in girls ($F_{1,614} = 3.44$) and in active children ($F_{1,614} = 3.07$) did not reach statistical significance.

Comparisons between populations are frequently complicated by substantial differences in the average level of maturation at a given age. Even within white populations, maturation is apparently delayed in small and economically under-privileged communities (Tanner, 1962; Wolanski, 1966, 1972). The group of children studied by the IBP working party in Toronto (Shephard *et al.*, 1968*c*) had a maturity rather comparable with that of the economically privileged Boston schoolchildren examined by Greulich & Pyle (1959). However, many other investigators have found their populations to be systematically less mature. The French Canadian children (Shephard *et al.*, 1975*d*) were on average six to eight months retarded relative to the Greulich–Pyle standards; as can be seen from the multiple regression equations, in their case maturation proceeded faster in a village (where there was some industry) than in a city of 150000 population.

Radiographic studies of primitive communities have generally shown quite large lags relative to the Boston standards (Bereni *et al.*, 1971; Blanco *et al.*, 1972; Clegg *et al.*, 1972; Michaut, Niang & Dan, 1972; Edgren *et al.*, 1974). The Finnish Lapps, for example, are at least one year behind the Greulich–Pyle standards (Edgren *et al.*, 1976). One recent study from the Fort Chimo region of Arctic Quebec (Eyman & Salter, 1976) found no retardation of hand and wrist maturation relative to a combination of Greulich–Pyle and Tanner–Whitehouse standards (Tanner *et al.*, 1975). The present author knows of no other completed studies of wrist maturity from the Canadian Arctic. However, general growth patterns (Rode & Shephard, 1973*d*) and the age of menarche (Hildes & Schaefer, personal communication) both suggest that development of the Canadian Eskimo is delayed relative to the white Canadian. The situation in the Fort Chimo area may be modified by admixture of white and North American Indian genes, together with the economic advantages of frequent air connections with Montreal. However, much still remains to be learnt about the course of maturation, and there have been reports (Garn, 1965) that in terms of dental age, American Indians, American Negroes, Formosans, and East Africans are all advanced relative to white North Americans.

Power functions of standing height

A theoretical basis for the standardization of performance data during the period of growth has been proposed by Scandinavian work physiologists (von Döbeln, 1966; Asmussen & Christensen, 1967; P. O. Åstrand & Rodahl, 1970). Variables such as leverage and stride length were considered proportional to standing height (H), muscle force (a function of muscle cross-section) as proportioned to the square of height, and volumetric variables (body mass, total body haemoglobin, heart volume and lung volumes) as proportional to the third power of height.

Unfortunately, many of the variables of interest to the work physiologist (such as aerobic power) also include the dimension of time. Von Döbeln (1966) made the ingenious but debatable suggestion that time (t) could be expressed as an equivalent length (L). His reasoning was as follows:

$$\text{acceleration} = \text{force/mass} = L^2/L^3$$
$$\text{acceleration} = L/t^2$$
$$\text{Thus } L/t^2 = L^2/L^3\text{, and } t \propto L$$

If this reasoning be accepted, then one can proceed to show that measures of the rate of working vary as L^2, or body mass$^{2/3}$:

$$\text{work} = \text{force} \times \text{distance} = L^2 \times L = L^3$$
$$\text{rate of working} = \text{work/time} = L^3/L = L^2$$

How well does this method of data standardization work in practice? For some variables, the agreement with theory is remarkably close. Asmussen took Åstrand's (1952) cross-sectional data on vital capacity, and calculated exponents of $H^{3.1}$ for boys and $H^{3.0}$ for girls. However, the concord may have been fortuitous. Firstly, it can be objected that Asmussen was using data on children who had grown through the privations of war and any subsequent 'catch-up' phase. Again, other authors have obtained rather different results. Thus, De Muth, Howatt & Hill (1965) demonstrated exponents of $H^{2.81}$ and $H^{2.82}$ for much larger samples of US boys and girls; here, one may presume that the potential for development of the vital capacity was not realized due to progressive inactivity during adolescence. In contrast, Canadian Eskimo boys exceeded the theoretical exponent, vital capacity varying as $H^{3.28}$ and $FEV_{1.0}$ as $H^{3.31}$, while in Eskimo girls the corresponding exponents were $H^{3.01}$ and $H^{2.50}$ (Rode & Shephard, 1973*a*).

For other variables, the discrepancies between theory and practice have been even wider. Asmussen found exponents of $H^{2.89}$ for the muscle strength and $H^{2.90}$ for the aerobic power of boys, rather than the theoretical curves of $H^{2.0}$. Again results have varied with the milieu. Thus, Bailey *et al.* (1974*b*) recently reported an exponent of $H^{2.50}$ for the aerobic power

Table 74. *A determination of growth spurts from longitudinal and cross-sectional data on Canadian Eskimo children*

	Age of maximum growth			
	longitudinal study		cross-sectional study	
Variable	Boys	Girls	Boys	Girls
Standing height	13.3	n.m.	13.3	n.m.
Body weight	13.8	12.8	> 17.0	12.3
Skinfold thickness	n.m.	n.m.	n.m.	n.m.
Grip strength	n.m.	n.m.	> 17.0	12.3
Leg extension strength	n.m.	n.m.	> 17.0	< 9.5
Aerobic power	13.8	n.m.	n.m.	13.0

From Shephard & Rode (1975).
n.m. = no clearly defined maximum.

of white Canadian boys participating in the longitudinal IBP study at Saskatoon.

Von Döbeln & Eriksson (1973) have suggested that departures from the theoretical exponents reflect a lack of normality in the children due to insufficient physical activity. There certainly seems some gradient of exponents with differences in the average level of habitual activity between populations. However, for the present, the basis of the power exponents remains rather insecure to permit their use in comparing children who have the same calendar age but are at differing levels of maturity.

Definition of growth spurts

The timing of growth spurts and their inter-relationships among different variables can be examined by fitting polynomial regressions of the type

$$Y = X_1(A) + X_2(A)^2 + X_3(A)^3 + X_4 \ldots$$

where Y is the variable of interest (for example, Δ standing height in a semi-longitudinal study or standing height in a cross-sectional study), A is the age of the individual, and X_1, X_2, X_3 and X_4 are constants. The type of information obtained by differentiating such curves is illustrated in Table 74. Unfortunately, the size of many primitive communities is rather small to permit accurate definition of maximum slopes. Our data from Igloolik suggest that in terms of standing height there is little disagreement between the differentiated cross-sectional curve and the semi-longitudinally observed period of maximum growth. On the other hand, the older boys as seen in the cross-sectional survey are heavier than would be predicted

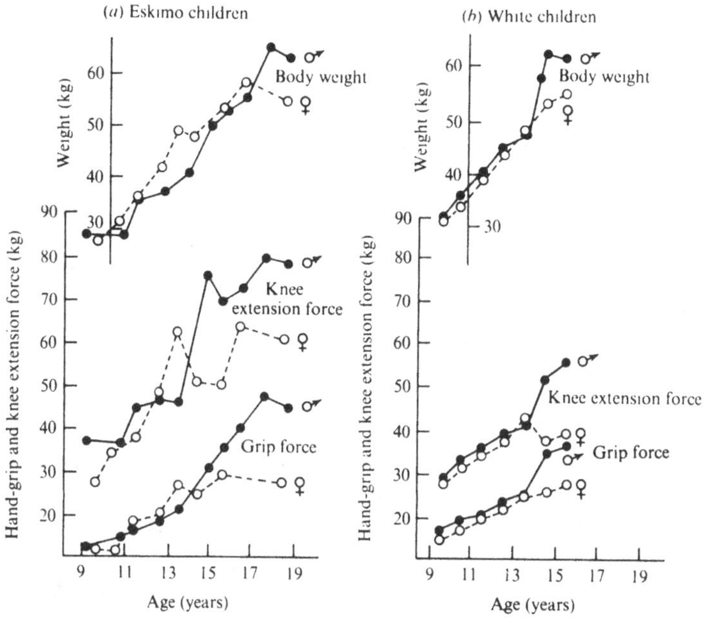

Fig. 10. Cross-sectional data for growth of body weight, knee extension force and hand-grip force. Data for Igloolik Eskimo children (Rode & Shephard, 1973*d*) and white children of Edmonton, Alberta (Howell *et al.*, 1965).

from current semi-longitudinal growth curves. Both types of survey show the absence of any clear age for a maximum rate of increase of skinfold thickness. The cross-sectional study also brings out the feature, well-recognized in white communities, that boys continue to develop their strength into the late teens, whereas girls reach a peak of development relatively early in puberty. It can further be seen that the height spurt precedes the weight and strength spurts, casting added doubt on the theoretical basis of the Scandinavian dimensional analysis.

One final feature of the growth curves is that all spurts occur earlier in the girls than in the boys. For this reason, there is a brief pre-pubertal period when the girls are heavier than the boys (Fig. 10). In a typical white population such as the Edmonton schoolchildren studied by Howell, Loiselle & Lucas (1965), the advantage held by the girls is scarcely significant, but in the Eskimo community of Igloolik (where the girls probably have greater encouragement to reach their potential), female subjects are heavier and stronger than the boys between the ages of eleven and fourteen years.

The secular trend and the genetic basis of growth

In the past, many authors have attributed the distinctive growth patterns of various populations to their genetic characteristics. Thus, the short adult stature of the Lapps and the Eskimos was attributed to their Mongoloid origin (Bryn, 1932), and a racial basis was suggested for the fact that French Canadian schoolchildren living in Montreal were shorter than their English-speaking contemporaries in other parts of North America (Demirjian, Jenicek & Dubuc, 1972). The Fels Institute has even proposed the use of parent-specific standards of stature (Garn, 1965).

However, over the past few decades many primitive groups have shown a rapid secular trend to an increase of adult stature and other supposedly genetically-determined distinctive features. The phenomenon can be shown rather crudely by a simple plot of adult heights against age (Fig. 11), although such graphs augment the true secular trend with certain effects of ageing (compression of inter-vertebral discs and increasing kyphosis). Lewin *et al.* (1970) and Skrobak-Kaczynski & Lewin (1976) were able to find precise measurements on Lapps born in 1885–1905, and thus to calculate true secular trends relative to their descendants born in the period 1945–1950. The trend was most dramatic for the Pasvik Skolts (1.4 cm/decade in the men, 1.3 cm/decade in the women), with a smaller change in the Suenjel Skolts (0.8 cm/decade in men, 0.6 cm/decade in women). In some populations such as the north coastal Alaskan Eskimos, part of the increase in stature has been attributable to hybridization (Jamison, 1970), but in the Lappish group Skrobak-Kaczynski & Lewin (1976) were careful to select only subjects with no history of racial admixture for at least four generations. They concluded that the secular trend was rapidly obliterating morphological differences between the Lapps and the surrounding Scandinavian populations, that the change in physical characteristics was an expression of improved living conditions, and that the current plasticity of traditional anthropological measurements cast serious doubt on much of the work on hominid taxonomy.

Garn (1965) concluded more cautiously that while improved nutrition and higher standards of medical care would bring some southern European and African populations into the North American ranges of height and weight, this was 'probably not true for the peoples of Japan, the Philippines, Malaysia and Indonesia, or for Mayan Indians of Central America'.

There is as yet little unequivocal evidence on how the secular trend is affecting working capacity. Within a given population, the short individuals commonly have a small absolute aerobic power, but a good maximum oxygen intake relative to body weight; the same usually holds for muscular strength. One might thus anticipate that the secular trend would lead to

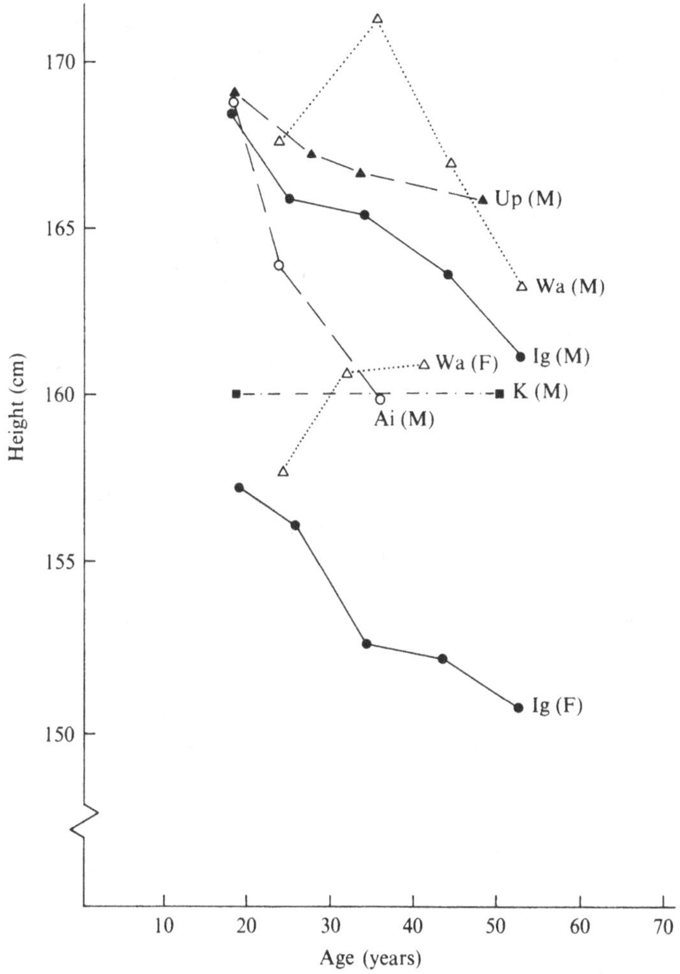

Fig. 11. The apparent 'secular trend' in the standing height of circumpolar populations. Ai = Ainu; Ig = Igloolik Eskimos; K = Kautokeino Lapps; Up = Upernavik Eskimos; Wa = Wainwright Eskimos. (From Shephard, 1975*a*, by permission of the editor, *Circumpolar Peoples*, and Cambridge University Press.)

an increase of absolute and a decrease of relative working capacity. With regard to white subjects, a comparison of Robinson's data (1938) and more recent statistics for US adults might give some support to such a view. Equally, within the Igloolik Eskimo population (page 107), acculturation is associated with both an increase of standing height and a decrease of relative aerobic power. The typical adult at Igloolik is shorter than the Eskimos of Wainwright and Upernavik, but he has a much higher relative aerobic power and a somewhat higher absolute aerobic power than the other two populations. However, in all of these comparisons, it is difficult to disentangle the relative effects of secular trends in body size and parallel trends to a diminution of habitual activity.

Body composition

Height and weight

The secular trend towards a greater standing height does not seem to be accompanied by proportionate increases of bone breadth and body weight – in consequence, the body form has become progressively thinner, a process that has been termed 'gracilization' (Skrobak-Kaczynski & Lewin, 1976; Suchý, 1971). It is less clear how the relative proportions of body fat and lean tissue have been changed by this phenomenon.

Current indications are that environment may exert its most profound effects on body composition during the late intra-uterine period and in the first six months of extra-uterine development. Subcutaneous fat is almost absent from infants that are born two months prematurely, but at term there are already well-marked sexual differences, with particularly thick skinfolds in the infants born of hyperglycaemic mothers (Pařízková, 1963). During the first six months of life, over-feeding can increase the total number of fat cells in the body, whereas subsequently it merely leads to an over-loading of existing cells (Knittle & Hirsch, 1968); thus, the present trend for primitive communities to replace prolonged breast-feeding by early weaning and the use of artificial baby foods (Sayed *et al.*, 1976*b*) may have permanent consequences for the fitness and working capacity of the rising generation in such populations.

The differences of growth curves between primitive and more-developed communities can be illustrated by cross-sectional data from Canada (Fig. 12). At all ages, the tallest children are to be found in the industrial cities of Anglophone Canada. The Eskimo children are much smaller, showing a later growth spurt and a lesser adult height. French Canadian children occupy an intermediate position, those from the large industrial metropolis (Montreal) being somewhat larger and more advanced than those from smaller urban centres and rural areas.

Most other studies have confirmed this pattern. Thus, in India, urban boys develop both their height and growth spurts one year earlier than those from rural areas, with the urban group attaining a 5 cm taller final stature and several kilograms heavier final body weight than the rural population. Baker and his associates (personal communication) commented that Sherpas living at 3500–4500 m showed no adolescent growth spurt. On the other hand, growth continued into the third decade of life. A disparate report from Tanzania (C. T. M. Davies *et al.*, 1974) found children in Dar Es Salaam and surrounding villages were short and light,

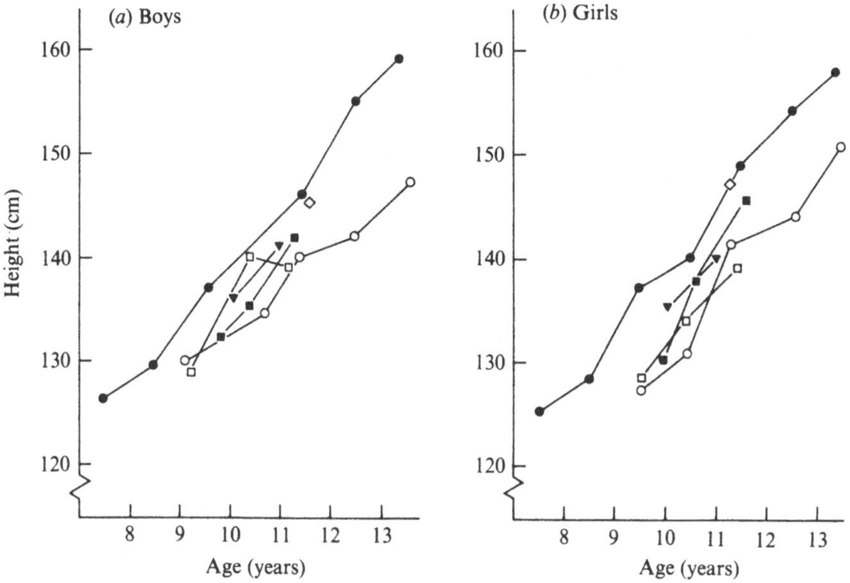

Fig. 12. Standing height growth curves for (*a*) boys and (*b*) girls. Cross-sectional data for Canadian Anglophones (Edmonton ●, Howell *et al.* 1965; Toronto ◇, Shephard *et al.*, 1968*c*); Francophones (metropolitan ▼ Demirjian *et al.*, 1971; urban ■, rural □, Shephard *et al.*, 1975*a*); and Eskimos (○, Rode and Shephard, 1973*d*).

but in this population there were no significant urban/rural differences and the timing of growth spurts was much as in UK samples.

Recent studies of the Canadian Eskimos (Sayed *et al.*, 1976*b*) suggest that their characteristic height and weight is apparent even in early childhood. Over the first five years of life, 70% of Eskimo boys and 77% of the girls lie below the 25th percentile for the US population of the same age. However, in primitive groups as in urban and metropolitan children, the girls commence their pubertal spurt earlier than the boys, and there is thus a period when the girls are taller than boys of the same age.

Cross-sectional curves for the growth of body-weight are illustrated in Fig. 10. Again, the pubertal spurt occurs earlier in the girls, and there is a stage when they weigh more than the boys. In the more-developed countries, a substantial part of the pubertal weight gained by the girls is fat, and they only exceed the weight of the boys for a very short time. However, in more active and primitive populations such as the Canadian Eskimo, the girls appear to realize their potential for muscular development more fully; they are thus heavier than the boys for about three years (age eleven to fourteen years). The importance of nutrition to the rate of maturation is illustrated by cross-sectional data from seven Latin American and seven Asian countries (Frisch & Revelle, 1969); the weight spurt

Table 75. *Weight (W) and weight/height ratio (W/H) for populations of various more-developed countries*

Population	Age (yr)	Boys *W* (kg)	Boys *W/H* (kg/cm)	Girls *W* (kg)	Girls *W/H* (kg/cm)	Author
Belgium	6	20.5	0.176	—	—	Hebbelinck & Borms (1969)
	8	24.5	0.194	—	—	Hebbelinck & Borms (1969)
	10	30.0	0.219	—	—	Hebbelinck & Borms (1969)
	12	35.5	0.244	—	—	Hebbelinck & Borms (1969)
Strong	14.7	56.0	0.344	—	—	van Uytvanck & Vrijens (1966)
Weak	14.8	47.8	0.293	—	—	van Uytvanck & Vrijens (1966)
Canada	10	36.0	0.252	33.7	0.240	Shephard *et al.* (1968*c*)
	11	38.6	0.264	38.0	0.254	Shephard *et al.* (1968*c*)
	11	37.7	—	36.2	—	Howell & MacNab (1968)
	11–12	37.0	0.253	40.0	0.272	G. R. Cumming & Cumming (1963)
	12	40.5	0.274	42.9	0.275	Shephard *et al.* (1968*c*)
Francophone						
Metropolitan	6	20.2	0.178	19.9	0.177	Demirjian *et al.* (1972)
	8	24.8	0.197	25.5	0.206	Demirjian *et al.* (1972)
	10	30.3	0.224	30.9	0.228	Demirjian *et al.* (1972)
	12	38.6	0.265	38.8	0.263	Demirjian *et al.* (1972)
	14	47.9	0.294	49.5	0.314	Demirjian *et al.* (1972)
	16	58.6	0.343	51.4	0.320	Demirjian *et al.* (1972)
Rural	10–12	32.5	0.238	31.0	0.230	Shephard *et al.* (1975*a*)
Urban	10–12	31.5	0.229	35.3	0.251	Shephard *et al.* (1975*a*)
Czechoslovakia	10.9	37.1	0.257	—	—	Šprynarova (1966)
	11.5	34.2	0.242	—	—	Seliger & Pachlopková (1967)
Rural	12	38.7	0.261	39.7	0.265	Seliger (1970)
	15	54.7	0.327	53.9	0.332	Seliger (1970)
Urban	12	38.9	0.261	39.7	0.263	Seliger (1970)
	15	58.0	0.342	54.5	0.333	Seliger (1970)
France	8	24.4	0.196	23.8	0.193	Sempé (1964)
	10	29.4	0.219	29.2	0.218	Sempé (1964)
	12	35.7	0.248	37.6	0.258	Sempé (1964)
	14	44.7	0.287	46.3	0.300	Sempé (1964)
Germany	11	38.7	0.344	38.5	0.266	Rutenfranz (1964)
	11	—	—	35.5	0.246	Hollman *et al.* (1967)
Holland	12–13	42.0	0.268	—	—	Kemper (1973)
Hungary	14	39.8	0.265	—	—	Bugyi (1972)
	18	59.5[a]	0.358[a]	—	—	Bugyi (1972)
	20	76.5[a]	0.442[a]	—	—	Bugyi (1972)
Israel	9–10	33.2	0.241	33.2	0.241	Bar-Or & Zwiren (1973)
Japan	12	40.0	0.268	43.3	0.289	Matsui *et al.* (1971)
	15	53.3	0.327	49.2	0.318	Matsui *et al.* (1971)
	18	58.2	0.349	48.7	0.309	Matsui *et al.* (1971)
Sweden	11	36.7	0.249	34.1	0.241	P. O. Åstrand (1952)
Rural	11	39.0	0.265	42.0	0.278	Adams *et al.* (1961*a*)
Urban	11	36.0	0.245	36.0	0.247	Adams *et al.* (1961*a*)

Table 75. (*cont.*)

Population	Age (yr)	Boys W (kg)	Boys W/H (kg/cm)	Girls W (kg)	Girls W/H (kg/cm)	Author
UK	6	20.5	0.175	20.4	0.178	Tanner *et al.* (1966)
	8	25.0	0.198	25.1	0.201	Tanner *et al.* (1966)
	10	30.3	0.221	31.1	0.228	Tanner *et al.* (1966)
	10	—	0.231	—	0.231	Holt (1948)
	11	—	0.241	—	0.241	Holt (1948)
	11.5	33.4	0.236	33.8	0.239	Provis & Ellis (1955)
	12	—	0.253	—	0.253	Holt (1948)
	12	37.7	0.256	40.5	0.271	Tanner *et al.* (1966)
	14	48.8	0.304	51.0	0.320	Tanner *et al.* (1966)
	16	59.6	0.346	55.8	0.344	Tanner *et al.* (1966)
USA	8	25.4	0.199	26.0	0.205	Reed & Stuart (1959)
	10	31.6	0.229	32.5	0.235	Reed & Stuart (1959)
	10	35.8	0.254	35.3	0.252	Rodahl *et al.* (1961)
	10.5	29.5	0.215	—	—	Robinson (1938)
	10–11	—	—	37.0	0.259	Wilmore & Sigerseth (1967)
	11	35.2	0.244	35.7	0.247	Stuart & Meredith (1946)
	11	46.0	0.303	44.0	0.297	Adams *et al.* (1961*b*)
	12	39.4	0.265	42.0	0.278	Reed & Stuart (1959)
	12	44.8	0.294	45.3	0.294	Rodahl *et al.* (1961)
	12–13	—	—	49.0	0.308	Wilmore & Sigerseth (1967)
	14	50.8	0.312	51.3	0.320	Reed & Stuart (1959)
	16	60.5	0.350	54.6	0.335	Reed & Stuart (1959)

[a] After working in heavy industry.

occurred 2 to 5 years earlier in boys and 1.2 years earlier in girls from countries with a daily per-capital food supply greater than 2300 kcal. The percentage of the mature weight attained at the time of the growth spurt was also greater in countries that had a low caloric intake.

Weights and weight/height ratios for representative world populations are summarized in Tables 75 and 76. These must be interpreted with some caution. We have noted already the dimensional argument for use of W/H^3; other authors, equally, have proposed W/H^2 rather than W/H. Garn (1965) instanced a study of Guamese children that concluded the population were relatively slender; however, they were also much shorter than the reference population in the United States, and if data were expressed as W/H^3 rather than W/H it could be argued that the Guamese had a relatively normal body build. All populations show an increase of W/H ratios as the children become older and taller; however, W/H^3 falls dramatically, and Forbes (1974) has commented that the fitting of a polynomial curve does not help in describing the relationship between body weight and standing height.

Table 76. *Weight (W) and weight/height ratio (W/H) for less-developed populations*

Population	Age (yr)	Boys W (kg)	Boys W/H (kg/cm)	Girls W (kg)	Girls W/H (kg/cm)	Author
Canada						
Eskimo	6	21	0.189	21.3	0.192	de Peña (personal communication)
	8	25	0.208	25.1	0.210	de Peña (personal communication)
	10	31	0.238	31.0	0.237	de Peña (personal communication)
	12	33	0.248	40.3	0.281	Rode & Shephard (1973*d*)
	14	44	0.291	50.6	0.335	Rode & Shephard (1973*d*)
	16	52	0.325	53.6	0.349	Rode & Shephard (1973*d*)
North American Indian	8	27.3	0.213	—	—	M. Singh, Howell & MacNab (1968)
	10	35.1	0.255	—	—	M. Singh *et al.* (1968)
	12	39.7	0.270	—	—	M. Singh *et al.* (1968)
	14	47.6	0.298	—	—	M. Singh *et al.* (1968)
Easter Island	10–11	36	0.257	33	0.236	Ekblöm & Gjessing (1968)
Egypt	8	22.6	0.188	—	—	McDowell, Taskar & Sarhan (1970)
	10	26.8	0.207	—	—	McDowell *et al.* (1970)
	11.5	30.3	0.223	—	—	McDowell *et al.* (1970)
Ethiopia						
Public school	10	26.3	0.197	—	—	Areskog *et al.* (1969)
	13	31.8	0.224	—	—	Areskog *et al.* (1969)
Private school	10	25.5	0.194	—	—	Areskog *et al.* (1969)
	13	34.8	0.236	—	—	Areskog *et al.* (1969)
Gambia	6	17.0	0.157	16.7	0.156	McGregor *et al.* (1961)
India	8	18.8	0.155	—	—	McDowell *et al.* (1970)
	10	22.5	0.179	—	—	McDowell *et al.* (1970)
	11.5	25.9	0.194	—	—	McDowell *et al.* (1970)
Italian Alps	10–13	35.9	0.252	35.3	0.248	Steplock *et al.* (1971)
Jamaica	6	17.8	0.163	18.0	0.165	Ashcroft *et al.* (1965*a*)
	8	22.4	0.187	21.6	0.180	Ashcroft *et al.* (1965*a*)
African	12	39.1	0.262	45.0	0.291	Ashcroft, Heneage & Lovell (1966*b*)
	14	50.0	0.307	51.6	0.321	Ashcroft *et al.* (1966*b*)
	16	59.6	0.346	54.0	0.336	Ashcroft *et al.* (1966*b*)
Rich	14	48.8	0.300	50.4	0.313	Ashcroft *et al.* (1966*b*)
Poor	14	38.6	0.255	45.3	0.286	Ashcroft *et al.* (1966*b*)
Afro-European	12	42.1	0.278	45.0	0.292	Ashcroft *et al.* (1966*b*)
	14	51.8	0.319	50.3	0.316	Ashcroft *et al.* (1966*b*)
	16	58.9	0.343	52.8	0.326	Ashcroft *et al.* (1966*b*)
Chinese	12	37.6	0.257	38.9	0.264	Ashcroft *et al.* (1966*b*)
	14	51.0	0.318	46.4	0.300	Ashcroft *et al.* (1966*b*)
	16	55.9	0.337	48.9	0.314	Ashcroft *et al.* (1966*b*)
European	12	42.4	0.282	42.4	0.277	Ashcroft *et al.* (1966*b*)
	14	50.6	0.311	51.8	0.324	Ashcroft *et al.* (1966*b*)
	16	60.6	0.354	53.5	0.329	Ashcroft *et al.* (1966*b*)

Table 76. (*cont.*)

Population	Age (yr)	Boys W (kg)	Boys W/H (kg/cm)	Girls W (kg)	Girls W/H (kg/cm)	Author
Malaya						
Chinese	15	48.0	0.294	—	—	Thinkaran *et al.* (1974)
Indian	15	43.4	0.270	—	—	Thinkaran *et al.* (1974)
Malay	14	43.6	0.281	—	—	Thinkaran *et al.* (1974)
New Guinea	6	18	0.171	18	0.171	Sinnett & Whyte (1973*b*)
	10	27	0.216	27.5	0.218	Sinnett & Whyte (1973*b*)
	14	37	0.268	37	0.270	Sinnett & Whyte (1973*b*)
St Kitts–Nevis	6	19.2	0.168	18.7	0.164	Ashcroft *et al.* (1965*b*)
	8	23.5	0.188	23.4	0.187	Ashcroft *et al.* (1965*b*)
	10	28.2	0.210	28.5	0.209	Ashcroft *et al.* (1965*b*)
	12	33.3	0.232	36.0	0.244	Ashcroft *et al.* (1965*b*)
	14	41.7	0.269	45.7	0.291	Ashcroft *et al.* (1965*b*)
	16	46.5	0.290	47.5	0.297	Ashcroft *et al.* (1965*b*)
Tanzania	8	22.8	0.191	22.4	0.186	C. T. M. Davies *et al.* (1974)
	10	25.5	0.200	26.7	0.207	C. T. M. Davies *et al.* (1974)
	12	30.9	0.227	30.1	0.222	C. T. M. Davies *et al.* (1974)
	14	38.5	0.263	34.4	0.242	C. T. M. Davies *et al.* (1974)
	16	43.9	0.289	—	—	C. T. M. Davies *et al.* (1974)
Tunisia	8	20.5	0.174	—	—	Lowenstein & Connell (1974)
	10	24.3	0.191	—	—	Lowenstein & Connell (1974)
	11.5	27.4	0.205	—	—	Lowenstein & Connell (1974)
	12	38.5	0.265	40.8	0.276	Pařízková *et al.* (1972)
Tunis						
Poor	8	21.5	0.178	—	—	Lowenstein & Connell (1974)
Rich	8	23.5	0.187	—	—	Lowenstein & Connell (1974)

In primitive groups, as in more-developed countries, the weight spurt usually lags behind the height spurt by six months to a year (Rode & Shephard, 1973*d*; C. T. M. Davies *et al.*, 1974), with inevitable effects on the *W*/*H* ratio. Amongst the boys the general trend to an increase of *W*/*H*, accelerated at puberty, reflects largely the development of the body musculature. For this reason, the *W*/*H* ratio differs very much (Van Uytvanck & Vrijens, 1966; H. H. Clarke & Borms, 1968) between robust boys (average leg extension force 107 kg) and weaker boys of the same age (average leg extension force 76 kg). The largest gain of *W*/*H* ratio is usually registered between twelve to fourteen years (Reed & Stuart, 1959; Tanner, Whitehouse & Takaishi, 1966) or fourteen to sixteen years (Demirjian *et al.*, 1972), although where heavy work is demanded in late adolescence (Bugyi, 1972), gains may continue through to the twentieth year. The *W*/*H* ratio for the girls temporarily surpasses that of the boys during the pubertal spurt (around the twelfth year); in the more-developed countries, gains from twelve to sixteen years are smaller than in the boys, and increases in the girls reflect mainly the accumulation of body fat.

In younger children (six to eight years), the *W/H* ratios are remarkably similar in many more-developed countries, as shown by statistics from Belgium, Francophone Canada, France, UK, and the USA. Comparisons of older age groups are complicated by differences in the timing of growth spurts. Thus, the Californian students studied by Wilmore & Sigerseth (1967) were relatively much heavier than other US children seen earlier by Robinson (1938 – a group from an orphanage) and by Reed & Stuart (1959); however, the heaviness of the Californian sample reflected early maturation rather than obesity, and they had a high aerobic power.

Social class no longer has an effect on *W/H* ratios in England, although W/H^2 drops progressively with employment of the mother and an increase in the number of siblings (Topp *et al.*, 1970). The effects of maternal employment and family size are more marked in fourteen- to seventeen-year-old children than in those aged five to eight years, perhaps partly because nutritional deficiencies have more impact during the adolescent growth spurt and partly because fourteen- to seventeen-year-olds are more likely to be members of completed families than are the younger age group. Three urban/rural comparisons (Adams *et al.*, 1961*a*; Seliger, 1970; Shephard *et al.*, 1975*a*) have disclosed quite small differences of *W/H* ratio between the two types of environment. In Sweden, rural children were slightly heavier than the urban group; in Czechoslovakia the older urban boys were slightly heavier than the rural population, and in Quebec the urban girls were slightly heavier than the rural sample; this last effect was not due to maturation, since the rural girls were slightly more mature than those from the urban centre. Metropolitan children from Montreal (Demirjian *et al.*, 1972) were heavier than either the urban (Trois Rivières) or rural (Champlain) groups from French Canada. The similarities of body build between Montreal and British children and the substantial differences of *W/H* ratios within the Quebecois community are both arguments against the distinctive and genetically determined growth pattern that Demirjian *et al.*, (1972) have ascribed to Canadian children of French origin.

Among primitive and less-developed communities such as Ethiopia, Gambia, India, New Guinea, St Kitts–Nevis and Tunisia, the young (six- to ten-year-old) children are light and generally have low *W/H* ratios. However, this is not an invariable rule, and both Canadian Eskimos and Canadian Indians are at this age heavier, with a higher *W/H* ratio than their contemporaries in the more-developed countries. Because of the later pubertal growth spurt in many primitive communities, the discrepancy in *W/H* ratios is generally larger for twelve- to fourteen-year-olds than for younger children; thus, the boys of Ethiopia, India, St Kitts–Nevis and

Tunisia have very low weights and *W/H* ratios at this stage of development, while even the Canadian Eskimos and Indians lose much of their advantage over the white boys. Among the girls, the Canadian Eskimos retain a high relative weight and *W/H* ratio through the pubertal period, but figures for New Guinea and St Kitts–Nevis are again low. Within Tunisia, the twelve-year-old girls studied by Pařízková, Merhautová & Prokopec (1972) were much heavier than those seen by Lowenstein & Connell (1974). The Czech authors noted 'Les enfants examinés à Tunis fréquentaient le lycée de la capitale et appartenaient à des familles ayant un niveau de vie assez élévé. . .' In the post-pubertal period, the Canadian Eskimo boys are a little lighter than their white counterparts, presumably because the Eskimos carry less fat. However, female Eskimos have an absolute weight that is well up to European standards, with a high *W/H* ratio.

Areskog (1969) examined differences between relatively wealthy Ethiopian boys who were attending a private school and those within the public school system. Somewhat surprisingly, at the age of ten both groups had a low *W/H* ratio relative to white standards, with the public school boys taller and heavier than those at the private school. However, by the age of thirteen, the private school boys had an advantage of 6 cm and 3 kg relative to the public school boys. We have noted above the greater weight of wealthy twelve-year-olds in Tunis (Pařízková *et al.*, 1972) relative to children from other parts of the same country. Lowenstein & Connell (1974) compared high and low income children from Tunis with very poor south Tunisian families living on the edge of the Sahara desert. Within the city of Tunis, there was a threefold gradient of protein intake between high and low income families, and the growth of the poor children was further retarded by a variety of diseases, including upper respiratory and middle-ear infections, together with a 21% family history of tuberculosis. In southern Tunisia, supplementation of breast-feeding did not begin until eighteen months of age, suggesting that growth may also have been retarded by nutritional deficiencies during infancy. After weaning, 60–75% of the caloric intake of the south Tunisians came from wheat, giving a likely deficiency in specific nutrients such as zinc. Perhaps more important in terms of assessing the relative importance of genes and environment, at the age of ten years the wealthy Tunisians were 6 cm taller and 5 kg heavier than those from the southern oases, but were only 1 cm shorter and 1 kg lighter than a US sample.

The African children in Jamaica and other parts of the Carribean have interested IBP and other investigators, since their ancestors were transported as slaves from West Africa, mainly in the eighteenth century. Young children (six to eight years of age) living in rural areas near Lawrence Tavern, Jamaica (page 99), had a diet deficient in both proteins and calories (Ashcroft *et al.*, 1965*a*). While they were short and light relative to European and North American samples, they were rather comparable with children from other parts of the Carribean, and were taller and heavier than those living in Gambia (McGregor, Billewicz & Thompson, 1961) and other parts of central Africa. As in Tunisia, environment played a major role in determining the size of the Jamaican children. At fourteen years of age, for example, boys from fee-paying secondary schools in Kingston (Ashcroft *et al.*, 1966*b*) were 8 cm taller and 10 kg heavier than their contemporaries from rural areas, but only about 3 cm shorter and 2 kg lighter than the US series of Stuart & Meredith (1946). Within the more privileged group of Kingston children (Ashcroft *et al.*, 1966*b*), African, Afro-European and European subjects did not differ significantly from each other with respect to height and weight, but the Chinese of both sexes were lighter and shorter than those from the other racial groups. Since the Chinese were relatively prosperous members of the Kingston community, Ashcroft *et al.* suggested that their characteristic body build had a genetic basis. The African children of St Kitts–Nevis (Ashcroft *et al.*, 1965*b*) were of similar size and build to those measured in rural Jaimaca. In both locations, the girls were closer in size to their white counterparts than were the boys. However, as yet it remains unclear whether there is a racial difference in the relative size of the two sexes, or whether there are environmental factors favouring the growth of girls in the West Indies.

Thinkaren *et al.* (1974) examined a small sample of twenty-six Malaysian children; they were said to be genetically pure representatives of the three main ethnic groupings, all living under similar urban conditions in Petaling Jaya. Here, the Chinese sample were heavier and a little taller than the Malays and Indians; although of approximately the same height as the Jamaican sample, the Malayan Chinese boys were substantially (5 kg) lighter. Ashcroft *et al.* (1966*b*) drew attention to the fact that their Jamaican sample was heavy relative to Hong Kong Chinese of equivalent socio-economic status, and they speculated that this might reflect racial admixture in the Jamaican sample.

Skinfold thicknesses

There have been several very extensive studies of skinfold thicknesses in the more-developed countries. For example, Tanner & Whitehouse (1962) have reported figures for 24 000 London (UK) schoolchildren aged five to sixteen years. The number of skinfolds examined by different authors has ranged from one to ten. Where possible, the summarized data (Table 77) is the average for the three folds recommended to the IBP ($\Sigma\bar{3}$: triceps, subscapular and suprailiac); however, in a few instances, figures are only available for one ($\Sigma\bar{1}$) or two ($\Sigma\bar{2}$) of the three folds.

The statistics are remarkably similar for most industrialized nations. The boys show a small increase, from 6–7 mm at age seven to 8–10 mm at maturity. Three studies from the United States, including the 'total community' survey of Tecumseh, Michigan (Montoye, Epstein & Kyelsberg, 1964), have shown rather thicker folds than in most other nations. The over-nutrition of the US groups was borne out by a comparison among children of Italian ancestry (H. B. Young, 1965). Where families had remained resident in Palermo or had migrated to Rome, the children had quite lean skinfolds, but where the parents or grandparents had migrated to Boston, USA, average values were 2–3 mm larger.

In Czechoslovakia, urban boys had slightly more subcutaneous fat than those from rural areas, this being particularly true of a high-school sample in Prague (Pařízková *et al.*, 1972). However, among French Canadian boys, the urban sample was thinner than the rural (Shephard *et al.*, 1975*a*). Although Francophones from Montreal (Jenicek & Demirjian, 1972) appear to be fatter than a random sample of Toronto schoolchildren (Shephard *et al.*, 1968*c*), this is an artefact due to the use of only two folds by Jenicek & Demirjian (1972); data for the triceps and the subscapular folds are closely similar in the two communities.

A study of 3975 boys and 3903 girls, southern Chinese living in Hong Kong, illustrates the importance of economic status; at all ages, children from high income families had more subcutaneous fat than those from middle and low income families (Fry *et al.*, 1965).

Sexual differences in both the average skinfold thickness and the distribution of body fat are established at an early age (Pařízková, 1963; Abbie, 1967). The sex difference for the three IBP folds is no more than 0.5–1.0 mm in the young child (Tanner & Whitehouse, 1962; Pařízková, 1963), but it increases rapidly with the onset of puberty, to reach a differential of 5–6 mm in late adolescence. The difference between US and European samples seems smaller for the girls than for the boys, and indeed the fattest group of girls come from Norway (Andersen & Ghesquière, 1972). In both Czechoslovakia and French-speaking Canada, girls from an urban environment have slightly thicker skinfolds than those from a rural milieu (Seliger, 1970; Shephard *et al.*, 1975*a*).

Among less-developed and primitive communities, skinfolds are generally thinner than in the more-developed nations, equally for boys and for girls (Table 78). In some groups such as the Canadian Eskimo (Rode & Shephard, 1973*d*), a high level of physical activity and the need for heat dissipation without soaking the clothing in sweat may favour a reduction of subcutaneous fat, but generally the main responsible factor seems a lower standard of living than that current in an industrialized community, with a less ready availability of 'rich' foods. Thus Abbie (1967) describes the Australian Aborigine's diet as 'lean meat protein and a varying amount

Table 77. *Skinfold thickness (mm) – children of more-developed countries*

Population	Age (yr) 7–8	9–10	11–12	13–14	15–16	17–18	Author
Boys							
Australia ($\Sigma\bar{3}$)	6.3	—	—	—	—	—	Roche & Cahn (1962)
Belgium ($\Sigma\bar{3}$)	—	—	7.2	7.4	7.4	—	Simons *et al.* (1974)
Canada ($\Sigma\bar{3}$)							
Anglophone	—	7.9	7.3	7.7	—	—	Shephard *et al.* (1968*c*)
Anglophone	—	6.9	8.6	8.0	—	—	Bailey *et al.* (1974*b*)
Francophone							
Metropolitan ($\Sigma\bar{2}$)	7.1	7.4	8.5	8.8	8.6	8.3	Jenicek & Demirjian (1972)
Rural	—	—	7.7	—	—	—	Shephard *et al.* (1975*a*)
Urban	—	—	7.0	—	—	—	Shephard *et al.* (1975*a*)
Czechoslovakia ($\Sigma\bar{3}$)							
Prague	4.8	5.3	7.0	8.2	8.5	—	Pařízková (1963)
Prague	—	—	9.5	—	—	—	Pařízková *et al.* (1972)
Rural	—	—	7.5	—	7.6	—	Seliger (1970)
Urban	—	—	8.7	—	8.2	—	Seliger (1970)
Holland ($\Sigma\bar{2}$)	—	—	—	8.1	7.6	7.6	Bink & Wafelbakker (1969)
Hong Kong (Chinese, $\Sigma\bar{3}$)							
High income	6.1	6.4	7.4	7.0	7.9	8.8	Fry *et al.* (1965)
Middle income	5.9	5.8	6.1	6.5	7.2	7.8	Fry *et al.* (1965)
Low income	5.3	5.5	5.8	6.2	7.1	7.8	Fry *et al.* (1965)
Italy ($\Sigma\bar{3}$)							
Boston	—	—	9.0	9.1	9.1	10.2	C. M. Young (1965)
Rome	—	—	6.8	6.9	6.5	6.5	C. M. Young (1965)
Norway ($\Sigma\bar{3}$)	—	7.0	—	7.0	—	—	Andersen & Ghesquière (1972)
UK							
($\Sigma\bar{2}$)	6.4	6.6	7.3	7.5	7.5	—	Tanner & Whitehouse (1962)

Table 77. (*cont.*)

Population	Age (yr) 7–8	9–10	11–12	13–14	15–16	17–18	Author
USA							
($\Sigma\bar{2}$)	7.0	8.2	9.0	—	—	—	F. E. Johnston, Hamill & Lemeshow (1972)
($\Sigma\bar{3}$)	—	—	—	11.7	12.1	9.2	Novak (1963)
($\Sigma\bar{2}$)	—	—	—	9.5	10.0	10.0	Montoye *et al.* (1964)
Girls							
Australia ($\Sigma\bar{3}$)	8.2	—	—	—	—	—	—
Canada ($\Sigma\bar{3}$)							
Anglophone	—	9.5	9.5	12.0	—	—	Shephard *et al.* (1968*c*)
Francophone							
Metropolitan ($\Sigma\bar{2}$)	8.7	9.3	10.0	11.5	12.7	—	Jenicek & Demirjian (1972)
Rural	—	—	9.3	—	—	—	Shephard *et al.* (1975*a*)
Urban	—	—	10.0	—	—	—	Shephard *et al.* (1975*a*)
Czechoslovakia ($\Sigma\bar{3}$)							
Prague	5.8	7.2	9.2	12.0	13.5	—	Pařízková (1963)
Prague	—	—	12.4	—	—	—	Pařízková *et al.* (1972)
Rural	—	—	8.7	—	11.5	—	Seliger (1970)
Urban	—	—	10.0	—	12.9	—	Seliger (1970)
Hong Kong (Chinese, $\Sigma\bar{3}$)							
High income	7.1	7.6	8.8	10.8	12.0	13.3	Fry *et al.* (1965)
Middle income	6.5	6.8	7.5	9.7	11.4	11.8	Fry *et al.* (1965)
Low income	6.0	6.4	6.8	9.0	11.3	11.9	Fry *et al.* (1965)
Italy ($\Sigma\bar{3}$)							
Rome	—	—	9.3	9.8	11.3	11.3	C. M. Young (1965)
Norway ($\Sigma\bar{3}$)	—	9.0	—	13.7	—	—	Andersen & Ghesquière (1972)
UK ($\Sigma\bar{2}$)	7.4	8.3	9.3	10.5	12.1	—	Tanner & Whitehouse (1962)
USA ($\Sigma\bar{3}$)	—	—	—	12.9	14.1	15.0	Novak (1963)

Table 78. *Skinfold thickness (mm) – children of primitive populations*

Population	Age (yr)						Author
	7–8	9–10	11–12	13–14	15–16	17–18	
Boys							
Australia ($\Sigma\bar{3}$)							
Aborigines	5.6	4.9	5.1	5.1	5.2	6.9	Abbie (1967)
Canada							
Eskimo ($\Sigma\bar{3}$)	—	4.1	5.5	4.6	4.9	6.0	Rode & Shephard (1973*d*)
Cook Islands ($\Sigma\bar{2}$)							
Polynesians	6.2	7.1	8.5	8.8	10.6	14.0	Fry (1960)
Dominica ($\Sigma\bar{2}$)	5.6	5.8	—	—	—	—	Robson, Bazin & Soderstrom (1971)
Ethiopia							
Public school ($\Sigma\bar{3}$)	—	4.7	—	5.2	—	—	Areskog *et al.* (1969)
Private school ($\Sigma\bar{3}$)	—	5.4	—	6.7	—	—	Areskog *et al.* (1969)
Private school ($\Sigma\bar{2}$)	4.8	4.9	5.1	5.2	4.5	—	Eksmyr (1971)
Guatemala ($\Sigma\bar{2}$)	4.8	—	—	—	—	—	Malina *et al.* (1974)
Lapps ($\Sigma\bar{10}$)				6.7			Andersen *et al.* (1962)
Libya ($\Sigma\bar{2}$)	5.3	5.3	5.5	6.0	—	—	Ferro-Luzzi & Ferro-Luzzi (1962)
Malaya ($\Sigma\bar{2}$)							
Muar	5.7	5.5	6.2	6.2	—	—	Wadsworth & Lee (1960)
Morocco ($\Sigma\bar{2}$)	5.2	5.3	5.7	5.9	6.7	7.1	Ferro-Luzzi & Ferro-Luzzi (1962)
New Guinea ($\Sigma\bar{3}$)							
Biak	4.6	4.3	5.6	4.8	—	—	Jansen (1963)
Nubuai	3.5	3.3	4.3	—	—	—	Jansen (1963)
Serong	4.0	3.7	3.7	3.8	—	—	Jansen (1963)
Tanzania ($\Sigma\bar{2}$)	—	—	—	—	—	5.5	Robson (1964)
($\Sigma\bar{1}$)	—	—	4	4	4	—	C. T. M. Davies *et al.* (1974)

Table 78. (*cont.*)

Population	Age (yr)						Author
	7–8	9–10	11–12	13–14	15–16	17–18	
Tunisia ($\Sigma\bar{2}$)	4.8	4.9	5.5	—	—	—	Lowenstein & Connell (1974)
Tunis ($\Sigma\bar{3}$)	—	—	8.8	—	—	—	Pařízková *et al.* (1972)
Girls							
Australia ($\Sigma\bar{3}$)							
Aborigines	6.0	7.9	7.1	7.8	9.5	12.0	Abbie (1967)
Canada ($\Sigma\bar{3}$)							
Eskimo ($\Sigma\bar{3}$)	—	5.6	6.4	7.9	11.1	9.7	Rode & Shephard (1973*d*)
Cook Islands ($\Sigma\bar{1}$)							
Polynesians	9.5	9.4	11.4	12.2	14.7	18.9	Fry (1960)
Dominica ($\Sigma\bar{2}$)	6.2	6.3	—	—	—	—	Robson *et al.* (1971)
Ethiopia ($\Sigma\bar{2}$)							
Private school	5.7	6.0	6.7	8.0	8.5	—	Eksmyr (1971)
Guatemala ($\Sigma\bar{2}$)	5.4	—	—	—	—	—	Malina *et al.* (1974)
Lapps ($\Sigma\overline{10}$)			9.3				Andersen *et al.* (1962)
New Guinea ($\Sigma\bar{3}$)							
Serong	4.0	3.7	3.7	8.4	—	—	Jansen (1963)
Nubuai	3.5	3.3	4.3	4.6	—	—	Jansen (1963)
Biak	4.6	4.3	5.6	5.1	—	—	Jansen (1963)
Tunisia ($\Sigma\bar{3}$)							
Tunis	—	—	10.0	—	—	—	Pařízková *et al.* (1972)

of vegetables, but almost completely lacking in sugar and in animal fats'. The importance of nutritional status, previously noted in the Hong Kong Chinese (Fry *et al.*, 1965) is also well brought out by the comparison between poor Tunisian children (Lowenstein & Connell, 1974) and the wealthy group attending a Tunis high school (Pařízková *et al.*, 1972).

Relatively thin skinfolds were found in two samples of Ethiopians attending private schools (Areskog *et al.*, 1969; Eksmyr, 1971); these observations have been linked with the thin skinfolds of older Tanzanian boys (Robson, 1964) to generate the hypothesis that Africans have a different distribution of body fat to white people. While this may yet prove to be true, a further factor that needs to be considered in Ethiopia is membership in the Coptic church. About 80% of upper-class families are reputed to belong to this organization, and at least for the faithful there is a strong tradition of periodic fasting.

Fry (1960) obtained rather large skinfold readings on a sample of 221 Rarotongan children. He used a non-standard form of skinfold caliper, but it is difficult to dismiss his readings on this count, since they were confirmed by soft-tissue radiographs. Presumably, food is more plenteous and activity levels are lower in Rarotonga than in many primitive communities.

Percentage of body fat

A number of authors have quoted their results as percentages of body fat (Tables 79 and 80); in some instances, the information has been obtained by underwater weighing, but in other series the numbers have been derived by a manipulation of skinfold readings. Problems of converting skinfold and body density measurements to estimates of body fat have already been noted (Tables 24 and 25). In children, such conversions are even more hazardous, for there is little guarantee that formulae established on adult cadavers will be appropriate for younger children. Equal uncertainties arise in the application of $^{40}K^{+}$ (Myhre & Kessler, 1966; Burmeister *et al.*, 1972) and deuterium oxide (Shephard *et al.*, 1973*b*) methods.

Studies from Czechoslovakia (Pařízková, 1974) and the United States (Heald *et al.*, 1963) agree in showing that boys have a substantial decline of body fat in the post-pubertal period. At a first inspection, this data seems at variance with the previously noted constant or slightly increasing thickness of subcutaneous fat. However, if account is taken of the very rapid pubertal gain in total body weight, it can be seen that a diminution in fat percentage is compatible with a small increase in the absolute fat content of the body. In the girls, the growth of fat tissue more closely

Table 79. *Body fat percentage – children of more-developed countries*

Population	Age (yr) 7–8	9–10	11–12	13–14	15–16	17–18	Author
Boys							
Canada							
Anglophone	—	—	15.6	—	—	—	Shephard *et al.* (1968*c*)
Francophone							
Rural	—	—	19.1	—	—	—	Shephard *et al.* (1975*a*)
Urban	—	—	17.7	—	—	—	Shephard *et al.* (1975*a*)
Czechoslovakia							
Cross-sectional	14.5	14.0	16.0	14.0	11.5	10.5	Pařízková (1974)
Longitudinal	—	—	17.5	15.5	15.0	11.3	Pařízková (1974)
	—	—	13.9[a]	10.4[a]	8.3[a]	8.4[a]	Pařízková (1974)
Denmark	—	20	18	—	—	—	Friis-Hansen (1965)
Germany	—	—	—	28.6[b]	—	—	Burmeister *et al.* (1972)
Hungary	—	—	—	12.1	10.0	9.4	Bugyi (1972)
Italy							
Liguria	—	—	18.0[c]	—	—	—	Di Prampero (personal communication)
Sardinia	—	—	15.0[c]	—	—	—	Di Prampero (personal communication)
USA	—	—	23.4	20.3	16.1	12.1	Heald *et al.* (1963)
	—	—	—	—	{13.0[d] 16.6[d]}	—	Myhre & Kessler (1966)
	—	18.7	—	—	—	—	Wilmore & McNamara (1974)
Girls							
Canada							
Anglophone	—	—	20.7	—	—	—	Shephard *et al.* (1968*c*)
Francophone							
Rural	—	—	23.2	—	—	—	Shephard *et al.* (1975*a*)
Urban	—	—	23.7	—	—	—	Shephard *et al.* (1975*a*)
Czechoslovakia							
Cross-sectional	19.0	20.0	22.0	18.0	24.0	22.0	Pařízková (1974)
Germany	—	—	—	38.6[b]	—	—	Burmeister *et al.* (1972)
Italy							
Liguria	—	—	22.5[c]	—	—	—	Di Prampero (personal communication)
Sardinia	—	—	24.8[c]	—	—	—	Di Prampero (personal communication)

[a] Boys attending special sports school.
[b] Institutionalized children; $^{40}K^+$ method.
[c] Estimates from skinfold readings.
[d] Lower value, underwater weighing; $^{40}K^+$ method.

Table 80. *Body fat percentage – children of primitive populations*

Population	Age (yr) 9–10	11–12	13–14	15–18	Author
Boys					
Canada					
Eskimo	—	—	—	12.7[a]	Shephard *et al.* (1973*b*)
East Africa	7[b]	—	—	—	Di Prampero & Cerretelli (1969)
Italy					
Alps	—	16.6[b]	—	—	Steplock *et al.* (1971)
Tunisia	—	12.0[b]	—	—	Di Prampero (personal communication)
Girls					
Canada					
Eskimo	—	—	—	19.4[a]	Shephard *et al.* (1973*b*)
East Africa	8[b]	—	—	—	Di Prampero & Cerretelli (1969)
Italy					
Alps	—	23.3[b]	—	—	Steplock *et al.* (1971)
Tunisia	—	18.0[b]	—	—	Di Prampero (personal communication)

[a] Estimated by use of deuterated water. [b] Estimated from skinfold thickness.

parallels the general growth curve, and the percentage of fat remains constant or even increases slightly in the post-pubertal period.

The densitometric readings generally confirm the conclusions drawn from the skinfold data. Canadian boys and girls have approximately the same proportions of body fat as those from Europe, but US children are fatter. The small urban/rural differences in the Quebec population coincide with those demonstrated by skinfold calipers. The main discrepancy is for the Canadian Eskimos; despite very low skinfold readings, deuterium dilution suggests a substantial body fat content (page 78).

Maximum oxygen intake

More-developed countries

Despite the difficulties of defining maximum effort in the child (pages 181–4), much of the data on the working capacity of children is based on oxygen intake measurements during bicycle ergometer or treadmill exercise to exhaustion (Tables 81 and 82).

In the more-developed countries, the dominant impression is of the uniformity of the data, both from one population to another and from one age group to another, typical readings being 48–50 ml/kg min STPD in the boys and 38–42 ml/kg min STPD in the girls. Despite theoretical objections (page 187), body weight seems to prove a good basis of standardization. Further, in practical terms, it indicates the power available to the child for its principal activities of walking and running.

The three longest series of cross-sectional data (Robinson, 1938; P. O. Åstrand, 1952; G. R. Cumming, 1967) all show rather low scores for the youngest boys, but this probably reflects a lack of cooperation from the youngest students rather than a true physiological difference. Most authors are agreed that the aerobic power of the girls deteriorates either at or just before puberty. The loss of working capacity cannot be attributed entirely to an increase in the percentage of body fat; it reflects also an alteration of attitudes towards physical activity – as the growing girl becomes older, she is conditioned to believe that running and other forms of vigorous physical activity are inappropriate to her sex.

In some countries, average results have differed quite widely from one report to another, reflecting the non-random nature of the samples tested. Thus, in Japan values of 52.7 ml/kg min and 40.7 ml/kg min have been quoted for the aerobic power of twelve-year-old boys with figures of 44.9 and 34.9 ml/kg min for twelve-year-old girls. The first sample (Ikai *et al.*, 1970) were 'normals' from the Tokyo district, while the second sample – also apparently from Tokyo – was chosen to provide children conforming in size to the 1967 Japanese Statistical Report of School Health (Nakagawa & Ishiko, 1970). Sampling problems seem particularly acute in Japan, due to the rapid urbanization of what were previously rural districts. Thus, Yoshizura & Ushihisa (1967) found an 8% lower $\dot{V}_{O_2(max)}$ and a 5% lower physical endurance in urban areas, but other authors (Ikai, 1967; Matsui, 1968) found no significant urban/rural gradients. In Canada (Shephard *et al.*, 1974*a*) and Czechoslovakia (Seliger, 1970), the urban/rural gradient apparently is slightly in favour of the urban children, partly because they walk more, and partly because they have more facilities for recreation.

Two anomalous reports seem the very high figures for Swedish children (P. O. Åstrand, 1952) and the very low results encountered in Philadelphia (Rodahl *et al.*, 1961). P. O. Åstrand (1952) did not draw a completely random sample; teachers were asked to select 'the best', 'rather good' and 'poor' members of their classes for testing. Further, his report notes that exceptionally fat children did not participate, and that all subjects were well trained. Nevertheless, the superiority of the Swedish child at this point in history is confirmed by a comparison of Californian and Swedish children (Adams *et al.*, 1961*a*, *b*; Table 83). The group seen by Rodahl *et al.* (1961) came largely from a poor part of Philadelphia, and the data contained evidence of a trend towards improvement of aerobic power with an increase of social class. In confirmation of the low maximum oxygen intake readings, PWC_{170} scores were also poor (Table 83); never-

Table 81. *Maximum oxygen intake (ml/kg min STPD) of children in developed countries*

Population	Age (yr) 6	8	10		12		14		16	18	Author
Boys											
Canada											
	45	54	44		47		49		52	55	G. R. Cumming (1967)
	—	—	49.7		45.8		—		—	—	Shephard *et al.* (1968*c*)
Francophone											
Rural	—	—		49.4[c]			—		—	—	Shephard *et al.* (1974*a*)
Urban	—	—		60.6[c]			—		—	—	Shephard *et al.* (1974*a*)
Czechoslovakia											
	—	—	—		49.1		—		—	—	Šprynarova (1966)
	—	—		51.2[a]			—		—	—	Seliger & Pachlopniková (1967)
Rural	—	—	—		43.0			44.4		—	Seliger (1970)
Urban	—	—	—		47.0			45.8		—	Seliger (1970)
Germany	—	—	—		45		48		45	—	König *et al.* (1961)
Holland	—	—	—		50.9		50.7		50.9	50.8	Bink & Wafelbakker (1969)
Israel											
Arabs	—	—	—			50.5			—	—	Bar-Or (1973)
Jews	—	—	—			52.7			—	—	Bar-Or (1973)
Jews	—	—	50.0		—		—		—	—	Bar-Or & Zwiren (1973)
Italy											
Milan	—	—	—			45.3			—	—	Brandi, Brambilla & Cerretelli (1960)
Milan	—	48.0	48.3		49.0		50.3		51.8	52.2	Cerretelli, Aghemo & Rovelli (1963)
Sardinia	—		38[b]						39[b]		Di Prampero (personal communication)
Liguria	—		43[b]						45[b]		Di Prampero (personal communication)
Japan											
	—	—	54.4		52.7		51.8		52.3	52.3	Ikai *et al.* (1970)
	—	—	—		40.7		42.0		37.7	—	Nakagawa & Ishiko (1970)
	—	50	49.1		48.1		48.5		47.7	43.5	Ikai & Katagawa (1972)
	—	—	—		49.1		48.7		52.4	53.8	Matsui *et al.* (1971)

Table 81. (*cont.*)

Population	Age (yr) 6	8	10	12	14	16	18	Author
Norway								
Direct	—	—	54.3	56.4	—	—	—	Hermansen & Oseid (1971)
Indirect	—	—	45.2	49.1	—	—	—	Hermansen & Oseid (1971)
South Africa								
Bantu	—	—	—	—	—	40.8	—	Smit (1973)
White	—	—	—	—	—	46.2	—	Smit (1973)
Sweden	49	57	56	57	60	58	58	P. O. Åstrand (1952)
Switzerland	—	—	—	—	49.0	—	47.2	Howald *et al.* (1974)
USA								
	47	47	52	—	47	53	—	Robinson (1938)
	—	—	48		45	—	51	Morse, Schultz & Cassels (1949)
	—	—	50.9		—	—	—	S. R. Brown (1960)
	—	—	32	32	33	32	31	Rodahl *et al.* (1961)
	—	—	50	46	49	47	—	Kramer & Lurie (1964)
	—	—	53.3	—	—	—	—	Wilmore & McNamara (1974)
Girls								
Canada								
	52	49	40	42	38	39	44	G. R. Cumming (1967)
	—	—	36.8	38.3	—	—	—	Shephard *et al.* (1968*c*)
Francophone								
Rural	—	—	42.4[c]		—	—	—	Shephard *et al.* (1974*a*)
Urban	—	—	49.4[c]		—	—	—	Shephard *et al.* (1974*a*)
Czechoslovakia								
Rural	—	—	—	36.7	34.9		—	Seliger (1970)
Urban	—	—	—	38.3	35.8		—	Seliger (1970)
Israel	—	—	45	—	—	—	—	Bar-Or & Zwiren (1973)

Table 81. (*cont.*)

Population	Age (yr)							Author
	6	8	10	12	14	16	18	
Italy								
Milan	—	—	—	43.6		—	—	Brandi *et al.* (1960)
Milan	—	46.3		44.5	44.1	43.7	42.9	Cerretelli *et al.* (1963)
Sardinia	—		35[b]			33[b]		Di Prampero (personal communication)
Liguria	—		41[b]		—	—	—	Di Prampero (personal communication)
Japan								
	—	—	47.2	44.9	40.5	37.7	38.3	Ikai *et al.* (1970)
	—	—	—	34.9	28.9	26.7	—	Nakagawa & Ishiko (1970)
	—	43.4	40.8	39.5	35.0	35.7	34.0	Ikai & Katagawa (1972)
	—	—	—	42.0	39.2	38.3	37.8	Matsui *et al.* (1971)
Sweden	48	55	52	50	46	47	47	P. O. Åstrand (1952)
USA								
	—	—	29	30	34	23	19	Rodahl *et al.* (1961)
	—	—	—	28	—	—	—	Kramer & Lurie (1964)
	—	53.5	50.7	48.7	—	—	—	Wilmore & Sigerseth (1967)

[a] Boys attending special sports school. [b] Submaximum step test. [c] Predicted from submaximum data.

theless, the population studied does not seem generally representative of the US, and it may be that a proportion were poorly motivated and failed to reach a true maximum oxygen intake.

Data from the United States are particularly interesting in that figures for the boys extend back to 1938, to include the carefully documented sample from a Boston orphanage studied by Robinson (1938). Unfortunately, none of the samples are nation-wide, but, with the possible exception of Rodahl's data, there is little evidence that the aerobic power of young boys has deteriorated from 1938 to the present day. Particularly high figures have recently been reported from California, both in girls (Wilmore & Sigerseth, 1967) and in boys (Wilmore & McNamara, 1974). The boys seen by Wilmore & McNamara were a fairly full sample (44.6%) of eligible students at two schools in Davis. The authors commented that in the area studied, bicycling was extremely popular, due to a network of bicycle trails, that the city provided an excellent year-round recreation programme with a high participation rate, and that the boys pursued many endurance-oriented activities in their daily thirty-minute school physical programme. Unfortunately, education in general and physical education in particular is a local municipal responsibility, both in the US and in Canada; many school districts (including disadvantaged areas such as the part of Philadelphia studied by Rodahl *et al.*, 1961) have much less effective programmes for the development of endurance fitness.

Less-developed countries

The majority of maximum oxygen intake measurements on children from less-developed countries are no greater than those for more-developed countries, even when expressed per unit of body weight (Table 82). Although physical activity is sometimes greater in the primitive groups than in a more-developed society, the potential stimulus to the development of aerobic power in most instances seems to be offset by poor nutrition and/or anaemia. In a few instances, improvised laboratory facilities have also necessitated testing at higher than optimal environmental temperatures (see, for example, the series of Thinkaren *et al.*, 1974, who were tested at an *average* ambient temperature of 26 °C).

Three of the five circumpolar populations that have been studied (the Canadian Eskimos, the Kautokeino Lapps and the Nellim Lapps) show a rather larger aerobic power than that seen in children from temperate and tropical latitudes (Andersen *et al.*, 1962; Karlsson, 1970; Rode & Shephard, 1973*d*). These findings are in keeping with the characteristics of adults from the same circumpolar regions. In the Eskimo boys, the form of the growth curve is much as in groups with a smaller aerobic power (Rode & Shephard, 1973*d*); aerobic power develops in parallel with gains

Table 82. *Maximum oxygen intake (ml/kg min STPD) of children in less-developed populations*

	Age (yr)						
Population	8	10	12	14	16	18	Author
Boys							
Canada							
Eskimos	—	61.4	64.2	61.1	60.2	57.1	Rode & Shephard (1973*d*)
East Africa	—	49	—	—	—	—	Di Prampero & Cerretelli (1969)
Easter Island	—	37	46	46	45	45	Ekblöm & Gjessing (1968)
Greenland							
Eskimos	—	—	—	—	—40.1—		Lammert (1972)
Italy							
Alps	—	—42.9[a]—		—	—	—	Steplock *et al.* (1971)
Japan							
Ainu	—	—	—	—	—46.5—		Ikai *et al.* (1971)
Malaya							
Malayan	—	—	—	49[b]	—	—	Thinkaran *et al.* (1974)
Indian	—	—	—	47[b]	—	—	Thinkaran *et al.* (1974)
Chinese	—	—	—	44[b]	—	—	Thinkaran *et al.* (1974)
Temiar	—	—	—	46[b]	—	—	Thinkaran *et al.* (1974)
Norway							
Kautokeino Lapps	—	53	53	55	53	—	Anderson *et al.* (1962)
Nellim Lapps	—	—	—	—	52	—	Karlsson (1970)
Tunisia	—48[a]—				—44[a]—		Di Prampero (personal communication)
Girls							
Canada							
Eskimos	—	53.5	48.4	45.6	44.5	45.5	Rode & Shephard (1973*d*)
East Africa	—	41[a]	—	—	—	—	Di Prampero & Cerretelli (1969)
Easter Island	—	36	35	36	32	—	Ekblöm & Gjessing (1968)
Italy							
Alps	—	—38.5[a]—		—	—	—	Steplock *et al.* (1971)
Norway							
Kautokeino Lapps	—	51	48	44	42	42	Andersen *et al.* (1962)
Tunisia		—45—			—40—		Di Prampero (personal communication)

[a] Submaximum step test. [b] Temperature of testing 26 °C.

in body weight, and the power per unit of body weight thus remains fairly constant throughout the adolescent period. The Eskimo girls show a deterioration of aerobic power per unit of body weight at and following puberty (Rode & Shephard, 1973*c*) but this is of later onset and less severe than that seen in white Canadian girls (G. R. Cumming, 1967). The explanation for the better showing of circumpolar children is complex. One factor may be the prosperity of the nations of which they form a part. Both Canada and the Scandinavian countries are currently striving very hard to equalize health care delivery for city-dwellers and the indigenous peoples of the Arctic (Shephard & Itoh, 1976), and there is not the malnutrition and disease so prevalent in the tropical countries. Again, physical activity remains a feature of Arctic life. Games (Shephard, 1974*d*) must be played while wearing heavy clothing and walking through deep snow. The early adolescent boy still wishes to follow the life-style of his forebears, and at every opportunity he will be absent from school, sharing in the hard work of the hunt (in Canada) or of reindeer herding (in Lapland). Many of the teenage girls in the Canadian and Alaskan Arctic spend substantial periods at more southerly residential schools (Shephard & Itoh, 1976); here, they learn the patterns of inactivity characteristic of white girls, and their aerobic power deteriorates. But while they remain in an Arctic village, they can often be seen struggling through the snow with a heavy younger brother or sister perched on their shoulders in an amauti, and this traditional method of carrying infants undoubtedly helps to sustain their endurance fitness.

Physical working capacity

More-developed countries

The world literature contains sufficient data on the physical working capacity of children to merit discussion of this material in addition to the more readily interpreted measurements of maximum oxygen intake (Tables 83 and 84). The two sets of data can be compared roughly through use of the work scale of the Åstrand nomogram; to a first approximation, a physical working capacity of 13 kg m/kg min in a boy is equivalent to an aerobic power of 42 ml/kg min, while in a girl 10 kg m/kg min is equivalent to 33.5 ml/kg min.

Such conversions necessarily assume that the mechanical efficiency of ergometer exercise is the same in the child as in the adult (page 183), and that the child has been habituated to the laboratory (Shephard, 1969*b*). In some samples where submaximal bicycle ergometer data has been collected on the same occasion as a maximal bicycle ergometer test (for example, Seliger, 1970) there is a fair concordance of data. Other series show significant discrepancies. Thus, the Canadian IBP working party

Table 83. *Physical working capacity of schoolchildren in more-developed countries*

Population	Age (yr)	PWC_{170} (kg m/min)	PWC_{170}/kg	Author
Boys				
Canada	7–9	347	12.5	Howell & MacNab (1968)
	10–11	546	14.1	G. R. Cumming & Cumming (1963)
	10–11	460	13.0	Howell & MacNab (1968)
	11	474	13.9	G. R. Cumming & Cumming (1963)
	12–13	605	13.7	Howell & MacNab (1968)
	14–17	799	13.4	Howell & MacNab (1968)
Czechoslovakia	8	372	13.8	Máček, Vavra & Zika (1970)
	10	444	13.8	Máček *et al.* (1970)
	11	564[a]	16.5[a]	Seliger & Pachlopniková (1967)
	12	630	13.8	Máček *et al.* (1970)
	14	1002	17.4	Máček *et al.* (1970)
Rural	12	548	14.4	Seliger (1970)
	15	855	15.6	Seliger (1970)
Urban	12	558	14.4	Seliger (1970)
	15	914	15.6	Seliger (1970)
Finland	8	420	14.4	Olavi *et al.* (1965)
Germany	8–9	502[b]	17.2[b]	Rutenfranz (1964)
	8–10	450	—	Blumchen, Roskamm & Reindell (1966)
	10–12	732[b]	19.1[b]	Rutenfranz (1964)
	12	580	13.8	Mocellin *et al.* (1971*a*, *b*)
Holland	12	653[b]	16.8[b]	Bink & Wafelbakker (1969)
	12	504	12.0	Kemper (1973)
	14	894[b]	18.0[b]	Bink & Wafelbakker (1969)
	16	1064[b]	18.0[b]	Bink & Wafelbakker (1969)
Israel				
Arabs	13	552	13.4	Bar-Or (1973)
Jews	13	709	15.3	Bar-Or (1973)
Japan	12	552	13.8	Ishiko (1971)
Sweden	7–9	309	—	Bengtsson (1956)
	10–12	468	—	Bengtsson (1956)
Rural	11	624	16.0	Adams *et al.* (1961*a*)
Urban	11	569	15.8	Adams *et al.* (1961*a*)
Switzerland	14	906	17.2	Howald *et al.* (1974)
	18	1147	17.2	Howald *et al.* (1974)
USA	8	282	9.7	Rodahl *et al.* (1961)
	10	330	9.3	Rodahl *et al.* (1961)
	11	650	14.1	Adams *et al.* (1961*b*)
	12	472	10.6	Rodahl *et al.* (1961)
Girls				
Canada	7–9	276	10.3	Howell & MacNab (1968)
	10–11	349	10.1	Howell & MacNab (1968)
	10–12	417	11.2	Shephard *et al.* (1968*c*)
	11	497	11.6	G. R. Cumming & Cumming (1963)

Table 83. (*cont.*)

Population	Age (yr)	PWC_{170} (kg m/min)	PWC_{170}/kg	Author
	12–13	434	9.6	Howell & MacNab (1968)
	14–17	454	8.4	Howell & MacNab (1968)
Czechoslovakia	8	354	10.8	Máček *et al.* (1970)
	10	432	12.0	Máček *et al.* (1970)
	12	486	10.8	Máček *et al.* (1970)
	14	558	10.8	Máček *et al.* (1970)
Rural	12	401	10.2	Seliger (1970)
	15	600	11.4	Seliger (1970)
Urban	12	384	9.6	Seliger (1970)
	15	551	10.2	Seliger (1970)
Finland	8	306	11.4	Olavi *et al.* (1965)
	10	408	12.6	Olavi *et al.* (1965)
	12	618	15.0	Olavi *et al.* (1965)
Germany	9	414[b]	14.8[b]	Rutenfranz (1964)
	9–11	319	8.9	Hollmann *et al.* (1967)
	10–12	518[b]	15.0[b]	Rutenfranz (1964)
Sweden	7–9	309	—	Bengtsson (1956)
	11	468	—	Bengtsson (1956)
Rural	11	518	12.3	Adams *et al.* (1961*a*)
Urban	11	423	11.8	Adams *et al.* (1961*a*)
USA	7–9	380	12.5	Wilmore & Sigerseth (1967)
	8	262	9.3	Rodahl *et al.* (1961)
	10	294	8.4	Rodahl *et al.* (1961)
	10–11	470	12.7	Wilmore & Sigerseth (1967)
	11	488	11.1	Adams *et al.* (1961*b*)
	12	368	8.1	Rodahl *et al.* (1961)
	12–13	620	12.7	Wilmore & Sigerseth (1967)

[a] Students attending special gymnastics school.
[b] Continuously increasing work load.

(Shephard *et al.*, 1968*c*) made a random selection of children from the same schools as those studied by Howell & MacNab (1968) for CAHPER, yet the IBP investigators obtained a 10% higher figure for the PWC_{170}. It is possible that by emphasizing the right of parents to refuse a test, the IBP team recruited a segment of the population who were of above average fitness, although the majority of the parents concerned did not consider their children as being unusually active for their age (page 11). It should also be stressed that the IBP studies were restricted to a major metropolis (Toronto), whereas the CAHPER field samples were drawn from metropolitan, urban and rural areas. However, the main reason for the discrepancy between the CAHPER and the IBP data seems that in the IBP series the children were 'habituated' by a previous visit to the laboratory.

Resting pulse rates dropped from 88 to 81 beats/min in the boys and from 97 to 83 beats/min in the girls and a proportionate modification of exercise responses was suggested by a 10% increase of predicted maximum oxygen intake from visit one to visit two.

As with the measurements of maximum oxygen intake, the PWC_{170} of the boys shows a rather regular development with body weight, although some series (for example, Bink & Wafelbakker, 1969; Seliger, 1970) have a suggestion of an increase per unit of body weight at or around the pubertal growth spurt. Many series of girls show a decrease of PWC_{170} per unit of body weight at puberty; however, this is not an inevitable accompaniment of maturation. Wilmore & Sigerseth (1967) found no decrease in girls receiving a vigorous physical education programme, and a random sample of Czechoslovakian girls (Seliger, 1970) showed a small increase of PWC_{170}/kg from age twelve to fifteen. The Czechoslovakian children did not show the same urban/rural gradient revealed by measurements of maximum oxygen intake. Whether this indicates a true difference in responses to maximum and submaximum work, or whether results have been complicated by some technical difficulty such as a difference of mechanical efficiency between urban and rural samples has yet to be explored.

The PWC_{170} data provide some support for the suggestion of a higher level of fitness in Europe than in North America. The comparison between Sweden and California (Adams *et al.*, 1961*a*, *b*) has already been noted. Again, Howell & MacNab (1968) tested a random sample of young Canadians; even if their data are scaled upwards by 10%, they scarcely match what Howald *et al.* (1974) considered a representative sample of Swiss youth. However, some of the very high values from Europe, particularly those of Rutenfranz (1964) and of Bink & Wafelbakker (1969) are due to a test protocol with a continuously increasing work load. Bink & Wafelbakker (1969) did not specify their source of subjects, but Rutenfranz (1964) demonstrated that his sample was substantially taller and heavier than an unselected Bavarian population.

Less-developed nations

Data from Ethiopia (Areskog *et al.*, 1969) showed extremely low PWC_{170} values, equally in the poorer boys attending a public school, and in more-privileged groups able to attend a fee-paying private school. The temperature of the laboratory was not quoted, but may have exceeded the optimum figure. Nevertheless, the poor working capacity of the Ethiopian children is consistent with their low body weight and the poor levels of maximum oxygen intake found in adults (Table 84).

M. Singh *et al.* (1968) examined a random sample of 125 Canadian Indian

Table 84. *Physical working capacity, less-developed populations*

Population	Age (yr)	PWC_{170} (kg m/min)	PWC_{170}/kg (kg m/ kg min)	Author
Boys				
Canada				
Indians	7	240	10.0	M. Singh *et al.* (1968)
	9	359	12.0	M. Singh *et al.* (1968)
	11	478	12.6	M. Singh *et al.* (1968)
	13	618	14.8	M. Singh *et al.* (1968)
	15	685	13.0	M. Singh *et al.* (1968)
Ethiopia				
Public school	9–11	315	12.0	Areskog *et al.* (1969)
	12–14	379	11.9	Areskog *et al.* (1969)
Private school	9–11	252	9.9	Areskog *et al.* (1969)
	12–14	357	10.2	Areskog *et al.* (1969)

boys aged seven to fifteen years and attending schools on Indian reserves in the province of Alberta. Each student performed a single twelve-minute progressive submaximal test in three four-minute stages; the results are thus best compared with the unadjusted data of Howell & MacNab (1968) for a random sample of Canadian children, which reputedly included some Canadian Indians but no Eskimos. The youngest Indian students had a poorer PWC_{170}/kg than their randomly selected Canadian counterparts, but this difference was no longer statistically significant in the older age groups.

Muscle strength and anaerobic power

Growth curves

Data on muscle force and anaerobic power are relatively sparse, particularly for young children. The general form of growth curves can be illustrated by cross-sectional data for white children from Edmonton, Alberta, and Canadian Eskimo children from Igloolik (Fig. 10). Curves for the boys tend to follow those for weight rather closely, although there may also be some augmentation of body weight by fat, particularly in white boys. The strength spurt lags at least a year behind the height spurt, and there is thus a sense in which boys outgrow their strength just prior to puberty. White girls show little increase of strength after puberty. As we have already noted for aerobic power, this is largely an expression of social conditioning, with an inactive life-style and failure to realize their potential for development. In terms of muscle strength, the Canadian Eskimo girls develop for longer than the white girls and probably approach more closely to their potential.

Table 85. *Hand-grip strength (kg) of various populations*

Population	Age (yr) 8	10	12	14	16	18	Author
Boys							
Belgium	13.5	17.5	23.5	—	—	—	Hebbelinck & Borms (1969)
Canada							
Edmonton	14.9	19.4	23.4	34.5	—	—	Howell *et al.* (1965)
Toronto	—	19.0	23.3	—	—	—	Shephard *et al.* (1968*c*)
Francophone							
Rural	—	———19———		—	—	—	Shephard *et al.* (1975*a*)
Urban	—	———19———		—	—	—	Shephard *et al.* (1975*a*)
Eskimo	—	14.8	19.0	30.0	39.4	45.2	Rode & Shephard (1973*d*)
Indian	8.1	12.4	16.9	21.7	—	—	M. Singh *et al.* (1968)
Denmark	17.8	23	28	35	—	—	Asmussen *et al.* (1959)
Ethiopia							
Public school	—	13.8[a]	———16.8———		—	—	Areskog *et al.* (1969)
Private school	—	13.0[a]	———19.2———		—	—	Areskog *et al.* (1969)
Hong Kong	—	—	—	—	—	45.6	Meshizuka & Nakanashi (1972)
Japan	13.5	—	28.4	—	—	49.4	Meshizuka & Nakanashi (1972)
Korea	—	—	—	—	—	44.5	Meshizuka & Nakanashi (1972)
	11.3	16.3	21.8	31.3	39.2	42.3	Park (1972)
Taiwan	—	—	—	—	—	46.2	Meshizuka & Nakanashi (1972)
Thailand	7.8	—	21.5	—	—	38.5	Meshizuka & Nakanashi (1972)
USA (Oakland)	—	—	26	37	50	58	H. E. Jones (1949)
Vietnam	5.3	—	15.5	—	—	37.5	Meshizuka & Nakanashi (1972)
Girls							
Canada							
Edmonton	13.0	16.2	21.3	26.0	—	—	Howell *et al.* (1965)
Toronto	—	15.5	23.1	—	—	—	Shephard *et al.* (1968*c*)
Francophone							
Rural	—	———17———		—	—	—	Shephard *et al.* (1975*a*)
Urban	—	———17———		—	—	—	Shephard *et al.* (1975*a*)
Eskimo	—	11.0	20.4	24.7	29.0	27.8	Rode & Shephard (1973*d*)
Denmark	15.4	20	25	31	—	—	Asmussen *et al.* (1959)
Hong Kong	—	—	—	—	—	27.1	Meshizuka & Nakanashi (1972)
Japan	12.6	—	26.0	—	—	34.0	Meshizuka & Nakanashi (1972)
Korea	—	—	—	—	—	24.8	Meshizuka & Nakanashi (1972)
	9.4	14.0	19.4	24.2	26.5	24.8	Park (1972)
Taiwan	—	—	—	—	—	27.5	Meshizuka & Nakanashi (1972)
Thailand	7.7	—	21.8	—	—	26.7	Meshizuka & Nakanashi (1972)
USA (Oakland)	—	—	26	32	35	36	H. E. Jones (1949)
Vietnam	4.2	—	13.3	—	—	22.9	Meshizuka & Nakanashi (1972)

[a] Age range 9–11 yr.

Data on hand-grip force from various countries is presented in Table 85. H. H. Clarke (1966) has found that in the average subject who has not developed his arm muscles specifically through sport or work there is a correlation of 0.80 between hand-grip force and the summated muscular strength. Much higher average hand-grip readings have been reported from Oakland, California (Jones, 1947, 1949) and from Copenhagen (Asmussen, Heebøll-Nielsen & Molbech, 1959) than from random samples of Belgians, Canadians, and students in other parts of the US. However, it is not clear how far these differences simply reflect differences of sampling and test equipment, rather than true inter-population discrepancies. Within Canada, differences between urban and rural children, and between Anglophones and Francophones are both small (Shephard *et al.*, 1975*a*).

Among primitive groups, values show a general gradient with nutritional status, with particularly poor readings for Ethiopia and Vietnam. The grip strength of the Canadian Eskimo child is not particularly large, although he or she has a good leg extension force; the relative weakness of the hand may reflect the wearing of clumsy mittens for much of the year.

Anaerobic power

Italian children (Di Prampero & Cerretelli, 1969) show a progressive increase in the maximum rate of ascent of a staircase (Margaria test) from the age of eight to twenty years (in the boys, the velocity increases from 0.8 to 1.5 m/s, equivalent to an increase in oxygen intake of 88 to 165 ml/kg min, while in girls the increase is from 0.8 to 1.3 m/s, equivalent to an oxygen intake increase of 88 to 143 ml/kg min). In Sardinia (Di Prampero, personal communication), results are not much poorer (boys 145 ml/kg min at ages eight to thirteen and 160 ml/kg min at ages fourteen to nineteen; girls 130 and 125 ml/kg min for the same age groups). However, in Tunis (Di Prampero, personal communication), scores are appreciably poorer (boys 95 and 150 ml/kg min; girls 80 and 130 ml/kg min). East Africans (Di Prampero & Cerretelli, 1969) also score relatively poorly (0.5 to 1.1 m/s, 55 to 121 ml/kg min in the boys; 0.8 m/s, 88 ml/kg min with little age gain in the girls). These inter-population differences in anaerobic power become even larger if results are expressed per unit of fat-free mass. Their significance is uncertain. The African children have lower limbs that are the equal of Italians after correction for a lesser quantity of subcutaneous fat. It may be that although muscular dimensions are comparable, poor nutrition has reduced the high energy phosphate content of the muscles in African children; it is also difficult to exclude the possibility of differences in skill and motivation between the various populations.

Performance tests

The assessment of working capacity by performance tests, particularly gymnasium batteries of the AAHPER and CAHPER types, has been popular in children, presumably because most schools now have gymnasia and the minimal equipment needed for such studies.

Interpretation of data

Certain difficulties in the interpretation of data have already been noted (page 6). Much of the variance in results both within and between populations can be attributed to differences in body height and weight (G. R. Cumming & Keynes, 1967). Results are greatly influenced by motivation and thus the competitiveness of a given society. Large changes in test score can arise from seasonal changes in track conditions (Knuttgen & Steendahl, 1963) and familiarity with either a single test item (Knuttgen, 1961; Knuttgen & Steendahl, 1963) or the entire test battery (Drake *et al.*, 1968). Thus Rennie *et al.* (1970) measured the distance covered in fifteen minutes of running. Among the white investigators of the IBP team, distances bore a close relationship to maximum oxygen intake, but only four of thirteen male Alaskan Eskimos approached the distance that would have been predicted from their aerobic power. This could possibly be interpreted as a lack of endurance on the part of the Eskimos, but Rennie *et al.* preferred a simpler explanation: 'they failed to pace themselves. Most sprinted down the airstrip. . . within two laps [they] were too exhausted to recover sufficiently. In short, the test for endurance failed to achieve its purpose.'

Lavallée, Larivière & Shephard (1974) reviewed both previous studies and their own experience of the correlations between performance tests and laboratory measurements of fitness. Their sample of French Canadian boys showed a fair correlation ($r = 0.79$) between summated laboratory tests of muscle force and the speed of a 50 yard dash. Three hundred and six hundred yard times were correlated with the aerobic power per kilogram of body weight ($r = -0.52$, -0.58 respectively), but correlations with $\dot{V}_{O_2(max)}$ were much poorer over 880 yards and twelve minutes of running. As in the Eskimo sample, this reflected errors of pacing rather than problems of motivation, since the heart rates (199 ± 17/min) and blood lactates (89.6 ± 16.0 mg/dl) at the end of the 880 yard run generally reflected maximum effort. Lavallée *et al.* (1974) also tested a sample of ten-year-old French Canadian girls; in this group, motivation was more suspect, and the heart rate at the end of the 880 yard run averaged only 193 ± 17/min. Klimt *et al.* (1974) have reported heart rates in excess of 200/min with maximal blood lactate levels, when nine-year-old German children have participated in an 800 metre run.

Basic differences

Despite problems of interpretation, a number of authors have used performance tests to examine differences of fitness and working capacity between populations.

Cullumbine (1949–50) found a gradient of increasing times for a 100 yard dash from Sinhalese (11.7 s) to Ceylonese Tamil (13.3 s), Indian Tamil (16.2 s) and Ceylonese Moor (16.4 s), but he considered that a large part of these differences reflected variations in weight and size between the several populations.

Cluver, de Jongh & Jokl (1942) classified 9214 South African children as white, Bantu, and Asiatic. The Bantu were smaller than the whites and were commonly affected by malnutrition and parasitic infections, but nevertheless in the pre-pubertal period they excelled at both 100 and 600 yard runs, with the white boys performing better at the strength-demanding shot-put. After puberty, the white boys had the best scores on all tests, and white girls also excelled at the 100 yard run. The Asiatic children, a mixture of Chinese and Indian subjects, were inferior at all tests. Smit (1961) reported that Bantu subjects had an advantage in terms of flexibility, although the test battery used (Kraus–Weber) favours shorter subjects. A more recent study of 5962 South African students (Sloan, 1966*a*) distinguished white, coloured (mixed European and African stock) and 'pure' Bantu. The children were mainly post-pubertal (ages twelve to eighteen years), and at this stage of development the white boys performed the AAHPER tests better than the coloured students, who in turn performed better than the Africans. Among the girls, racial differences in performance were less marked. The white students were taller at all ages, and with the exception of female Bantus older than fifteen years, the whites were also heavier than the other groups. To answer possible criticisms that the inter-racial performance differences were simply an effect of body build, a covariance analysis was carried out on a representative sub-sample of 1050 students; even after elimination of the variance due to height and weight, significant differences of performance remained (Sloan & Hansen, 1969). The pooled data for all 5962 students was further compared with published results for British and US children (Sloan, 1966*b*). Although somewhat smaller, South African students under the age of seventeen excelled at all tests except the softball throw. Sloan attributed the greater scores of the South Africans to enjoyment of an outdoor life and an absence of television. The US students performed well on the softball throw, but on all other tests were inferior to the British group. The British and US populations studied were of rather comparable stature, but the US children were somewhat lighter than the British sample.

Ponthieux & Barker (1965) compared Negro and white children living in Texas. They found that the Negro children performed better on most items of the AAHPER test battery. Unfortunately, no anthropometric measurements were made, but it seems unlikely that the Negro children were larger than the whites. Much of the difference in performance probably reflects relative enthusiasm for the test, since in the US athletics has been a traditional pathway of upward social mobility for the Negro.

Japanese students performed better than US children on standing broad jump, vertical jump and softball throwing tests (Ministry of Education, Japan, 1959), despite their substantially smaller size. A comparison of performance scores for Asian children (Meshizuka & Nakanishi, 1972) showed the best results for the Japanese, equal and lower scores for students in Hong Kong, Taiwan and Korea, and the poorest results for the Philippines, Thailand and Vietnam. Average heights and weights showed a similar gradation (Japan > Korea, Taiwan and Hong Kong > Thailand and Philippines > Vietnam).

We may conclude that there are some inter-racial differences of performance that cannot be explained simply on the basis of differences in body size. However, it is by no means certain that such residual effects have a genetic basis, since existing studies have by no means ruled out differences in motivation, skill, and opportunity to practise the specific test requirements.

Environmental factors

Beran & Flores (1972) studied the performance of children from three different schools in the Philippines. Two were urban schools from the coastal plain, one representing fairly high income families who were able to buy meat and fish, and the other a lower income group, many of whom had fishermen for fathers. The majority of children at these two schools had an adequate protein intake. The third school was in rural upland, and here the diet was mainly root crops and vegetables. In consequence, the children were much lighter and shorter than those from the urban area. Nevertheless, the upland group scored better than the urban children on a number of tests, including the endurance run in the girls and the standing broad jump in the boys. The fitness scores showed little relation to either nutrition or body size. It was thus concluded that the working capacity of the upland children had been developed by walking long distances to school and by helping on family farms.

Simons, Beunen & Renson (1974) made a detailed analysis of environmental factors influencing the physical performance of Belgian schoolchildren. Height, weight, and skinfold readings all showed a small but steady increase with gains of socio-economic status, as measured by the educational level of the father and mother, and the occupation of the father. Farmers' sons tended to a mesomorphic body build, whereas the sons of small-businessmen were endomorphic. On the whole, there was a positive correlation between socio-economic status and performance scores such as vertical jump, bent arm hang, and running speed. However, gradients of body build could account for some of these observations. The sons of small-businessmen scored particularly well on the static arm strength, perhaps because they were heavy and thus muscular. The mesomorphic farmers' sons were marked by poor eye–limb coordination, limb speed and vertical jump. Simons *et al.* (1974) were unable to explain these differences in terms of the amounts of athletic exercise obtained by the various groups; however, they noted that a wider variety of sports were open to students from the upper social classes. A further variable of possible significance was that farm children tended to come from larger families.

Seasonal variations

Seasonal differences of performance were examined by Knuttgen & Steendahl (1963). There was a marked general improvement in the scores of sixteen-year-old Danish high school boys over the course of an academic year, with deterioration during vacation periods. The greatest increases of endurance occurred in the autumn and the spring, when outdoor physical education was practical. Other test items, such as pull-ups, sit-ups, and standing broad jump, improved during the winter, when classes were held in the gymnasium. The authors attributed these changes to a training effect from the physical education programme, with detraining during the vacation months. However, it is also difficult to rule out a contribution from recent practice of the test procedures.

Long-term trends

Espenschade & Meleney (1961) compared the performance of Californian children in 1934 and 1958, drawing their samples from the same school. The main change over the 24-year interval was an increase of height and weight in both sexes (Table 86). The performance of the girls had deteriorated on the 50 yard dash, ball throw and broad jump, but grip strength showed no change, and jump and reach had increased slightly. The boys also ran more slowly in 1958 than in 1934, and the broad jump had deteriorated by an average of 12 cm (5 inches). On the other hand, the ball throw, jump and reach, and grip strength had all improved slightly.

Interpretation of these potentially interesting comparisons was unfortunately complicated by the development of urban over-crowding between 1934 and 1958; it is thus probable that the more recent students had less

Table 86. *Changes in the motor performance of thirteen-year-old children in Oakland, California (difference of test scores, 1934 and 1958)*

	Girls	Boys
Height (cm)	+2.3	+5.6
Weight (kg)	+2.7	+4.7
50 yd dash (s)	+0.5	+0.2
Ball throw (cm)	−61	+610
Standing broad jump (cm)	−12.7	−12.7
Jump and reach (cm)	+2.5	+5.6
Hand-grip force (kg)	No change	+2.2

Based on data of Espenschade & Meleney (1961).

opportunity to practise running than did those educated before World War II.

Activity patterns

In a number of studies of children, the suspicion has arisen that differences of fitness between populations have reflected differences of habitual activity. Thus, it was suggested that rural schoolchildren in French Canada were less fit than an urban sample because the school bus delivered the rural group directly to their homes, while many of the urban children had to walk two to three kilometres to school every day (Shephard *et al.*, 1974*a*). Equally, in Czechoslovakia the daily walk of many rural children was curtailed by the provision of transport to and from school (Seliger, 1970).

Formal studies of activity patterns in children have been relatively few. Stefanik, Heald & Mayer (1959) compared fourteen obese and fourteen normal boys attending a vacation camp in the US. The obese boys had a lower food intake, but also participated much less than the normal boys in the various active pursuits at the camp. Bullen, Reed & Mayer (1964) made parallel observations on athletic girls, also at a US summer camp. Ishiko, Ikeda & Enomoto (1968) noted that the proportion of obese children was increasing in Japan; in Kanagawa prefecture, there were three times as many obese pupils in urban districts as in rural and mountainous districts, with a particular preponderance of overweight children in private schools and in managers' families. A study from northern Sweden (Samuelsson, 1971) again found more obesity in urban than in rural children, despite the lower caloric intake of the former group. The obese Japanese children (Ishiko *et al.*, 1968) were questioned as to their taste in physical activities. This coincided with their pattern of performance. Thus, they had a normal throwing ability, and enjoyed swimming and games such as dodge-ball and baseball. On the other hand, they scored poorly on running, jumping and chinning tests, and expressed a firm dislike for both running and gymnastics. Durnin (1966) found no differences in the activity patterns of thin and normal Scottish children. There were few plump boys in his series; however, he also noted that the plump girls devoted less time than their thinner counterparts to exercise, this being true both for 'very heavy' exercise at school and also for 'moderate' exercise in their leisure hours.

Durnin & Passmore (1967) found that the Scottish girls spent more time than the boys on 'moderate' exercise and less time on 'heavy' and 'very heavy' activities, both at school and in their leisure time. The sex difference was well illustrated by the energy cost of walking; this was 3 kcal/min in nine- to eleven-year-old boys (equivalent to a walking speed of 6.4 km/hr in an adult), but in girls of the same age it was only 2 kcal/min. At the age of fourteen, there was still an 0.7 kcal/min sex difference in the average cost of walking, with total daily expenditures of 2800 kcal for the boys and 2300 kcal for the girls.

Klimt (1966) used telemetry to study activity patterns in German preschool children. During spontaneous play, heart rates of 160–180/min were interspersed with brief rest pauses. The spontaneously chosen leisure activities of older children produced pulse rates of 150–160/min for much longer periods of time. Laboratory studies (Oseid *et al.*, 1969) indicate that boys ten to twelve years of age can sustain 70–80% of their maximum oxygen intake (corresponding to a pulse rate of 170–180/min) for an hour or more without exhaustion. Biochemical studies have shown that children have less muscle phosphofructokinase than adults (Eriksson *et al.*, 1973*a*). They are thus less able to produce lactate from glycogen. This is a disadvantage in very heavy work, but it may decrease their liability to fatigue in more moderate effort. Seliger *et al.* (1974) made 24-hour pulse counts on twelve- and sixteen-year-old Czechoslovakian boys. During the waking period, the average counts (~ 90/min) were significantly higher in the twelve-year-old boys than in the sixteen-year-olds (~ 80/min). When sleeping, the lowest readings were 48–51/min in the younger boys, and 36–49/min in the older group. The corresponding average daily energy expenditures were 2500 and 3000 kcal, significantly different in absolute terms, but rather similar when expressed as a percentage of the basal metabolic rate. The main difference between the twelve- and the sixteen-year-old boys was that the latter showed an increase in the daily period of sitting at the expense of sleep.

Comment has already been made upon Engström's (1972) study of active pursuits in 2000 fourteen-year-old Swedish children. The boys in this sample were devoting an average of 5 hours per week to vigorous leisure activities, and the girls an average of 3.5 hours. Participation was strongly influenced by the mass media, school and parents, and it is thus disturbing to note that while the students rated gymnastics as their most popular subject, the teachers placed it at the bottom of a list of eight subjects.

The children of most more-developed countries currently spend a large part of their leisure time in watching television. One study from Quebec (Shephard *et al.*, 1975*b*) showed urban boys spending thirty hours, rural boys twenty hours, urban girls twenty-two hours, and rural girls sixteen hours before the television each week, with little difference in summer and winter viewing times. One consequence of the television broadcasts is a wide discrepancy between expressed interests in particular sports and active participation. Almost all of the boys and most of the girls in the Quebec sample, for example, expressed an interest in both ice-hockey and baseball. However, of forty boys, only eighteen were playing ice-hockey regularly, and only twelve were playing baseball, while out of forty girls, only two were playing ice-hockey and three baseball.

The adolescent athlete

Training in adolescence

There is some evidence from animal experiments that training is most effective during the period of growth, and that it becomes progressively harder to improve physical working capacity as a person ages. The number of fat cells in the body is determined at a very early stage of extra-uterine development, and thereafter the fat content of the tissues can be modified only through a change in the loading of existing fat cells. Equally, hyperplasia and splitting of muscle cells may occur in the young animal, but in later life muscle power can only be improved by increasing the size and modifying the chemical constituents of existing fibres. It is thus of interest to examine the training responses of the adolescent, and to determine whether an increase of vigorous activity at an early stage in life can modify the normal course of growth, potentiating the development of both static and functional dimensions concerned with physical working capacity.

Short-term training, seasonal effects

One factor that tends to counteract the inherent trainability of the child is a relatively high level of habitual activity. The spontaneous patterns of activity in some communities may be enough to hold the majority of children at or close to the limits imposed by constitutional factors.

There is considerable disagreement regarding the responses to short-term training and seasonal variations in the physical working capacity of children. Mocellin & Wasmund (1973) found that when a group of eight- to ten-year-old German children trained by running 1000 metres twice a week for seven weeks, there was a 12–18% improvement of running speed; however, this was due largely to an improvement of coordination, with only an insignificant 2–3% augmentation of aerobic power.

Bar-Or & Zwiren (1973) compared the responses of nine- to ten-year-old Israeli children to a normal games and calisthenics programme with the changes induced by two to four sessions per week of endurance-type interval training. The interval-trained students gained no advantage in terms of maximum oxygen intake, although the boys did show some reduction of heart rate during submaximal work, and as in the series of Mocellin & Wasmund (1973) there was an improvement of running speed. Kemper (1971) and Johnson (1970) have also commented on the lack of physiological response to an increase in the number of physical education sessions per week.

With regard to seasonal effects, Adams *et al.* (1961*a*) found an improvement of PWC_{170} in Swedish children during the school holidays particularly in students with a low initial level of fitness. However, other reports from Scandinavia (P. O. Åstrand, 1961; Knuttgen & Steendahl, 1963) have suggested a gain of working capacity over the school year, with deterioration during the summer vacation (page 225). G. R. Cumming & Cumming (1963) found little difference of PWC_{170}/kg between children living in a poor district of Winnipeg (where few owned skates or bicycles) and public school students from more favoured areas of the city. They suggested that the existence of opportunities for exercise meant little unless accompanied by a vigorous and supervised programme of endurance effort. Working capacity scores were higher in a private school that had superior facilities for recreation and allocated additional time for sports and games. However, even at this institution there was little evidence of a gain in PWC_{170}/kg over the course of the school year (G. R. Cumming, Goulding & Baggley, 1969). A further study from the Winnipeg area (G. R. Cumming *et al.*, 1967) examined six boys and six girls aged thirteen to sixteen years over the course of a week's intensive track training. Although the group were running as much as 70 km per day, there was little improvement of $\dot{V}_{O_2(max)}$ over the week; on the other hand, the boys showed a progressive decrease of the heart rate during submaximum work. One reason for the lack of change of aerobic power in this group may have been two months of preliminary conditioning; all students commenced the study with a high level of aerobic power, averaging 3.96 l/min STPD in the boys, and 2.85 l/min STPD in the girls.

Cross-sectional comparisons

There is no strong evidence that child athletes exceed the fitness of their contemporaries by a wider margin than would be anticipated in adults.

Competitors in swimming events commonly reach a peak of performance during adolescence. Cunningham & Eynon (1973) found that Ontario male swimmers aged ten to sixteen years had an aerobic power of 56.6 ml/kg min, only slightly below the figure in many international-class swimmers (Shephard *et al.*, 1973*a*). Cunningham & Eynon speculate that the main difference between the juveniles and world-class competitors may be in the area of skill, rather than of physiological characteristics. P. O. Åstrand *et al.* (1963) studied thirty of the best Swedish girl swimmers; their average aerobic power was 54 ml/kg min, with a reading of 60 ml/kg min in the top performer. The same girls were re-examined a mere five years after stopping swimming. At this stage, their aerobic power had declined to 37 ml/kg min, a figure lower than that for untrained women of the same age (Eriksson *et al.*, 1971); on the other hand, vital capacity and heart volume were still large in relation to standing height. Does this indicate initial selection, or a lasting consequence of twenty-six hours training per week? Longitudinal studies of swimmers (page 231) are somewhat in favour of a causal relationship between training and the increase of heart and lung dimensions.

Sobolova *et al.* (1971) examined boys of thirteen, fourteen and fifteen who were attending a special swimming school in Czechoslovakia. At least ten hours of training per week were provided by specialized trainers. The initial PWC_{170}/kg was high (21.0 kg m/kg min) and it increased further (to 22.8 kg m/kg min) at the age of fifteen. Vital capacity also increased relative to normal age and height standards.

In the Zdar highlands of Czechoslovakia, some 9.4% of the children engage in competitive sports, 43% of the competitors being cross-country skiers. Cross-sectional studies suggest a faster growth of PWC_{170}/kg in the skiers than in other athletes, the difference being apparent at the age of fourteen in the boys, and fifteen in the girls (Daněk, 1974).

Longitudinal surveys

Cross-sectional comparisons between average children and athletes enrolled in special programmes unfortunately confound effects due to initial selection with the response to the programme itself. More satisfactory information can be gained from longitudinal surveys.

An ambitious project is currently under way in French Canada (Jéquier, 1974), with annual examination of many fitness-related variables in some 546 schoolchildren, drawn equally from an urban area (Trois Rivières) and a rural community (Pont Rouge). Half of the children in both communities have been given an additional hour of physical education five days per week, while the remainder have continued with the normal forty-minute periods of physical education once weekly. The children were enrolled at the age of six, and the majority were eleven- to twelve-year-olds in 1977, with a cumulative drop-out rate of less than 10%. This study will continue until the children reach the age of seventeen, and should then provide a categoric answer to the effects of added exercise upon growth and development.

Kemper *et al.* (1974) reported a study where twelve- and thirteen-year-old Dutch boys were given five versus three lessons of physical education per week. Over the year of study, the only gains in the more active group were increased scores on performance-type tests and a larger augmentation of explosive strength than that seen in the control group.

A more widely known study is that of Ekblöm (1969). He took thirteen eleven-year-old Swedish boys; seven continued with the regular school physical education programme (two 45-minute sessions per week, mainly ball games with some calisthenics), and the remaining six were given two additional 45–60-minute periods of endurance-type activity per week. At the end of six months, one of the active students stopped training, and two of the inactive commenced to train. All students were then followed for a further twenty-six months, changes of physiological status being evaluated according to the dimensional scheme of von Döbeln (1966). Over the thirty-two months of observation, body weight increased 45% in the active group and 28% in the reference group, these changes being slightly more and slightly less than predicted from H^3 respectively. Vital capacity increased by 54% and 34%, while heart volume increased by 49 and 36% respectively; for both variables the active group exceeded the H^3 prediction, while the reference group conformed to it. Ekblöm commented that similar enhanced development of heart and lung volumes had been described in girl swimmers (P. O. Åstrand *et al.*, 1963), although in the latter population it is difficult to rule out the influence of initial selection. The maximum oxygen intake of Ekblöm's two groups increased by 60 and 38% respectively, both groups departing quite markedly – and by rather similar amounts – from the H^2 prediction. One might thus conclude from the Swedish study that static dimensions were enhanced by the added exercise. However, it is plain that the main effects of activity were upon the rate of increase of height and weight, and it is conceivable that the apparent gains of vital capacity and heart volume represent no more than an advancement of the normal pubertal growth spurt.

A longitudinal survey of twenty-nine Swedish girl swimmers commenced when the group was thirteen years old (Engström *et al.*, 1971). Over the first three years of observation, the vital capacity of the group increased more than predicted from H^3, with the largest gains occurring in the girls who trained the hardest. The explanation suggested was a change in breathing pattern and/or a gain in strength of the chest muscles, rather than a simple anatomical growth of the chest, since the functional residual capacity and the total lung capacity did not show parallel gains.

A conclusive answer to this question of training and growth should emerge from the Quebec study, where a much larger series of children are being followed over a longer portion of the growth curve.

Heredity and the adolescent athlete

The age of the adolescent athlete is such that his parents are usually still alive but have adopted sedentary habits. There is thus a possibility of examining the parents for their genetic potential, independent of superimposed training effects. Ness *et al.* (1974) compared the parents of eleven ten-year-old prospective girl swimmers with the parents of nine other girls not selected for competitive swimming and thirteen contemporaries with no interest in competition. The swimmers' fathers had somewhat larger values of functional residual capacity (FRC) and total lung capacity (TLC) than the other two classes of parent, but showed no advantage of pulmonary diffusing capacity or aerobic power.

9. Age and working capacity

This chapter will examine the normal ageing of working capacity, and will consider what changes are induced by sustained athletic participation and by maintenance of a primitive life-style.

Normal ageing in more-developed countries

The normal ageing of working capacity has been the subject of several reviews (P. O. Åstrand, 1973; Roskamm, 1973; Shephard, 1973*b*). In the context of the IBP human adaptability project, it is sufficient to discuss the technical considerations of sample definition and measurement techniques, together with the usual course of changes in maximum oxygen intake, body fat, muscle strength, and vital capacity.

Technical considerations

In western society, such minor ailments as varicose veins or a 'touch' of bronchitis are considered almost a normal accompaniment of old age. Up to 50% of a 65-year-old population may have some diagnosable medical condition, and it becomes a difficult decision which individuals to exclude from the sample, particularly as many of the ailments in question can modify working capacity. For many purposes, it may be appropriate to include within the group all individuals who are sufficiently healthy to continue with their normal daily work.

Partly because of medical rejections and partly because caution increases with age, the proportion of a population volunteering for testing is progressively attenuated in the older age groups. We have observed this phenomenon equally when drawing a random sample of white subjects in Saskatoon (Bailey *et al.*, 1974*a*) and when testing the whole Eskimo community of Igloolik (Shephard, 1974*c*). In the Eskimo survey (page 14) 77% of male and 65% of female subjects under the age of twenty volunteered for testing. However, percentages dropped to 41% of the men and 48% of the women over the age of forty. Older volunteers in more-developed communities have tended to be the health-conscious and often the more physically fit members of the population. Sample attenuation may thus lead to an underestimate of the true rate of ageing in cross-sectional studies.

A longitudinal experiment might be thought to avoid such problems. However, in practice an extended series of observations leads to a progressive attrition of the sick and the poorly motivated. A further major source of difficulty is a substantial decrease in habitual activity over the period of study. We have already noted this effect in the general, sedentary population (page 119, Table 47). Among those who are initially vigorous and athletic, the change tends to be even more marked. It is thus almost inevitable that in a longitudinal study the true course of ageing is exaggerated by a detraining effect. Unfortunately, most recruits have an initial interest in physical activity, and the ultimate error in estimates of ageing may be greater than in a cross-sectional study where the majority of those studied – both young and old – are inactive.

Many of these factors operated in the study of Dehn & Bruce (1972). The annual decrease in aerobic power was 0.28 ml/kg min in their cross-sectional data, but 0.94 ml/kg min in a brief (2.3 year) longitudinal study. Presumably as a result of selection, the population for the longitudinal study had a relatively high initial aerobic power (37.8 ml/kg min at 52 years of age). If the rate of loss were to continue at 0.94 ml/kg min per year, then the aerobic power would drop below the minimum compatible with life (around 5 ml/kg min) at 87 years of age, a prediction at variance with the experience of many nonogenarians!

Whether a cross-sectional or a longitudinal approach is used, the results of maximum tests may be biassed downwards in older subjects; the person tested is often reluctant to exert himself maximally, while the attending physician halts an increasing proportion of tests due to such indications as myocardial ischaemia (anginal pain, ventricular extrasystoles and depression of the ST segment of the e.c.g.), cerebral ischaemia (confusion and incoordination) and peripheral vascular ischaemia (intermittent claudication). Sidney (1973) found that some 80% of sedentary middle-aged Canadians who passed the initial medical screening were able to continue treadmill walking to a satisfactory oxygen plateau. Many of the remaining 20% were also able to complete the treadmill test after seven weeks of light training. G. R. Cumming & Borysyk (1972) found more difficulty in realizing a satisfactory plateau of oxygen consumption when elderly Manitoban subjects were asked to pedal a bicycle ergometer to exhaustion. Unfortunately, in their series, there was little correlation between the various subsidiary criteria of maximum effort. Maximum heart rate, with 95% confidence limits of 143 to 191, varied widely. Blood lactate levels were greater than 72 mg/dl in 78% of subjects, but again there was a large scatter, presumably because of inter-subject variability in the relative volumes of muscle and blood. The respiratory gas exchange ratio exceeded 1.12 in 54% of subjects, but it was not closely related to lactate production, presumably because it was recorded in an unsteady state with a variable degree of over-ventilation. At exhaustion, only 43% of subjects had an oxygen consumption less than 90% of that predicted for a given work load. Presumably, the effort of Cumming & Borysyk's subjects was in many instances limited by weakness of the leg muscles; if so, their difficulties could have been lessened by (*a*) use of treadmill walking rather than bicycle ergometry, and (*b*) arrangement of a brief pre-test period of training for those who were extremely weak.

If submaximum predictions of maximum oxygen intake are performed in the elderly, difficulty arises from uncertainty regarding the maximum heart rate. S. M. Fox (1969) summarized ten studies of older populations that showed average heart rates ranging from 149/min to 177/min at the age of 65. Most recent studies of sedentary individuals who have continued to an oxygen plateau rather than subjective exhaustion or a limit imposed by the observer suggest that a pulse rate of at least 170/min should be attained by the 65-year-old. Some authors have also found ageing to cause a small decrease in the heart rate at a given level of submaximal work (Kilbom & Åstrand, 1971; Cotes *et al.*, 1973*c*), but others have found no change (C. T. M. Davies, 1972*b*) or an increase (I. Åstrand *et al.*, 1973). The pattern of response may change with intensity, a decrease of heart rate being observed at light loads, but an increase in heavier work. In men, Cotes *et al.* (1973*c*) suggested that the heart rate at an oxygen consumption of 1 l/min STPD could be predicted from age (*A*, yr) and a skinfold estimate of fat-free weight (*F*, kg) according to the equation:

$$f_{h(1.0)} = 67.5 - 0.39(A) + 3071/F \quad (\text{CV } 13\%)$$

The net effect of ageing on population predictions of $\dot{V}_{O_2(max)}$ is fortunately relatively small. Sidney (1973) compared direct and indirect treadmill measurements of maximum oxygen intake in average sedentary 65-year-old Torontonians. The systematic error of the Åstrand nomogram (when using published age correction) was 13% in the men and 3% in the women. I. Åstrand *et al.* (1973) retested the original athletic population on whom the nomogram was based after the elapse of twenty-one years; findings were much as in the series of Sidney (1973), predictions for the 42-year-old women being without systematic error, while results for the 46-year-old male athletes underestimated the directly measured maximum by 10%. Heart rates at a given submaximal oxygen consumption were generally increased at re-examination (thirty of thirty-one men, twenty-six of thirty-five women) but the relationship between this increase and the decrease of maximum oxygen intake showed a wide scatter.

Maximum oxygen intake

The general course of the decline in aerobic power of urban populations is evident from cross-sectional data on 6633 non-athletic men and 653 women, collected by the present author in 1966 (Figs. 7 and 8). More recently, G. R. Cumming & Borysyk (1972) collected 252 direct measurements of $\dot{V}_{O_2(max)}$ on 50- to 60-year-old men. The weighted average for this grouped data (32.4 ml/kg min STPD) coincides closely with the present author's line for 'all the world' at age 55. Men apparently lose 11–14 ml/kg min of aerobic power between the ages of 25 and 55, with a steeper fall (6–7 ml/kg min) between 55 and 65. Women remain relatively constant in aerobic power from age 25 to 35, lose 8–14 ml/kg min from 35 to 55, and a further 7 ml/kg min from 55 to 65.

There are naturally problems in interpreting such composite data, but fortunately several individual studies are also available. Bink (1962) examined 500 Dutch workers, Shephard (1969*a*) 505 men and 156 women in the Toronto area, P. O. Åstrand (1973) 350 Swedish men and women, and Bailey *et al.* (1974*a*) 558 men and 672 women living in Saskatoon. The total loss of aerobic power in the male subjects over four decades was 1 l/min in Åstrand's series (31% loss) and 1.15 l/min (39%) in Bink's data. The Toronto step test results showed a decline of 20.2 ml/kg min STPD from age 25 to 65 (total change 43% or 0.51 ml/kg min per year), any discrepancy from the absolute changes reported by Åstrand and Bink reflecting mainly an accumulation of body fat. Bicycle ergometer tests on the Saskatoon population suggested a loss of 13.6 ml/kg min from age 25 to 65 (total change 37% or 0.34 ml/kg min). Dehn & Bruce (1972) collected 700 miscellaneous observations from seventeen published cross-sectional studies, and computed an annual loss of 0.40 ml/kg min; however, their own cross-sectional data for men aged 40 to 69 years had a somewhat shallower slope (0.28 ml/kg min per year), possibly because the sample included a large proportion of active individuals.

With regard to female subjects, Åstrand's cross-sectional data showed a loss of only 0.4 l/min (18%) over four decades; one must suspect that the older participants in this study were of above average fitness. Cumulative losses were much greater in the Toronto series (14.1 ml/kg min, 37%) and in Saskatoon (11.6 ml/kg min, 38%), both of the latter being rather comparable with the ageing changes observed in Canadian men.

Seliger (personal communication) made a cross-sectional study of more than 2000 Czech subjects, both men and women, aged 18, 25, 35 and 45 years. In this sample, the 25-year-old groups were already substantially less fit than the 18-year-old subjects, and perhaps for this reason there was a relatively small further decline of aerobic power from the twenty-fifth to the forty-fifth year (3.7 ml/kg min, 9.5%, in the men; 4.3 ml/kg min, 13.4% in the women).

We have noted already the large decrement of aerobic power that Dehn & Bruce (1972) found in a brief longitudinal study of men in the Seattle area (0.94 ml/kg min per year). The slope became even more dramatic when inactive men (1.62 ml/kg min per year) were distinguished from active men (0.56 ml/kg min per year). The inactive men gained some weight, while the active men became lighter; however, there was still a twofold difference between the two segments of the population (respective values 1.32 and 0.65 ml/kg min) when adjustments were made for the weight changes. Dehn & Bruce drew support for these startling figures from a more extended (twelve to fifteen year) longitudinal study of healthy German men (Hollmann, 1965), and (perhaps with rather less justification) from long-term changes in the aerobic power of former US champion runners (Dill, Robinson & Ross, 1967); both of the latter sets of data also yielded slopes of about 1.5 ml/kg min per year, although in the case of the former runners inactivity must have compounded the true effect of ageing.

Asmussen & Mathiasen (1962) re-examined physical education students after the elapse of twenty-five years. Again, changes in activity patterns may have helped boost the loss of aerobic power to 0.57 ml/kg min per year in the men and 0.70 ml/kg min per year in the women. I. Åstrand *et al.* (1973) made a somewhat similar repetition of tests on sixty-six physical education students twenty-one years after their initial evaluation. A few of the original sample were excluded because they had subsequently developed medical problems. The majority of the subjects were said to have remained fairly active, and most trained for three to four weeks prior to re-examination. Nevertheless, the arobic power of the men had declined by 0.8 l/min (19.6%, 0.64 ml/kg min per year), while that of the women had decreased 0.63 l/min (22.2%, 0.44 ml/kg min per year). Dill *et al.* (1964) reported data on the deterioration in performance of two famous physiologists. The maximum oxygen intake of Dill declined by 0.66 l/min (25.1%) from 1929 to 1961, while that of Talbott dropped by 1.42 l/min (36.1%) from 1928 to 1962. Both investigators had remained lean (weights of 72 and 74 kg respectively) so that on a per kilogram basis, their losses would have been approximately 0.29 and 0.56 ml/kg min per year.

We must thus conclude that longitudinal data generally indicate somewhat more rapid ageing of aerobic power than cross-sectional studies. However, not all authors have found the very steep decline of function seen by Dehn & Bruce (1972) in their brief longitudinal study. It is most unlikely that such rapid ageing could be sustained over the adult life span; indeed, the true rate of ageing is probably much closer to that shown by cross-sectional studies. This is fortunate, since all existing studies of ageing among primitive peoples are based on a cross-sectional approach.

Body composition

The typical sedentary male population shows an increase of body weight, commencing in the fourth decade of life and reaching its peak at about 45 years of age (Table 23). This is mainly body fat, as can be seen from the parallel increase of skinfold readings. Between the ages of 45 and 65, there is a constant or a decreasing body weight, but skinfold thicknesses do not change. It must thus be presumed that at this stage the lean tissue of the body is diminishing. A further factor, particularly in the final decade, is a progressive attrition of the sample through the selective mortality of overweight individuals.

Young Canadian women (20 to 30 years) are already heavy relative to actuarial standards. The Toronto sample shows a decrease of both body weight and skinfold thicknesses in the next decade (Table 23). Although not duplicated in every report, the present author regards this as a reasonable finding for North American society; the young 'working' girl is extremely inactive, but has a compulsory augmentation of her physical curriculum with the birth of two or three active children. In subsequent decades, the children learn to care for themselves, and the mother then shows a further augmentation of both body weight and skinfold thicknesses.

Very similar findings could be tabulated for many other more-developed countries. Much of the large-scale data collection has been for life assurance purposes, and it could be argued that this represents a privileged segment of the community. In practice, this does not seem a serious objection. In Toronto, young men of 25 and women of 35 come close to meeting actuarial standards. Wyndham, Watson & Sluis-Cremer (1970) made a detailed comparison between South African life assurance statistics and data for white miners and army personnel. When account was taken of the shoes normally worn by life insurance candidates, there was a fair concurrence of the data. Miners of height 175 cm (5 ft 9 in) for example, were 0.9 kg (2 pounds) lighter than the assured group at age 25, and by gaining 10.0 kg (22 pounds) over the age span 25 to 55 were finally 1.8 kg (4 pounds) heavier than those assured. How far such differences reflected socio-economic factors, and how far the high proportion of Afrikaans-speaking individuals in the mining group was not explored. Kemsley, Billewicz & Thomson (1962) related the weight of British subjects to that observed at age 25. Men were 2.3 kg (5 pounds) heavier from 30 to 39 years, 4.5 kg (10 pounds) heavier from 40 to 49 years, and 2.3 kg (5 pounds) heavier from 50 to 64 years. Women showed a more continuous gain of weight, the increments being 2.3 kg (5 pounds; 30 to 39 years), 4.5 kg (10 pounds; 40 to 44 years) and 6.8 kg (15 pounds; 45 to 64 years). Khosla & Lowe (1967) reported the heights and weights of 5000 men employed at an electrical engineering company in Birmingham, UK. Readings were obtained at the time of mass-radiography, and encompassed all grades of staff from works manager to serviceman. The body weight of this series was almost identical at ages 20 to 24 and 60 to 64, but the older men were on average 7.6 cm (3 in) shorter. Thus, at a fixed height, there was a weight gain of about 4.5 kg (10 pounds) from age 22 to 62, and the index W/H^2 increased from an average of 3.29 to 3.60. As in the series of Kemsley *et al.* (1962), the relation of body weight to age was quadratic, all grades of staff reaching a peak of body weight around their 35th year. At all ages, monthly staff were heavier than those paid weekly, who in turn were heavier than those paid on an hourly basis; however, this was due largely to differences in standing height, and when data were expressed as weight at a fixed height or as W/H^2 there were no trends with socio-economic status.

Bjelke (1971) used a mail survey to collect detailed statistics on the interactions of height, weight and age for a randomly selected sample of the Norwegian population (8638 men, 79% of respondents; 10331 women, 82% of respondents) aged from 37 to 71 years. In the men, weight at a given height increased by only 1.0–1.4 kg over this age span, but in the women the gain was 6.7–8.1 kg. Much of the data was analyzed as W/H^2 ratios. Perhaps because muscularity and obesity are confounded in all weight/height relationships, no gradation of W/H^2 was seen with social class. However, the ratio was negatively correlated with both habitual physical activity and cigarette smoking. Among the female subjects, W/H^2 was greater for married than for single women, with particularly large ratios in the multiparous.

North American samples have generally shown a greater body fat than those from Europe, particularly in the US. Norris, Lundy & Shock (1963) described the anthropometric characteristics of a group of men living in the Baltimore area. They seemed typical US citizens – 'busy, successful people, active in community life, engaged in sedentary work, and for the most part did not engage in strenuous physical activities'. Between the ages of 25 and 65, there was a weight gain of over 8 kg, although the older subjects were 4.4 cm shorter than the 25-year-olds; 6.3 kg of the additional weight was gained between 25 and 35 years of age. Even the young subjects were quite obese, and there was surprisingly little change of body density between 25 years (specific gravity 1.022, equivalent to 33% body fat as calculated by the formula of Brozek *et al.*, 1963*a*; see Table 25), and 65 years (specific gravity 1.019, 34.3% body fat). It seems unlikely that sedentary subjects developed muscle or bone between 25 and 65 years, and some systematic error in the body density measurements must be suspected.

C. M. Young (1965) summarized other US studies of skinfold thickness and body density. Brožek *et al.* (1963*b*) reported data for 854 railroad clerks and 382 switchmen aged 40 to 59 years. Averages for the triceps and subscapular folds were slightly greater than found in Toronto (16.3 mm in clerks, 15.5 mm in switchmen). C. M. Young *et al.* (1963) studied a population of university women volunteers in New York. Again, the US citizens were a little fatter than the Toronto sample, the average readings for twelve skinfolds being 17.2 mm at age 16 to 30, with a slight decrease to 16.6 mm at age 30 to 40, and subsequent recovery to 21.0 mm at age 40 to 50 and 22.1 mm at age 60 to 70. A second cross-sectional study of US women (Wessel *et al.*, 1963) suffered from the fact that those volunteering for examination, particularly in the older age groups, were lighter than in the general population.

Several authors (C. M. Young, 1965) have reported body densities of 1.055 to 1.075 (10.9–19.0% body fat) in young US men (age about 20 years). The density in these studies dropped progressively to about 1.035 at age 65 (27.4% fat). Myhre & Kessler (1966) compared body density and potassium 40 estimates of body composition in a 'fairly representative' group of 100 mid-western males aged 15 to 87 years. Body density readings suggested an increase from 12.7% fat at 19.9 years to 23.4% at 71.1 years. The $^{40}K^{+}$ readings indicated a much larger increase, from 14.9% to 36.2% body fat, and the authors suggested that in the older subjects the body potassium figures may have been distorted by an increase in the ratio of connective tissue to muscle. Data for young women (C. M. Young, 1965) showed a body density around 1.040 (25.2% body fat), with a progressive decrease to 1.005 (40.5% body fat) at the age of 65 years. Skinfold readings are lower, and body densities higher in some European data. Thus Pařízková (1963) found average skinfold readings (ten sites) of 10.8 and 12.0 mm in Czechoslovak men aged 30 and 60 years respectively, while her readings for women were 14.0 and 16.0 mm at the same ages. The body density of 16- to 17-year-old boys was 1.070; this dropped to 1.053 at age 30, and 1.048 at age 60, the latter being equivalent to 21.9% fat. For 17- to 18-year old girls, readings averaged 1.043, dropping to 1.039 at age 30, and 1.033 at age 60, the last reading corresponding to 28.2% fat. The more recent Czechoslovakian IBP studies (Seliger, personal communication) are in fair agreement with Pařízková's observations. Between the ages of 25 and 45, the body fat in a large random sample of men increased from 15.5% to 16.9%, while in the women the increase was from 20.1% to 25.8% over the same age span. On the other hand, skinfold estimates of body fat from Milan (Steplock *et al.*, 1971) are more comparable with the US figures (Italian males, 18% at age 20, rising to 28% at age 60; Italian females, 26% at age 20, rising to 32% at age 50).

An alternative approach to the measurement of body fat is to examine soft-tissue radiographs. Comstock & Livesay (1963) exploited this method when studying age changes in the body composition of 40 000 people living in Muscogee County, Georgia. The films were taken as part of an anti-tuberculosis programme, and for this reason the sample was biassed somewhat towards Negroes and the poorer segment of the white popula-

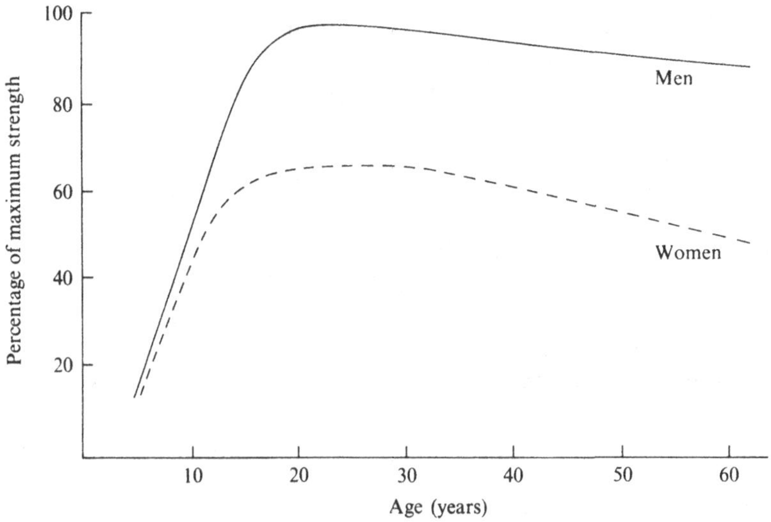

Fig. 13. The relationship of hand-grip strength to age. Based on data of author (Toronto) and of Howell *et al.* (Edmonton). (Illustration from Shephard, 1972*b*, by permission of C. C. Thomas, Publishers, Springfield, Illinois.)

tion. Among the men, the thickness of the subcutaneous fat was slightly greater in whites than in Negroes (5.0 and 4.6 mm respectively). Readings for the white men increased from an average of 4.6 mm at age 20 to 24 to 6.4 mm at 60 to 64, with the big gain of fat occurring in the fourth decade; in the Negroes, the increase was much less (from 4.6 mm at 20 to 24 to 5.0 mm at 60 to 64 years). Among the female subjects, the white were slightly thinner than the Negroes (average readings 6.7 and 7.2 mm). The white women showed a relatively uniform increase of skinfolds from 6.1 mm at age 20 to 24 to 8.6 mm at 60 to 64, while in the Negroes the gain was from 6.4 to 8.5 mm. Both in white and in Negro subjects, the greatest proportion of the 'obese' (those with a trapezius fat layer greater than 10 mm) was found in the higher socio-economic groups; however, at all levels in society, the proportion of obese subjects increased with age.

Muscle force and anaerobic power

The general course of ageing of muscle strength in the more-developed countries can be illustrated by cross-sectional data for hand-grip force (Fig. 13). In the present author's series of Canadians (Shephard, 1969*a*), the hand-grip force of the men remained relatively constant at an average of 54 kg from age 25 to age 45; thereafter, it declined by 20% to a figure of 44 kg at age 65. Unpublished data for male Canadian physical education students (Pimm & Shephard, in preparation) are marginally higher (average 58.2±7.7 kg). Our only systematic data for young Canadian women relate to relatively unfit physical education majors (aerobic power 40.2 ml/kg min, mean skinfold thickness 13.3 mm). This group had an average grip strength of 36.0±5.7 kg. The average result for female departmental store employees was 31.2 kg at age 45, dropping to 27.5 kg at age 65 (Shephard, 1969*a*).

Asmussen & Heebøll-Nielsen (1962) reported that Danish men reached a peak hand-grip force of 55–65 kg at the age of 30, with a subsequent decline, 20% at 60 years of age. Danish women had 55–65% of the strength of the men, with peak readings (35–40 kg) a little lower than the female Toronto physical education students. Seliger (personal communication) found no change in the hand-grip force of a large sample of Czechoslovakians between the ages of 25 and 45 years (respectively values 49.7 and 49.9 kg in the men, 30.0 and 30.0 kg in the women).

The main statistics on the Margaria test of anaerobic power have come from the Milan laboratory. In city-dwelling Milanese, there was a decrease from a vertical speed equivalent to an oxygen consumption of 170 ml/kg min at age 20 to about 100 ml/kg min at age 60. For the women, the decline was from 145 ml/kg min at age 20 to about 80 ml/kg min at age 60.

Vital capacity

The world literature contains an enormous number of prediction formulae describing the rate of ageing of vital capacity and related lung volumes (Shephard, 1971*d*), particularly with respect to the more-developed countries. The commonest statistical treatment has been to fit a multiple regression equation based on standing height and age to data collected over the span 20 to 60 or 20 to 70 years of age (Table 87). This is slightly erroneous, since most subjects reach a peak of vital capacity at about the age of 24 (Anderson *et al.*, 1968); however, in practice the error from assuming a linear deterioration of vital capacity from the age of 20 does not seem too serious.

Cole (1974) has argued for the existence of an age/height interaction. There seems some problem in the placing of the decimal in his published report; it is unlikely that a man of 170 cm would lose 39.4 ml/yr due to age alone, and a further 95.2 ml/yr due to the age/height interaction (Table 88); presumably, the interaction should be 9.5 ml/yr. Nevertheless, the underlying concept seems sound, and it is readily explained on the basis that all subjects suffer an equivalent proportional loss, but that tall subjects with a large vital capacity have a larger absolute change. Since populations differ in average height, it may be desirable to make comparisons in terms of percentage changes as well as in absolute units.

Most recent prediction equations yield rather comparable figures for

Table 87. *Partial regression coefficients describing the effect of age on vital capacity*

Population	Age coefficient (ml decrease/yr)	Author
Male		
African & Indian (Guyana)	24.0	Miller *et al.* (1970)
Asian	42.0	Oscherwitz *et al.* (1971)
Aymara Indians	37.5	Donoso *et al.* (1974*a*)
Bhutanese	20.0	Cotes & Ward (1966)
Chinese	17.6	Da Costa (1971)
Irish	31.0	Sidor & Peters (1973)
Italian	21.0	Sidor & Peters (1973)
Negro	18.0	Abramowitz *et al.* (1965)
Negro	27.0	Oscherwitz *et al.* (1971)
Negro	17.5	Rossiter & Weill (1974)
White	16.0	Abramowitz *et al.* (1965)
White	36.0	Oscherwitz *et al.* (1971)
White	23.0	Rossiter & Weill (1974)
Female		
African & Indian (Guyana)	20.0	Miller *et al.* (1970)
Negro	7.0	Abramowitz *et al.* (1965)
White	11.0	Abramowitz *et al.* (1965)

Table 88. *A comparison of cross-sectional and longitudinal assessments of the age, height and age/height coefficients for deterioration of forced vital capacity*

Survey	Age (ml/yr)	Height (ml/cm)	Age/height[a] (ml/yr per cm)
1957 Cross-sectional	−39.4	53.6	−0.56
1966 Cross-sectional	−40.4	50.2	−0.58
25–34 yr Longitudinal	−17	not stated	0.24
55–64 yr Longitudinal	−45	not stated	0.98

Based on the data of Cole (1974).

[a] The absolute size of this coefficient is here presented as quoted by the author. It seems too large by a factor of 10.

the average 20-year-old (Shephard, 1971*d*). Assuming a young man of 170 cm height, two Canadian equations predict a vital capacity of 5012 and 5048 ml BTPS, a South African equation gives 4925 ml, the three most popular equations from the US 4870, 4860 and 4610 ml, two from Scandinavia 4960 and 5179 ml, two from France 5040 and 5340 ml, and one from the UK 4930 ml. In the case of a 160 cm young woman, the Canadian equation predicts a vital capacity of 3360 ml, two Scandinavian equations 3690 and 3483 ml, a British equation 3600 ml, and two US equations 3420 and 3400 ml. Such comparisons suggest the European may have a slight advantage of vital capacity over US populations, presumably because of greater muscular strength. The ageing of vital capacity is also relatively consistent in different more-developed countries. Where multiple regression equations of the type

$$\mathrm{VC} = a(H) - b(A) + c$$

have been fitted, the coefficient b has amounted to 17.4 and 22.6 ml/yr in two series of Canadian men, 22.0 ml/yr in a US series, 20.0 and 33.2 ml/yr in two Scandinavian series, and 31.0 ml/yr in a South African series. Taking the average of the six estimates (24.4 ml/yr), the vital capacity of a typical healthy white male of 170 cm height should drop from 5001 to 4026 ml BTPS between the ages of 20 and 60 years. The deterioration of the female subjects proceeds somewhat more slowly in absolute terms (average loss of 18.3 ml/yr in five studies); however, when account is taken of the smaller initial vital capacity, the relative effect (decrease 3515 to 2783 ml BTPS between the ages of 20 and 60 years) is approximately the same (20% loss) as in the men.

Most studies to date have been cross-sectional. The apparent rate of deterioration is thus modified by the smoking habits and air pollution experience of successive generations. There is also a progressive elimination of those with a low vital capacity from the older age groups. Cole (1974) compared cross-sectional and longitudinal data for two age and occupation random-stratified samples of the community of Staveley, Derbyshire. The younger longitudinal group unfortunately covered the age span 25 to 34 when the steady rate of deterioration had not fully developed, while the older age group covered the span where loss was most rapid (55 to 64). The most that can be learnt from Cole's comparison is thus that longitudinal and cross-sectional studies give results that are not in gross disagreement with each other.

Ageing of athletes

Maximum oxygen intake

There have been reports that the deterioration of maximum oxygen intake proceeds more slowly in the continuing athlete than in the sedentary subject (for example, Hollmann, 1965; Dehn & Bruce, 1972). However, in both of the comparisons cited, the apparent rate of ageing of the 'control' group seems to have been exaggerated by an increase of body weight and by a decrease of habitual activity. Thus, the rates of deterioration cited for the continuing athletes were 0.70 and 0.56 ml/kg min per year, much as many cross-sectional studies have indicated for the sedentary population.

The longitudinal data of Dill *et al.* (1967) showed a slope of 0.67 ml/kg min per year for champion runners who were still active. The longitudinal data of Asmussen & Mathiasen (1962) and of I. Åstrand *et al.* (1973) also refers to fairly active subjects; nevertheless, losses in the men were 0.57 and 0.69 ml/kg min per year respectively, while in the women the corresponding figures were 0.70 and 0.44 ml/kg min per year.

Saltin & Grimby (1968) made a cross-sectional comparison between continuing orienteers and those who had formerly been active in this sport. To the age of 65, the loss was 0.34 ml/kg min per year in the active group, and 0.40 ml/kg min per year in those who were now inactive.

We may thus conclude that the longitudinally measured rate of functional loss in the male athlete is about 0.60 ml/kg min and that part of the discrepancy from cross-sectional comparisons reflects a progressive diminution of training schedules even in the continuing athlete. In absolute terms, the rate of loss seems similar in active and in sedentary individuals, but because the athlete starts with a large working capacity, his relative loss is substantially smaller.

Body composition

The expectation might be that the continuing athlete would avoid middle-aged obesity. This is borne out in the cross-sectional study of orienteers (Saltin & Grimby, 1968). Their average weight at age 25 was 70 kg. In those who remained active, the weight was still 70 kg at age 45, but there was a surprisingly large loss of tissue over the next two decades (weights 64 and 63 kg at ages 55 and 65 respectively). The inactive former orienteers, on the other hand, were 5 kg heavier than their active contemporaries at age 45, 12 kg heavier at 55 and 7 kg heavier at 65 years. Montoye *et al.* (1957) have also documented the rapid weight gain of athletes when they abandon training.

In the series of Asmussen & Mathiasen (1962), the men gained an average of 5.5 kg from age 24 to 50, while the women gained 2 kg from age 23 to 51. The physical education students of I. Åstrand *et al.* (1973) showed a similar picture, the men gaining 2.5 kg, and the women losing 1.5 kg from age 22 to 43.

Irrespective of activity patterns, the serum cholesterol levels of the middle-aged and older Swedish orienteers (Saltin & Grimby, 1968) were relatively high (at age 65, 286 and 266 mg/dl in the active and inactive groups respectively). However, the inactive subjects carried a heavier burden of serum lipids (corresponding values 1.10 and 1.85 mmol/l at age 65).

Some authors (for example, C. T. M. Davies, 1972*b*) have argued that athletes maintain a large heart, and other dimensional characteristics. Studies of detrained female swimmers (Eriksson *et al.*, 1971) suggest this is true over short periods such as five years. However, the data of Saltin & Grimby (1968) shows that the heart volume of inactive former orienteers dropped to 835 ml (11.1 ml/kg) at age 45, compared with 1047 ml (15.0 ml/kg) in those who had remained active. Over the next two decades, heart volumes diminished even in those who claimed to be remaining active, to reach a figure of 830 ml (13.2 ml/kg) at the age of 65. The apparent *rise* of heart volume from age 20 to 50 in the study of C. T. M. Davies (1972*b*) presumably reflects a sampling problem, with inclusion of more active subjects in his older age groups; this suspicion is borne out by the absence of any decrease in $^{40}K^{+}$ estimates of muscle mass with ageing.

Muscle force

The decrease of hand-grip force over twenty-five years was at least as great in the physical education graduates of Asmussen & Mathiasen (1962) as others have seen in the sedentary population.

	Right hand (Age)		Left hand (Age)	
	23–24 years (kg)	50–51 years (kg)	23–24 years (kg)	50–51 years (kg)
Men	54.3	44.2	48.2	38.9
Women	36.7	26.6	32.7	23.0

Vital capacity

The decrease of vital capacity seen by Asmussen & Mathiasen (1962) was quite small (in the men 0.2 litres, 7.7 ml/yr, in the women 0.25 litres, 8.9 ml/yr). I. Åstrand *et al.* (1972) were not able to detect any decrease of vital capacity in their physical education graduates over twenty-one years. While this may indicate some protective effect of physical activity on the lungs, two other considerations are that relatively few physical education students smoke, and there have been substantial improvements in the design of spirometers between the initiation and the completion of these two studies.

Table 89. *The ageing of aerobic power in primitive populations*

Population	Age span	Change of aerobic power (l/min STPD)		(ml/kg min STPD)		(ml/kg min per year)	Author
Male		Young	Older	Young	Older		
Cold regions							
Alaskan Eskimos	20–60	—	—	50	37	0.31	Rennie *et al.* (1970)
Canadian Eskimos	25–55	3.67	2.56	54.3	37.0	0.58	Rode & Shephard (1971)
Kautokeino Lapps	25–55	3.45	2.80	53	43	0.33	Andersen *et al.* (1962)
Skolt Lapps	25–65	—	—	50	25	0.65	Karlsson (1970)
Warm regions							
Bantu	30–55	2.8	2.3	46.5	39.5	0.28	Di Prampero & Cerretelli (1969)
Nilo–Hamitics	30–55	2.9	2.6	48	44	0.18	Di Prampero & Cerretelli (1969)
Pascuans	25–65	2.9	1.5	41	22	0.48	Ekblöm & Gjessing (1968)
Sardinians	25–45	2.25	2.4	36.5	30.5	0.30	Di Prampero (personal communication)
Tunisians	25–55	2.6	2.2	41	32	0.30	Di Prampero (personal communication)
Mountainous regions							
Aymara Indians	25–45	2.75	2.68	45.4	44.0	0.07	Donoso *et al.* (1974*a*)
Italian Alps	25–50	2.9	2.8	42	35	0.28	Steplock *et al.* (1971)
Tarahuma Indians	30–52	—	—	63	50	0.59	Aghemo *et al.* (1971)
Gurkhas	25–31.5	2.50	2.42	41.5	38.2	0.51	Ramaswamy *et al.* (1966)
Indians at 3100 m	26–36	2.35	2.37	43.5	42.0	0.16	Ramaswamy *et al.* (1966)
Indians at 3500 m	24–36	2.37	2.16	41.9	38.6	0.29	Ramaswamy *et al.* (1966)
Female							
Cold regions							
Canadian Eskimos	25–55	2.53	2.13	44.3	33.5	0.36	Rode & Shephard (1971)
Kautokeino Lapps	22–40	2.4	2.0	40	30	0.56	Andersen *et al.* (1962)
Skolt Lapps	25–55	—	—	33	24	0.30	Karlsson (1970)
Warm regions							
Nilo–Hamitics	25–35	2.1	2.2	42	40	0.20	Di Prampero & Cerretelli (1969)
Pascuans	25–55	1.8	1.8	32	28	0.13	Ekblöm & Gjessing (1968)
Tunisians	25–55	2.0	1.8	38	26	0.40	Di Prampero (personal communication)
Mountainous regions							
Italian Alps	25–50	2.0	2.3	35	31	0.16	Steplock *et al.* (1971)

Primitive groups

Maximum oxygen intake

Problems in the interpretation of cross-sectional studies (page 233) are even more acute for less-developed than for more-developed countries. In the primitive environment, attenuation of the ageing sample by disease and natural hazards is much more severe, and indeed in some communities only exceptional individuals survive to the age of 50. There are sometimes so few elderly subjects in a primitive settlement that it becomes necessary to calculate ageing over the span 25 to 55 years, or even 25 to 45 years; because of the usual shape of the ageing curve (Figs. 7 and 8), this in itself may favour the primitive group. Precise calculations are further complicated in communities where true ages are unknown (page 184).

In cold regions (Table 89), the overall loss of aerobic power (average of four studies, 0.47 ml/kg min per year in the men; average of three studies 0.41 ml/kg min per year in the women) seems much as reported for cross-sectional studies in developed countries. The form of the curve in the Igloolik population is also rather similar to that described for white Canadians; the aerobic power of the Eskimo is well sustained until about the age of 40, but societal customs dictate that grandparents can take their pick of the meat from returning hunters, and there is thus a sudden drop in physical activity and a parallel deterioration of aerobic power in the fifth decade of life.

In warmer regions, the aerobic power of the younger generation tends to be lower than in the circumpolar groups. Habitual activity is less than in the cold and mountainous regions at all ages, and probably for this reason the tropical populations show less deterioration with age (average for men, 0.31 ml/kg min per year, five studies; average for women 0.24 ml/kg min per year, three studies).

In mountainous regions, the rate of loss of aerobic power is very variable, rapid in the Tarahumara Indians who give up their endurance running, and very slow in other groups who must climb as a part of their daily life.

Body composition

The influence of poor nutrition upon the 'normal' increase of body fat is well illustrated by the data of Frisch & Revelle (1969) for South American and Asian countries. In only two of eleven nations (Table 90) was the daily energy supply greater than 2500 kcal; in these countries (Uruguay and Chile), there was an increase of body weight from 22 to 62 years, much as in more-developed nations. The diet of the remaining nine nations provided an average of only 2127 kcal/day; here, the men lost an average

Table 90. *The relationship between calorie supply and weight change from age 22 to 62 years*

Nation	Calorie supply per day (kcal)	Change in weight: Men (kg)	Change in weight: Women (kg)
Uruguay	3030	+3.0	+10.3
Chile	2610	+10.4	+7.2
Malaya	2400	−1.6	−1.2
Japan	2360	−2.2	−2.0
Venezuela	2330	+2.1	−0.5
Colombia	2280	−7.6	+0.2
Thailand	2120	−3.6	−5.2
Ecuador	2100	0.0	+0.5
Bolivia	2010	−2.0	−4.9
Vietnam	1944	−1.5	−4.1
NE Brazil	1600	−1.5	+1.4

Based on data of Frisch & Revelle (1969).

of 2.0 kg, and the women an average of 1.8 kg between the ages of 22 to 62.

The contribution of socio-economic factors to the development of body fat was further demonstrated by Ashcroft *et al.* (1966*a*). They found that in rural areas, Jamaican men lost 4.5 kg (10 pounds) of body weight from age 25 to 55, and women gained only 0.5 kg (1 pound) over the same period. However, in an urban environment, the men gained 2.7 kg (6 pounds), and the women 8.2 kg (18 pounds). An equally dramatic example was provided by Evans & Prior (1969); they compared body weights and weight/height indices for Polynesians living at subsistence level (coconut, taro and fish) on Pukapuka with the town-dwellers of Rarotonga (who were eating mainly European-type food). On Pukapuka, the men gained 2.90 kg (6.4 pounds) from 25 to 65 years, and the women lost 0.99 kg (2.4 pounds). However, on Rarotonga, the men gained 4.67 kg (10.3 pounds), and the women 14.47 kg (31.9 pounds)!

Skinfold readings are summarized in Table 91. Some groups such as the Australian aboriginal men (Abbie, 1967) and the high altitude Ethiopians (Andersen, 1971), have shown a small gain of subcutaneous fat over the adult span; the small group of five elderly Eskimo women seen by Rode & Shephard (1971) were also quite obese. However, in other populations, there has been a loss of subcutaneous fat during adult life; in the New Guinea group studied by Sinnett & Whyte (1973*b*) there was an associated decrease of average body weight (15 kg in the men, 13 kg in the women).

Body fat as estimated from skinfold readings (Table 92) shows relatively

Table 91. *The effect of age on the skinfold readings of primitive populations*

Population	Skinfold readings ($\Sigma\bar{N}$, mm) 25 yr	40 yr	55 yr	Author
Male				
Australian				
Aborigines ($\Sigma\bar{2}$)	7.3	9.0	9.6	Abbie (1967)
Canada				
Eskimos ($\Sigma\bar{3}$)	6.0	5.9	5.7	Rode & Shephard (1971)
Ethiopia ($\Sigma\bar{3}$)				
High altitude	6.8	7.4	7.9	Andersen (1971)
Low altitude	6.0	5.8	6.5	Andersen (1971)
New Guinea ($\Sigma\bar{4}$)	5.1	4.6	4.4	Sinnett & Whyte (1973*a*)
Biak ($\Sigma\bar{3}$)	5.3	5.0	—	Jansen (1963)
Female				
Australia				
Aborigines ($\Sigma\bar{2}$)	10.7	—11.2—		Abbie (1967)
Canada				
Eskimos ($\Sigma\bar{3}$)	9.3	11.3	24.8	Rode & Shephard (1971)
New Guinea ($\Sigma\bar{4}$)	6.5	4.2	4.1	Sinnett & Whyte (1973*a*)
Biak ($\Sigma\bar{3}$)	7.9	—5.8—		Jansen (1963)
Mappia ($\Sigma\bar{3}$)	8.3	—3.4—		Jansen (1963)

Table 92. *The effect of age on estimates of body fat of primitive populations*

Population	Body fat (%) 25 yr	40 yr	55 yr	Author
Male				
East Africans	12	15	11	Di Prampero & Cerretelli (1969)
Italian Alps	17[a]	24[a]	19[a]	Steplock *et al.* (1971)
Tarahumara Indians	11.1[b] 17.9[d]		14.7[c]	Aghemo *et al.* (1971)
Tunisians	18	20.5	22	Di Prampero (personal communication)
Female				
East Africans	18	18	18	Di Prampero & Cerretelli (1969)
Italian Alps	28[e]	34[e]	—	Steplock *et al.* (1971)
Tunisians	27	26	34	Di Prampero (personal communication)

[a] Values estimated from graph – final values at age 50.
[b] Runners aged 29.8 yr.
[c] Former runners aged 52.3 yr.
[d] Non-runners aged 28.0 yr.
[e] Values estimated from graph.

Table 93. *The effects of age on anaerobic power – data from primitive populations*

Population	Age span (yr)	Anaerobic power (ml O_2/ kg min)	Annual loss (ml/kg min)	Author
Male				
Alaskan Eskimos	25–65	180→110	1.75	Rennie *et al.* (1970)
East Africans	20–50	120→110	0.33	Di Prampero & Cerretelli (1969)
Italian Alps	20–50	175→125	1.67	Steplock *et al.* (1971)
Sardinians	25–45	130→120	0.50	Di Prampero (personal communication)
Tunisians	25–55	142→85	1.90	Di Prampero (personal communication)
Female				
Italian Alps	20–50	145→80	2.17	Steplock *et al.* (1971)
Tunisians	25–55	120→45	2.50	Di Prampero (personal communication)

low readings for the poorly nourished East Africans and the active Tarahumara Indians, but higher readings for Tunisians and residents in the foothills of the Italian Alps.

Muscle force, anaerobic power

Rode & Shephard (1971) provided data on the ageing of hand-grip and leg extension force for the Canadian Eskimos. In the men, hand-grip force decreased from 52.0 kg at age 25 to 38.5 kg at age 55, while leg extension force dropped from 88.3 kg to 76.9 kg. In the women, the corresponding changes were 29.8 to 27.3 kg and 68.9 kg to 69.5 kg. It would appear that in this community the men lose grip strength at least as fast as among white subjects, but that leg extension strength is unusually well preserved – perhaps by the need to walk through deep snow for much of the year.

The annual loss of anaerobic power seems extremely variable from one population to another (Table 93); in some primitive groups the rate of deterioration is faster than in the white population, and in others (particularly where the young perform badly) it is slower.

Vital capacity

Ageing coefficients for a number of populations are summarized in Table 87. If attention is directed to comparative studies (for example, Abramowitz *et al.*, 1965; Oscherwitz *et al.*, 1971; Rossiter & Weill, 1974), there seems little difference in the rate of deterioration of vital capacity between white and other racial groups. The average loss for eight studies of non-white male populations is 25.5 ml/year, very close to the rate previously indicated for white citizens of the more-developed countries. We may thus conclude on present evidence that the ageing of vital capacity is uninfluenced by racial factors. However, there is insufficient data to decide whether the rate of deterioration is affected by primitive environmental conditions.

10. Epilogue

In this final chapter, we shall assess how far the HA-IBP project has realized its objectives with respect to the measurement of physiological work capacity, indicating what conclusions can be drawn from the research findings, and what major areas of uncertainty remain for subsequent study.

Unification of scientific endeavour

One fundamental premise of the HA-IBP project was that scientists from different nations and different disciplines would accept curtailment of personal freedom in order to undertake a joint and definitive study of questions relating to the broad field of human ecology. Many other world problems are becoming so vast that they require a similar united attack by the scientific community. The HA-IBP project thus provides a model for those who would examine both the possibilities and the handicaps of international and inter-disciplinary research.

The immediate impression left by HA-IBP was that an exciting degree of international cooperation had been realized. Friendly contacts were established between many widely-separated laboratories, with a very free exchange of both information and scientific staff. In the specific context of the measurement of human physical performance, agreement on a standard protocol was reached quite rapidly through a bench-level working party convened in Toronto. Unfortunately, no rapid and easy mechanism was discovered to make this working party completely representative of the scientific community, and perhaps for this reason other groups (usually with slightly different objectives) put forward their own plans for a national or international methodology at much the same time. There was inevitably some confusion, and a number of HA-IBP investigators chose to ignore all of the various recommendations for a common protocol. This was plainly less than an ideal situation, although one comforting feature of the several methodological studies was their demonstration that most techniques for the measurement of working capacity yielded relatively similar results. Possibly, the HA-IBP should have pushed more vigorously for adoption of its standardized procedures; in fact, the HA-IBP handbook merely described its protocol as 'recommendations'. However, many scientists remain rugged individualists, and it is uncertain that a longer period of discussion with trial by an obviously representative group of scientists would have yielded a more widely accepted protocol. The HA-IBP recommendations may themselves yet earn general acceptance as they are proven by further years of use. Certainly, existing experience has demonstrated their effectiveness in both the city laboratory and the remote field station.

At the inter-disciplinary level, HA-IBP project directors faced inevitable problems in regulating budgets and ensuring the collection of vital pieces of information. On occasion, one discipline would press for a vast allocation of funds, and it would later transpire that this was destined for the pursuit of some personal interest unrelated to the IBP investigation. On other occasions, a group would undertake to collect some piece of information vital to the study of a community and would subsequently neglect its responsibility, leaving a glaring gap in the array of data. The answers to such personal problems still elude both scientists and their financial supporters. The skills of the team leader are certainly crucial to success – the individual selected for this role must have a broad understanding of the various disciplines represented, and substantial talents for the guiding of human relationships. There may also be a scope for a more direct administrative stimulation of laggards who fail to deliver promised results, through such mechanisms as withholding a part of any financial support until a final report has been filed.

Sampling techniques

The HA-IBP handbook strongly recommended that exercise physiologists review their rather careless approach to population sampling. There is little evidence that the suggestion was heeded, and in many studies the relationship between the data collected and the characteristics of the population under study was somewhat tenuous. Ever-increasing control of human experimentation, coupled with a growing militancy of indigenous populations makes it unlikely that observations can be repeated at some later date, using a closer approach to random sampling of the communities concerned. If any further work is possible, it will be necessary to accept a substantial volunteer bias within the sample tested, and to develop methods of allowing for this distortion of the data. Nevertheless, HA-IBP has performed a useful service in alerting those concerned with measurements of work capacity to their neglect of an important area of experimental design.

Description of 'threatened' communities

An urgent justification for the HA-IBP project was the need to document the status of primitive communities before they became acculturated to a western life-style. Accomplishments of the IBP investigators were particularly impressive in this regard. Despite the vast distances separating many of the primitive populations from base laboratories, formidable barriers of language and culture, and physical handicaps such as the absence of a mains electrical supply, a substantial volume of results was accumulated safely and effectively, using the full range of modern technology.

Unfortunately, government financing of the research did not always move as fast as the directors of the HA-IBP project would have wished, and in some instances western society had already made deep in-roads into traditional life-patterns by the time that data could be collected. Nevertheless, a bench-mark was established for the working capacity of many indigenous populations. Future generations of scientists will thus have opportunity to examine the long-term impact of acculturation on such peoples. Some assessment of the more immediate effects of acculturation is already possible, since villagers within most communities show substantial inter-individual differences in their degree of acculturation. Comparisons can also be made with other groups of similar ethnic background, having greater or lesser contact with western society.

There is good evidence that acculturation leads to a loss of maximum oxygen intake, and a replacement of lean tissue by fat, with a corresponding reduction in the tolerance of physical exertion. However, in certain habitats such as hill-farming areas and the circumpolar regions, subjects who have preserved a traditional life-style still have above average values for working capacity. This finding was not always apparent in previous examinations of the communities concerned, and there are reasons to believe that earlier data may have been biassed downwards by problems of methodology and sampling, poor nutrition and endemic disease. Furthermore, it should be stressed that differences in average scores for the hunter–gatherer and the sedentary city-dweller are quite small relative to inter-individual variation. Thus most large samples of the urban populations contain occasional subjects who have a working capacity in excess of the average for the primitive group.

Work capacity of other populations

The HA-IBP project proposed the comparison of physiological work capacity between urban and rural populations. Relatively little data of this type was actually collected, the notable exception being the exhaustive study of Seliger and his colleagues in Czechoslovakia. Some investigators may have been deterred by practical problems such as current difficulties in obtaining informed consent. It has become at least as difficult to obtain a valid sample from a vast urban area as to examine all (or almost all) of a small, discreet, and cooperative primitive tribe. There was also a belief that adequate data for the urban population was already available. Certainly, some of the gaps in the HA-IBP project can be completed by reference to the world literature. However, we still cannot make valid comparisons of working capacity among the more-developed nations, since with one or two exceptions there have been no measurements on random samples of national populations. Within a given nation, fitness seems at least as great in urban as in rural areas. Many rural children get less exercise than their city counterparts because school buses collect them from the doors of their homes each day. Rural residents are often indigent poor or wealthy commuters rather than farmers, and in any event a large part of traditional heavy farm work is now mechanized, at least in the more-developed countries.

HA-IBP investigators did little to exploit the stated interest of the project in top athletes as indicators of human potential for variation. Again, there may have been practical difficulties in recruiting the necessary subjects. The present author contacted some 200 Canadian athletes prior to the Pan American games in Winnipeg, but was able to persuade only four of the group to attend for exercise testing. Subsequent invitations to our laboratory have been extended by colleagues who were themselves top national sportsmen, with a much greater response rate. Many HA-IBP investigators may have shared the handicap of not themselves being outstanding athletes. Certainly, a number had a medical orientation, and studies of the sportsmen have become increasingly the prerogative of Departments of Physical Education. Despite the lack of specific HA-IBP research, the world literature contains a modest amount of data concerning almost all classes of sportsman. By tabulating this information, I have been able to realize in large measure the objective of defining potential limits for the various determinants of physical performance. However, the figures must be interpreted with care, since many reports do not describe methodology or sampling techniques with the precision that one would wish. There plainly remains scope for the establishment of a world-wide data bank that would accumulate information based on an agreed methodology and restricted to athletes of national and international calibre.

Genetic factors

A number of small-scale twin studies were carried out under HA-IBP auspices. The findings were not particularly helpful, since some experiments showed that almost all of the variation in physiological work capacity was inherited, while other experiments indicated almost zero inheritance of such variation. The difficulty can be traced to the type of analysis adopted, with its inherent assumptions that error and environmental variance can be equated for monozygotic twins, dizygotic twins and the general population. Little further progress can be anticipated with this question until methods have been developed to estimate differences in the methodological error and the environmental variance for the three classes of subject.

A few comparisons were made between different ethnic groups living in the same environment, and between similar ethnic groups living in different environments. Again, effects on physiological work capacity were relatively small; indeed, in some instances, the statistical significance of observed differences was disputed between the various observers involved. However, even if larger differences of working capacity had been apparent, interpretation of the findings would still have been difficult. There seems no convenient method to isolate the effect of such extraneous variables as (i) differences of personality and constitution between migrants and those remaining in their place of birth, and (ii) persistent cultural differences between immigrants and long-term residents in a given habitat.

Growth and ageing

Because of the short term nature of the HA-IBP project, most of the information on the growth and ageing of physiological work capacity was collected on a cross-sectional rather than a longitudinal basis. As anticipated, maturation was somewhat delayed in the primitive populations, and adolescent females showed a lesser tendency to deterioration of working capacity than would have been anticipated in an urban environment. The high mortality rates among earlier decades of children left relatively few old people to examine in some primitive tribes; however, ageing of the residual population apparently proceeded at much the same rate as in white communities. It is doubtful whether more reliable information could be obtained by repeating surveys on a longitudinal basis. Any curves thus established would reflect the interaction of true growth and ageing with the effects of rapid cultural change.

Disease

The HA-IBP project expressed an interest in the assessment of disease as a selective agent. We have noted occasional references to the interaction of disease and physiological work capacity. Most HA-IBP investigators have been concerned with the acute effects of disease – for example, the reductions of working capacity associated with parasitic anaemia or chronic respiratory tuberculosis. Neglect of such variables may account for previous reports of low working capacity in primitive groups. In the modern setting, the sick individual receives both medical and social care, so that the disease process exerts relatively little evolutionary pressure. However, this was not the case in the recent past, when large fractions of many indigenous populations were destroyed by epidemics of imported disease. The selective action of such epidemics may have contributed to the high current fitness levels of some primitive peoples. Unfortunately, there seems no ready method of assessing the importance of disease relative to other selective pressures such as skill in hunting and an ability to adapt to extreme climatic conditions.

Activity patterns and ecological balance

Perhaps the most fascinating of the HA-IBP objectives was to measure activity patterns within a closed community, and where possible to chart energy flows with a view to predicting the potential of a population for survival as an independent unit. In practice, relatively few activity studies were undertaken. Some of the tools proposed for this phase of the HA-IBP project apparently did not function too well in remote areas. With primitive groups, there was the further difficulty that occupation did not follow a simple, set pattern, but varied widely with season and weather conditions. An observer had to accompany a single hunter or gatherer for many weeks in order to obtain reliable and representative data; thus, the costs of studying an entire community were often prohibitively large. The energy flow concept required even larger amounts of data; furthermore, the idea was somewhat out-dated by the time the HA-IBP project was finally initiated, since the majority of communities destined for study had already become heavily dependent upon the outside world. One fairly complete analysis was made for an Eskimo population; this was chosen for its primitive characteristics, yet at the time of data collection only about a third of its food requirements was met from local natural resources.

Implications for society

It would be gratifying if one could offer the populations studied some tangible rewards for their stoic patience in the face of a multiplicity of measurements. In some cases, investigators were able to offer immediate medical and dental care. However, the more general analysis of the HA-IBP data points to the dangers facing the communities rather than to easy practical remedies.

Civilization has brought a measure of environmental control to many primitive societies. There is now protection from the worst extremes of weather and provision of a relatively stable supply of basic foods, often coupled with the advantages of modern health care and education. On the other hand, administrative convenience has dictated that villagers be collected into relatively large settlements. Such population pressures have been enhanced by abandonment of the natural contraceptive effect of breast feeding and by greatly improved standards of neonatal care. In consequence, the previous precarious balance between community size and natural resources has been irreversibly destroyed. Perpetuation of the traditional life-style is no longer feasible. There are few opportunities for formal employment. Most communities are thus faced with an agonizing choice. They must either emigrate and become totally assimilated or remain in their traditional habitat and accept the frustrations of dependence on western society.

It can be argued that in many respects the city-dwelling population faces a problem of equal incompatibility with its environment. The pressures of air and water contamination are well recognized. Furthermore, air-conditioning and power appliances are continually narrowing the necessary range of human adaptability. The immediate decline of working capacity of the inactive man is well documented, although less is known about the impact of this change on subsequent health. Perhaps of greater long-term significance, prospects for adaptation could be further reduced through the emergence of individuals who are fertile, attractive to the opposite sex, and yet have a very limited capacity to respond to environmental challenge. This variety of *Homo sapiens*, the ultimate form of *Homo sedentarius*, would survive only while modern technology could maintain the constancy of his external environment. Such a luxury may no longer be practicable when non-renewable energy reserves have been exhausted. If the HA-IBP project has done no more than warn mankind of the ultimate dangers associated with life in an over-sheltered habitat, it may well have merited the substantial sums expended in completing its investigations.

IBP Human Adaptability section publications

J. S. Weiner (1969).
A guide to human adaptability proposals
Blackwell Scientific Publications, Oxford. 88 pp.
J. S. Weiner & J. A. Lourie (1969).
A guide to field methods
Blackwell Scientific Publications, Oxford. 652 pp.
S. Biesheuvel (ed.) (1969).
Methods for the measurements of psychological performance
Blackwell Scientific Publications, Oxford. 110 pp.
P. B. Eveleth & J. M. Tanner (1976).
Worldwide variation in human growth
Cambridge University Press, London. 498 pp.
G. A. Harrison (ed.) (1977).
Population structure and human variation
Cambridge University Press, London. 342 pp.
P. Baker (ed.) (1978).
The biology of high-altitude peoples
Cambridge University Press, London. 357 pp.
R. J. Shephard (1978).
Human physiological work capacity
Cambridge University Press, London. 303 pp.
J. S. Weiner
Components of human physiological function
Cambridge University Press, London. (In preparation.)
F. Milan (ed.)
The human biology of circumpolar populations
Cambridge University Press, London. (In press.)

A detailed guide to all the IBP projects contributing to the theme of this volume is given in *Human adaptability: a history and compendium of research in the International Biological Programme* (1977) by K. J. Collins & J. S. Weiner, published by Taylor & Francis Ltd, 10–14 Macklin Street, London WC2B 5NF. A collection of HA reports, reprints and archival material is held in the Library of the British Museum (Natural History), Cromwell Road, London SW7 5BD.

References

Abbie, A. A. (1967). Skinfold thickness in Australian Aborigines. *Archaeol. Physical Anthrop. Oceania* **2**, 207–19.

Abramowitz, S., Leiner, G. C., Lewis, W. A. & Small, M. J. (1965). Vital capacity in the Negro. *Amer. Rev. Resp. Dis.* **92**, 287–92.

Adams, F. H., Bengtsson, E., Berven, H. & Wegelius, C. (1961*a*). The physical working capacity of normal school children. II. Swedish city and country. *Pediatrics* **28**, 243–59.

Adams, F. H., Linde, L. M. & Miyake, H. (1961*b*). The physical working capacity of normal school children. I. California. *Pediatrics* **28**, 55–64.

Aghemo, P., Limas, F. P. & Sassi, G. (1971). Maximal aerobic power in primitive Indians. *Int. Z. angew. Physiol.* **29**, 337–42.

Akgün, N. & Özgönül, H. (1969). Spirometric studies in normal Turkish subjects aged 8 to 20 years. *Thorax* **24**, 714–21.

Allen, C., O'Hara, W. & Shephard, R. J. (1976). Changes in body composition during an Arctic winter exercise. In: *Circumpolar Health.* Ed: Shephard, R. J. & Itoh, S. Toronto, Ontario: University of Toronto Press.

Allen, J. A. (1877). The influence of physical conditions in the genesis of species. *Radical. Rev.* **1**, 108–40.

Allen, J. G. (1966). Aerobic capacity and physiological fitness of Australian men. *Ergonomics* **9**, 485–94.

American Association for Health, Physical Education and Recreation (1965). *Youth Fitness Test Manual.* (Revised edition.) Washington, DC: NEA Publications.

Andersen, K. L. (1964). Physical fitness – studies of healthy men and women in Norway. In: *International Research in Sport and Physical Education.* Ed: Jokl, E. & Simon, E. Springfield, Illinois: C. C. Thomas.

Andersen, K. L. (1966). Work capacity of selected populations, p. 15. In: *The Biology of Human Adaptability.* Ed: Baker, P. T. & Weiner, J. S. Oxford, UK: Clarendon Press.

Andersen, K. L. (1967*a*). The effect of physical training with and without cold exposure upon physiological indices of fitness for work. *Canad. Med. Ass. J.* **96**, 801–3.

Andersen, K. L. (1967*b*). The capacity of aerobic muscle metabolism as affected by habitual physical activity. In: *Physical Activity and the Heart.* Ed: Karvonen, M. J. & Barry, A. J. Springfield, Illinois: C. C. Thomas.

Andersen, K. L. (1967*c*). Ethnic group differences in fitness for sustained and strenuous muscular exercise. *Canad. Med. Ass. J.* **96**, 832–5.

Andersen, K. L. (1969). Racial and inter-racial differences in work capacity. *J. Biosoc. Sci. Suppl.* **1**, 69–80.

Andersen, K. L. (1971). The effect of altitude variation on the physical performance capacity of Ethiopian men. In: *Human Biology and Environmental Change.* Ed: Vorster, R. Malawi: IBP.

Andersen, K. L. & Ghesquière, J. (1972). Sex differences in maximal oxygen uptake, heart rate and oxygen pulse at 10 and 14 years in Norwegian children. *Human Biol.* **44**, 413–32.

Andersen, K. L. & Hart, J. S. (1963). Aerobic working capacity of Eskimos. *J. Appl. Physiol.* **18**, 764–8.

Andersen, K. L., Bolstad, A., Loyning, Y. & Irving, L. (1960). Physical fitness of Arctic Indians. *J. Appl. Physiol.* **15**, 645–8.

References

Andersen, K. L., Elsner, R. E., Saltin, B. & Hermansen, L. (1962). Physical fitness in terms of maximal oxygen intake of nomadic Lapps, Fort Wainwright, Alaska. *USAF Arctic Medical Laboratory, Tech. Rept.* AAL TDR-61-53.

Andersen, K. L., Shephard, R. J., Denolin, H., Varnauskas, E. & Masironi, R. (1971). *Fundamentals of Exercise Testing.* Geneva, Switzerland: WHO.

Anderson, T. W., Brown, J. R., Hall, J. W. & Shephard, R. J. (1968). The limitations of linear regressions for the prediction of vital capacity and forced expiratory volume. *Respiration* **25**, 140–58.

Andrew, G. M., Becklake, M. R. & Guleria, J. S. (1972). Heart and lung function in swimmers and non-athletes during growth. *J. Appl. Physiol.* **32**, 245–51.

Antel, J. & Cumming, G. R. (1969). Effect of emotional stimulation on exercise heart rate. *Research Quart.* **40** (2), 5–10.

Apfelbaum, M., Bostsarron, J. & Lacatis, D. (1971). Effect of caloric restriction and excessive caloric intake on energy expenditure. *Amer. J. Clin. Nutr.* **24**, 1405–9.

Archer, J., Hamley, E. J. & Robson, H. E. (1965). The fitness assessment of competitive cyclists. *Physical Education*, July 1965.

Areskog, N. H. (1971). Short term exercise and nutritional status in Ethiopian boys and young males. *Acta Paed. Scand. Suppl.* **217**, 138–41.

Areskog, N. H., Selinus, R. & Vahlquist, B. (1969). Physical work capacity and nutritional status in Ethiopian male children and young adults. *Amer. J. Clin. Nutr.* **22**, 471–9.

Ashcroft, M. T. & Lovell, H. G. (1964). Heights and weights of Jamaican children of various racial origins. *Trop. Geograph. Med.* **4**, 346–53.

Ashcroft, M. T. & Lovell, H. G. (1966). The validity of surveys of heights and weights of Jamaican school-children. *West Indian Med. J.* **15**, 27–33.

Ashcroft, M. T., Lovell, H. G., George, M. & Williams, M. (1965). Heights and weights of infants and children in a rural community of Jamaica. *J. Trop. Pediatr.* **11**, 56–68.

Ashcroft, M. T., Buchanan, I. C. & Lovell, H. G. (1965*b*). Heights and weights of primary school children in St. Christopher – Nevis – Anguilla, West Indies. *Trop. Med. Hygiene* **68**, 277–83.

Ashcroft, M. T., Ling, J., Lovell, H. G. & Miall, W. E. (1966*a*). Heights and weights of adults in rural and urban areas of Jamaica. *Brit. J. Prev. Soc. Med.* **20**, 22–6.

Ashcroft, M. T., Heneage, P. & Lovell, H. G. (1966*b*). Heights and weights of Jamaican school-children of various ethnic groups. *Am. J. Phys. Anthrop.* **24**, 37–44.

Ashcroft, M. T., Lovell, H. G., Miall, W. E. & Moore, F. (1967). Heights and weights of adults in rural and industrial areas of South Wales. *Brit. J. Prev. Soc. Med.* **21**, 159–62.

Asmussen, E. & Christensen, E. H. (1967). *Kompendium: Legemsövelsernes Specielle Teori.* Kobenhavns Universitets Fond til Tilvejebringelse af Läremidler, Copenhagen.

Asmussen, E. & Mathiasen, P. (1962). Some physiologic functions in physical education students reinvestigated after twenty-five years. *J. Amer. Geriatr. Soc.* **10**, 379–87.

Asmussen, E. & Molbech, S. V. (1959). Methods and standards for evaluation of the physiological working capacity of patients. *Comm. Test. Obs. Inst.* **4**, Hellerup, Denmark.

Asmussen, E., Heebøll-Nielsen, K. & Molbech, S. V. (1959). Description of muscle tests and standard values of muscle strength in children. *Comm. Test. Obs. Inst.* **5** (suppl.), Hellerup, Denmark.

References

Asmussen, E. & Heebøll-Nielsen, K. (1962). Isometric muscle strength in relation to age in men and women. *Ergonomics* **5**, 167–9.

Åstrand, I. (1960). Aerobic work capacity in men and women with special reference to age. *Acta Physiol. Scand.* **49**, Suppl. **169**, 1–92.

Åstrand, I. (1967). Degree of strain during building work as related to individual aerobic work capacity. *Ergonomics* **10**, 293–303.

Åstrand, I., Åstrand, P. O., Hallbäck, I. & Kilbom, A. (1973). Reduction in maximal oxygen intake with age. *J. Appl. Physiol.* **35**, 64.

Åstrand, P. O. (1952). *Experimental studies of physical working capacity in relation to age and sex.* Copenhagen, Denmark: Munksgaard.

Åstrand, P. O. (1961). *Fysiologiska synpunkter på skolungdomens fysika fostran; preliminär rapport till folksain.* Stockholm, Sweden: Central Gymnastic Institute.

Åstrand, P. O. (1964). Human physical fitness with special reference to sex and age. In: *International Research in Sport and Physical Education.* Ed: Jokl, E. & Simon, E. Springfield, Illinois: C. C. Thomas.

Åstrand, P. O. (1967). Commentary. *Canad. Med. Ass. J.* **96**, 730.

Åstrand, P. O. (1973). Physiological bases for sport at different ages. In: *Sport in the Modern World – Chances and Problems.* Ed: Grupe, O., Kurz, D. & Teipel, J. M. Berlin, Germany: Springer Verlag.

Åstrand, P. O. & Rodahl, K. (1970). *A Textbook of Work Physiology.* New York: McGraw-Hill.

Åstrand, P. O. & Ryhming, I. (1954). A nomogram for calculation of aerobic capacity (physical fitness) from pulse rate during submaximal work. *J. Appl. Physiol.* **7**, 218–21.

Åstrand, P. O. & Saltin, B. (1961*a*). Oxygen uptake during the first minutes of heavy muscular exercise. *J. Appl. Physiol.* **16**, 971–6.

Åstrand, P. O. & Saltin, B. (1961*b*). Maximal oxygen uptake and heart rate in various types of muscular activity. *J. Appl. Physiol.* **16**, 977–81.

Åstrand, P. O., Engström, L., Eriksson, B., Karlberg, P., Nylander, I., Saltin, B. & Thoren, C. (1963). Girl swimmers. *Acta Paed. Scand. Suppl.* **147**, 1–75.

Åstrand, P. O., Ekblöm, B. & Goldberg, A. N. (1971). Effects of blocking the autonomic nervous system during exercise. *Acta Physiol. Scand.* **82**, 18A.

Auchincloss, J. H. & Gilbert, R. (1973). Estimation of maximum oxygen uptake with a brief progressive stress test. *J. Appl. Physiol.* **34**, 525–6.

Auger, F. (1974). Poids et plis cutanés chez les esquimaux de Fort Chimo (Nouveau Québec). *Anthropologica* **16**, 137–58.

Bailey, D. A., Shephard, R. J., Mirwald, R. L. & McBride, B. A. (1974*a*). Current levels of Canadian cardio-respiratory fitness. *Canad. Med. Ass. J.* **111**, 25–30.

Bailey, D. A., Ross, W. D., Weese, C. & Mirwald, R. L. (1974*b*). A dimensional analysis of aerobic power in boys. *Proceedings of the Sixth International Symposium on Pediatric Work Physiology, Seč, Czechoslovakia.*

Bakács, T. (1972). *Urbanization and human health.* Budapest, Hungary: Akadémia Kiadó.

Baker, J. A., Humphrey, S. J. E. & Wolff, H. S. (1967). Socially acceptable monitoring instruments (SAMI). *J. Physiol.* **88**, 4P.

Baker, P. (1958*a*). A theoretical model for desert heat tolerance. *US Army Quartermaster Res. Eng. Command Tech. Rept.* E.P. **96**, 1–28.

Baker, P. (1958*b*). American Negro–white differences in heat tolerance. *US Army Quartermaster Res. Eng. Command Tech. Rept.* E.P. **75**, 1–19.

Baker, P. T. (1966). Ecological and physiological adaptation in indigenous South Americans. In: *The Biology of Human Adaptability*, pp. 275–303. Ed: Baker, P. T. & Weiner, J. S. Oxford, UK: Clarendon Press.

References

Baker, P. T. (1974). The work capacity of highland natives: some genetic and experimental considerations. In: *Man in the Andes: A Multidisciplinary Study of High Altitude Quechua Indians*. Ed: Baker, P. T. & Little, M. A. In preparation.

Baker, P. T. & Weiner, J. S. (1966). *The Biology of Human Adaptability*. Oxford, UK: Clarendon Press.

Balke, B. & Snow, C. (1965). Anthropological and physiological observations on Tarahumara endurance runners. *Amer. J. Phys. Anthropol.* **23**, 293–302.

Bang, H. O. & Dyerberg, J. (1972). Plasma lipids and lipoproteins in Greenlandic West Coast Eskimos. *Acta Med. Scand.* **192**, 85–94.

Banister, E. W. & Jackson, R. C. (1967). The effect of speed and load changes on oxygen intake for equivalent power outputs during bicycle ergometry. *Int. Z. angew. Physiol.* **24**, 284–90.

Barcroft, J. (1914). *The Respiratory Function of the Blood*. London, UK: Cambridge University Press.

Bar-Or, O. (1973). A comparison of responses to exercise and lung functions of Israeli, Arabic and Jewish 12 to 14 year-old boys. In: *Pediatric Work Physiology. Proceedings of the Fourth International Symposium*. Ed: Bar-Or, O. Natanya, Israel: Wingate Institute.

Bar-Or, O. & Zwiren, L. D. (1973). Physiological effects of increased frequency of physical education classes and of endurance conditioning on 9 to 10 year old girls and boys. In: *Pediatric Work Physiology. Proceedings of the Fourth International Symposium*. Ed: Bar-Or, O. Natanya, Israel: Wingate Institute.

Bar-Or, O., Skinner, J. S., Buskirk, E. R. & Borg, G. (1972). Physiological and perceptual indicators of physical stress in 41 to 60 year old men who vary in conditioning level and in body fatness. *Med. Sci. Sports* **4**, 96–100.

Bates, H. W. (1884). *The Naturalist on the River Amazon, a Record of Adventures, Habits of Animals, Sketches of Brazilian and Indian Life, and Aspects of Nature under the Equator, During Eleven Years of Travel.* (5th edition.) London, UK: John Murray.

Baugh, C. W., Bird, G. S., Brown, G. M., Lennox, C. S. & Semple, R. E. (1958). Blood volumes of Eskimos and white men before and during acute cold stress. *J. Physiol.* **140**, 347–58.

Beaudry, P. H. (1968). Pulmonary function of the Canadian Eastern Arctic Eskimo. *Arch. Env. Health* **17**, 524–8.

Behnke, A. R., Feen, B. G. & Weltham, W. C. (1942). The specific gravity of healthy men. *J. Amer. Med. Ass.* **118**, 495–501.

Bell, W. (1973). Distribution of skinfolds and differences in body proportions in young adult rugby players. *J. Sports Med. Fitness* **13**, 65–73.

Bengtsson, E. (1956). The working capacity in normal children evaluated by sub-maximal exercise on the bicycle ergometer and compared with adults. *Acta Med. Scand.* **154**, 91–109.

Beran, J. A. & Flores, V. A. (1972). Relationship between physical fitness, nutrition and size of selected Philippine children. In: *Proceedings of the ACSPFT and the ICSPFT – 1972*. Ed: Simri, U. Natanya, Israel: Wingate Institute.

Bereni, L., Heches, P., Blanchot, M. & Duvinage, J. F. (1971). Enquête preliminaire sur la croissance osseuse en milieu scolaire à Saint Louis. *Bull. Soc. Med. Afr. Noire lgue. Frse.* **16**, 559–63.

Berg, K. (1972). Body composition and nutrition of adolescent boys training for bicycle racing. *Nutr. Metabol.* **14**, 172–80.

Bergmann, C. (1847). Über die Verhältnisse der Wärmeökonomie des Thiere zu ihrer Grösse. *Gottinger Studies* **3**, 595–708.

Bharadway, H., Singh, A. P. & Malhotra, M. S. (1973). Body composition of the high altitude natives of Ladakh – a comparison with sea-level residents. *Human Biol.* **45**, 423–34.

Bink, B. (1962). The physical working capacity in relation to working time and age. *Ergonomics* **5**, 25–8.

Bink, B. & Wafelbakker, F. (1969). Physical working capacity of boys 12–18 years of age. *Acta Physiol. Pharm. Neerl.* **15**, 387–8.

Bjelke, E. (1971). Variation in height and weight in the Norwegian population. *Brit. J. Prev. Soc. Med.* **25**, 192–202.

Blanco, R. A., Acheson, R. M., Canosa, C. & Salomon, J. B. (1972). Sex differences in retardation of skeletal development in rural Guatemala. *Pediatrics* **50**, 912–15.

Bloomfield, J. & Sigerseth, P. (1965). Anatomical and physiological differences between sprint and middle-distance competitors at the university level. *J. Sports Med. Phys. Fitness* **5**, 76–81.

Bloomfield, J., Blanksby, B. A. & Elliott, B. C. (1973). Profiles of national level oarsmen. *Brit. J. Sports Med.* **7**, 353–9.

Blumchen, G., Roskamm, H. & Reindell, H. (1966). Herzvolumen und Körperliche Leistungsfähigkeit. *Kreislaufforschung* **55**, 1012–16.

Bobbert, A. C. (1960). Physiological comparison of three types of ergometer. *J. Appl. Physiol.* **15**, 1007–14.

Bonjer, F. H. (1966). Measurement of working capacity by assessment of the aerobic capacity in a single session. *Fed. Proc.* **25**, 1363–5.

Bonjer, F. H. (1968). Relationship between physical working capacity and allowable caloric expenditure. In: *International Colloquium on Muscular Exercise and Training.* Ed: Rohmert, H. Darmstadt, West Germany: Gertner Verlag.

Borg, G. (1971). The perception of physical performance. In: *Frontiers of Fitness.* Ed: Shephard, R. J. Springfield, Illinois: C. C. Thomas.

Borg, G. & Linderholm, H. (1967). Perceived exertion and pulse rate during graded exercise in various age groups. *Acta Med. Scand. Suppl.* **472**, 194–206.

Bottin, R., Deroanne, R., Petit, J. M., Pirnay, F. & Juchmes, J. (1966). Comparaison de la consommation maximum d'O_2 mesurée à celle predite en fonction de la fréquence cardiaque an moyen du nomogramme d'Åstrand. *Rev. Educ. Phys.* **6**, 1–6.

Bottin, R., Petit, J. M., Deroanne, R., Juchmes, J. & Pirnay, F. (1968). Mesures comparées de la consommation maximum d'O_2 par paliers de 2 ou de 3 minutes. *Int. Z. angew. Physiol.* **26**, 355–62.

Bouchard, C., Godbout, F., Landry, F., Mondor, J. C., Houde, P. & Levesque, C. (1973). Une etude sur le taux de generalité et de specificité de la puissance aerobique maximale lors d'efforts en Vita maxima. *First Canadian Congress for the Multidisciplinary Study of Sport and Physical Activity, Montreal.*

Brandi, G., Brambilla, I. & Cerretelli, P. (1960). Modificazioni respiratorie nel lavoro in ragazzi di 12–14 anni. *Medicina Sportiva* **14**, 30–7.

Brooke, J. D. & Davies, G. J. (1973). Comment on 'Estimation of the energy expenditure of sporting cyclists'. *Ergonomics* **16**, 237–8.

Brouha, L. (1943). The step test: a simple method of measuring physical fitness for muscular work in young men. *Res. Quart.* **14**, 31–6.

Brown, G. M. (1954). Metabolic studies of the Eskimo. In: *Cold Injury. Transactions of the Third Conference, Feb. 22–25, 1954, Fort Churchill, Manitoba.* Ed: Ferrer, M. I. New York: Josiah Macey Fdn.

Brown, G. M., Bird, G. S., Boag, L. M., Delahaye, D. J., Green, J. E., Hatcher, J. D. & Page, J. (1954). Blood volume and basal metabolic rate of Eskimos. *Metabolism* **3**, 247–54.

Brown, S. R. (1960). Factors influencing improvement in the oxygen intake of young boys. Ph.D. Dissertation, Urbana, Illinois: University of Illinois.

Brown, S. R. (1966). Specific fitness for middle distance running. *Track Technique*, 25 September 1966.

Brown, S. R., Pomfret, J. B. & Parsons, D. (1966). Physical fitness characteristics of Canadian champion swimmers. *Swimming Technique* **3**, 84–8.

Brožek, J., Grande, F., Anderson, J. T. & Keys, A. (1963*a*). Densitometric analysis of body composition: revision of some quantitative assumptions. *Ann. N.Y. Acad. Sci.* **110**, 113–40.

Brožek, J., Kihlberg, J. K., Taylor, H. L. & Keys, A. (1963*b*). Skinfold distribution in middle-aged American men: a contribution to norms of leanness–fatness. *Ann. N.Y. Acad. Sci.* **110**, 492–502.

Bryn, H. (1932). Norwegische Samen. *Mitt. Anthrop. Ges.* **62** (Wien). Cited by Skrobak-Kaczynski & Lewin (1974).

Bugyi, B. (1972). Zur Beurteilung des Korperbaus der industrïeleholinge und Jungarbeiter in der schweren Industrie (in Ungarn). *Glasnik Antropoloskog Drustva Jugoslavve* **8–9**, 61–3.

Bullen, B. A., Reed, R. B. & Mayer, J. (1964). Physical activity of obese and non-obese adolescent girls. Appraised by motion picture samples. *Amer. J. Clin. Nutr.* **14**, 211–23.

Burmeister, W., Rutenfranz, J., Sbresny, W. & Radny, H. G. (1972). Body cell mass and physical performance capacity (W_{170}) of school children. *Int. Z. angew. Physiol.* **31**, 61–70.

Buskirk, E. R. (1974). Work performance of newcomers to the Peruvian Highlands: a comparison with high altitude Quechna. In: *Man in the Andes: A Multidisciplinary Study of High Altitude Quechua Indians*. Ed: Baker, P. T. & Little, M. A. In preparation.

Buskirk, E. R., Harris, D., Mendez, J. & Skinner, J. (1971). Comparison of two assessments of physical activity and a survey method for calorie intake. *Amer. J. Clin. Nutr.* **24**, 1119–25.

Buzina, R. (1972). Nutrition, status, working capacity and absenteeism in industrial workers. *First International Symposium, Alimentation et Travail*. Paris, France: Masson et Cie.

Caplan, A. (1944). Critical analysis of collapse in underground workers on the Kolar Gold Field. *Trans. Inst. Mining Met.* **53**, 95–179. Cited by Ladell (1964).

Carey, P., Stensland, M. & Hartley, L. H. (1974). Comparison of oxygen uptake during maximal work on the treadmill and the rowing ergometer. *Med. Sci. Sports* **6**, 101–3.

Carrier, R., Landry, F., Potvin, R. & Bouchard, C. (1972). Comparison between athletes, normal and Eskimo subjects from the point of view of selected biochemical parameters. In: *Training – Scientific Basis and Application*. Ed: Taylor, A. W. Springfield, Illinois: C. C. Thomas.

Carter, J. E. L., Kasch, F. W., Boyer, J. L., Phillips, W. H., Ross, W. D. & Sucec, A. (1967). Structural and functional assessments on a champion runner – Peter Snell. *Res. Quart.* **38**, 355–65.

Caru, B., LeCoultre, L., Aghemo, P. & Limas, F. P. (1970). Maximal aerobic and anaerobic muscular power in football players. *J. Sports Med. Phys. Fitness* **10**, 100–3.

Cavalli-Sforza, L. L. & Bodmer, W. C. (1971). *The Genetics of Human Populations*. San Francisco: Freeman.

Cerretelli, P. & Radovani, P. (1960). Il massimo consumo di O_2 in atlete olimpionici di varie specialtà. *Boll. Soc. Ital. Sper.* **36**, 1871–2.

Cerretelli, P., Aghemo, P. & Rovelli, E. (1963). Morphological and physiological observations on school children in Milan. *Med. Dell Sport* **3**, 731–43.

Chan, O. L., Thinkaran, T., Nor, M., Sundsten, J. W., Duncan, M. T. & Klissouras, V. (1974). Work capacity of the Temiars – a primitive jungle tribe in Malaysia. *Twentieth World Congress of Sports Medicine, Melbourne.*

Christensen, E. H. (1932). Beitrage zur Physiologie Schwerer Körperlicher Arbeit. VI. Mittelung: Der Stoffwechsel und die respiratorischen Funktionen bei schwerer Körperlicher Arbeit. *Arbeitsphysiologie* **5**, 463–78.

Christensen, E. H. (1953). Physiological valuation of work in Nykroppa iron works. In: *Symposium on Fatigue.* London, UK: H. K. Lewis.

Clark, C. & Haswell, M. R. (1964). *The Economics of Subsistence Agriculture.* London: MacMillan.

Clarke, H. H. (1966). *Muscular Strength and Endurance in Man.* Englewood Cliffs, NJ: Prentice Hall.

Clarke, H. H. (1973). Characteristics of athletes. *Physical Fitness Research Digest, Washington* **3** (2), 1–20.

Clarke, H. H. & Borms, J. (1968). Differences in maturity, physical and motor traits for boys of high, average, and low gross and relative strength. *J. Sports Med. Fitness* **8**, 143–52.

Clarke, R. S. J., Hellon, R. F. & Lind, A. R. (1958). The duration of sustained contractions of the human forearm at different temperatures. *J. Physiol.* **143**, 454–73.

Clausen, J. P. (1973). Muscle blood flow during exercise and its significance for maximal performance. In: *Limiting Factors of Physical Performance.* Ed: Keul, J. Stuttgart, West Germany: Thieme.

Clausen, J. P., Trap-Jensen, J. & Lassen, N. A. (1970). The effects of training on the heart rate during arm and leg exercise. *Scand. J. Clin. Lab. Invest.* **26**, 295–301.

Clausen, J. P., Klausen, K., Rasmussen, B. & Trap-Jensen, J. (1971). Effects of selective arm and leg training on cardiac output and regional blood flow. *Acta Physiol. Scand.* **82**, 35–6A.

Clegg, E. J., Pawson, I. G., Ashton, E. H. & Flinn, R. M. (1972). The growth of children at different altitudes in Ethiopia. *Roy. Soc. (Lond.) Biol. Sci. Philosoph. Trans.* **264**, 403–37.

Cluver, E. H., de Jongh, T. W. & Jokl, E. (1942). *Manpower (South Africa)* **1**, 39. Cited by Sloan (1966*a*).

Cole, T. J. (1974). The influence of height on the decline in ventilatory function. *Int. J. Epidemiol.* **3**, 145–52.

Comstock, G. W. & Livesay, V. T. (1963). Subcutaneous fat determinations from a community-wide chest X-ray survey in Muscogee County, Georgia. *Ann. N.Y. Acad. Sci.* **110**, 475–91.

Consolazio, C. F., Matoush, L. O., Nelson, R. A., Torres, J. B. & Isaac, G. J. (1963). Environmental temperature and energy expenditures. *J. Appl. Physiol.* **18**, 65–8.

Cooper, K. H. & Zechner, A. (1971). Physical fitness in United States and Austrian Military personnel. A comparative study. *J. Amer. Med. Ass.* **215**, 931–4.

Cormack, R. S. & Heath, J. R. (1974). New techniques for calibrating the Lloyd–Haldane apparatus. *J. Physiol.* **238**, 627–38.

Costill, D. L. (1967). The relationship between selected physiological variables and distance running performance. *J. Sports Med. Phys. Fitness* **7**, 61–6.

Costill, D. L. & Winrow, E. (1970). A comparison of two middle-aged ultramarathon runners. *Res. Quart.* **41**, 135–9.

Costill, D. L., Hoffmann, W. M., Kehoe, F., Miller, S. J. & Myers, W. C. (1968). Maximal anaerobic power among college football players. *J. Sports Med. Phys. Fitness* **8**, 103–6.

Costill, D. L., Branam, G., Eddy, D. & Sparks, F. (1971). Determinants of marathon running success. *Int. Z. angew. Physiol.* **29**, 249–54.

Cotes, J. E. (1966). *Occupational Safety and Health Series. Rept. 6.* Geneva, Switzerland: ILO.

Cotes, J. E. (1974). Genetic factors affecting the lung. Genetic component of lung function. *Bull. Physio-pathol. Resp.* **10**, 109–17.

Cotes, J. E. & Malhotra, M. S. (1965). Differences in lung function between Indians and Europeans. *J. Physiol.* **177**, 17–18P.

Cotes, J. E. & Ward, M. P. (1966). Ventilatory capacity in normal Bhutanese. *J. Physiol.* **186**, 88–9P.

Cotes, J. E. & Woolmer, R. F. (1962). A comparison between twenty-seven laboratories of the results of analysis of an expired gas sample. *J. Physiol.* **163**, 36–7P.

Cotes, J. P., Davies, C. T. M., Edholm, O. G., Healy, M. J. R. & Tanner, J. M. (1969). Factors relating to the aerobic capacity of 46 healthy British males and females, ages 18 to 28 years. *Proc. Roy. Soc. Lond. B* **174**, 91–114.

Cotes, J. E., Saunders, M. J., Adam, J. R., Anderson, H. R. & Hall, A. M. (1973*a*). Lung function in coastal and highland New Guineans – Comparison with Europeans. *Thorax* **28**, 320–30.

Cotes, J. E., Berry, G., Burkinshaw, L., Davies, C. T. M., Hall, A. M., Jones, P. R. M. & Knibbs, A. V. (1973*b*). Cardiac frequency during submaximal exercise in young adults; relation to lean body mass, total body potassium and amount of leg muscle. *Quart. J. Exp. Physiol.* **58**, 239–50.

Cotes, J. E., Hall, A. M., Johnson, G. R., Jones, P. R. M. & Knibbs, A. V. (1973*c*). Decline with age of cardiac frequency during submaximal exercise in healthy women. *J. Physiol.* **238**, 24–5P.

Cotton, F. S. & Dill, D. B. (1935). On the relationship between the heart rate during exercise and that of the immediate post-exercise period. *Amer. J. Physiol.* **111**, 554–8.

Craig, A. B. (1968). Limitations of the human organism. Analysis of world records and Olympic performance. *J. Amer. Med. Ass.* **205**, 110–16.

Crockford, G. W. & Davies, C. T. M. (1969). Circadian variations in responses to submaximal exercise on a bicycle ergometer. *J. Physiol.* **201**, 94–5P.

Cruz-Coke, R., Donoso, H. & Barrera, R. (1973). Genetic ecology of hypertension. *Clin. Sci. Mol. Med.* **45**, 55s–65s.

Cullumbine, H. (1949–50). Relationship between body build and capacity for exercise. *J. Appl. Physiol.* **2**, 155–68.

Cumming, A. (1961). Programme of physiological research carried out at Base F., Argentine Islands. Report to the Falkland Islands Dependencies Survey.

Cumming, G. R. (1967). Current levels of fitness. *Canad. Med. Ass. J.* **96**, 868–77.

Cumming, G. R. (1970). Fitness testing of athletes. *Can. Fam. Physician*, August 1970, pp. 48–52.

Cumming, G. R. (1973*a*). Correlation of athletic performance and aerobic power in 12 to 17 year old children with bone age, calf muscle, total body potassium, heart volume, and two indices of anaerobic power. In: *Pediatric Work Physiology. Proceedings of the Fourth International Symposium.* Ed: Bar-Or, O. Natanya, Israel: Wingate Institute.

Cumming, G. R. (1973*b*). Exercise testing. Newsletter, American College of Sports Medicine, Madison, Wisconsin.

Cumming, G. R. (1974). Attempts at maximal exercise testing in 3–6 year old children. *Proceedings, Canadian Association of Sports Sciences, Edmonton, Alberta.*

Cumming, G. R. & Alexander, W. D. (1968). The calibration of bicycle ergometers. *Can. J. Physiol. Pharm.* **46**, 917–19.

Cumming, G. R. & Borysyk, L. M. (1972). Criteria for maximum oxygen uptake in men over 40 in a population survey. *Med. Sci. Sports* **4**, 18–22.

Cumming, G. R. & Cumming, P. M. (1963). Working capacity of normal children tested on a bicycle ergometer. *Canad. Med. Ass. J.* **88**, 351–5.

Cumming, G. R. & Friesen, W. (1967). Bicycle ergometer measurement of maximal oxygen uptake in children. *Can. J. Physiol. Pharm.* **45**, 937–46.

Cumming, G. R. & Keynes, R. (1967). A fitness performance test for school children and its correlation with physical working capacity and maximl oxygen uptake. *Canad. Med. Ass. J.* **96**, 1262–9.

Cumming, G. R., Goodwin, A., Baggley, G. & Antel, J. (1967). Repeated measurements of aerobic capacity during a week of intensive training at a youth's track camp. *Can. J. Physiol. Pharm.* **45**, 805–11.

Cumming, G. R., Goulding, D. & Baggley, G. (1969). Failure of school physical education to improve cardiorespiratory fitness. *Canad. Med. Assoc. J.* **101**, 69–73.

Cumming, G. R., Goulding, D. & Baggley, G. (1971). Working capacity of deaf and visually and mentally handicapped children. *Arch. Dis. Childh.* **46**, 490–4.

Cumming, G. R., Garand, T. & Borysyk, L. (1972). Correlation of performance in track and field events with bone age. *J. Pediatrics* **80**, 970–3.

Cunningham, D. A. & Eynon, R. B. (1973). The working capacity of young competitive swimmers, 10–16 years of age. *Med. Sci. Sports* **5**, 227–31.

Cureton, T. K. (1951). *Physical Fitness of Champion Athletes.* Urbana, Illinois: University of Illinois Press.

Cureton, T. K. (1972). What does maximum oxygen intake ($\dot{V}_{O_2(max)}$) measure? How is it interpreted? What are its principal limitations? In: *Proceedings of the ACSPFT and the ICSPFT – 1972.* Ed: Simri, U. Natanya, Israel: Wingate Institute.

Da Costa, J. L. (1971). Pulmonary function studies in healthy Chinese adults in Singapore. *Amer. Rev. Resp. Dis.* **104**, 128–31.

Damoiseau, J., Bottin, R., Petit, J. M. & Legros, R. (1966). Intéret des mesures de consommation maximum d'O_2 chez des joueurs de football. *Theorie de l'Education Physique* **4**, 1–10.

Daněk, K. (1974). Longitudinal observations of somatometric, spirometric and ergometric values in cross-country skiers aged 11–15. *Proceedings of the Sixth International Congress on Pediatric Work Physiology, Seč, Czechoslovakia.*

Daniels, J. & Oldridge, N. (1970). The effects of alternate exposure to altitude and sea level on world-class middle-distance runners. *Med. Sci. Sports* **2**, 107–12.

Davies, C. T. M. (1968). Limitations to the prediction of maximum oxygen intake from cardiac frequency measurements. *J. Appl. Physiol.* **24**, 700–6.

Davies, C. T. M. (1971). Body composition in children: a reference standard for maximum aerobic power output on a stationary bicycle ergometer. *Acta Paed. Scand. Suppl.* **217**, 136–7.

Davies, C. T. M. (1972*a*). Maximum aerobic power in relation to body composition in healthy sedentary adults. *Human Biol.* **44**, 127–39.
Davies, C. T. M. (1972*b*). The oxygen-transporting system in relation to age. *Clin. Sci.* **42**, 1–13.
Davies, C. T. M. (1973). The contribution of leg (muscle plus bone) volume to maximum aerobic power output: the effects of anaemia, malnutrition and physical activity. *J. Physiol.* **231**, 108P.
Davies, C. T. M. & Van Haaren, J. P. M. (1973). Maximum aerobic power and body composition in healthy East African older male and female subjects. *Amer. J. Phys. Anthropol.* **39**, 395–401.
Davies, C. T. M. & Van Haaren, J. P. M. (1974). Heart volume in relation to maximal aerobic power output in young anaemic and normal African subjects. *Proceedings of the Sixth International Symposium on Pediatric Work Physiology, Seč, Czechoslovakia.*
Davies, C. T. M., Tuxworth, W. & Young, J. M. (1970). Physiological effects of repeated exercise. *Clin. Sci.* **39**, 247–58.
Davies, C. T. M., Barnes, C., Fox, R. H., Ojikutu, R. O. & Samueloff, A. S. (1972). Ethnic differences in physical working capacity. *J. Appl. Physiol.* **33**, 726–32.
Davies, C. T. M., Mbelwa, D. & Doré, C. (1974). Physical growth and development of urban and rural East African children, aged 7–16 years. *Ann. Hum. Biol.* **1**, 257–68.
Davies, L. E. C. & Hanson, S. (1965). The Eskimos of the North-West passage: a survey of dietary composition and various blood and metabolic measurements. *Canad. Med. Ass. J.* **92**, 205–16.
Dehn, M. & Bruce, R. A. (1972). Longitudinal variations in maximal oxygen intake with age and activity. *J. Appl. Physiol.* **33**, 805–7.
Demirjian, A., Jenicek, M. & Dubuc, M. B. (1972). Les normes staturo-ponderales de l'enfant urbain canadien-français d'âge scolaire. *Can. J. Publ. Health* **63**, 14–30.
De Muth, R. G., Howatt, F. W. & Hill, M. B. (1965). The growth of lung function. *Pediatrics Suppl.* **35** (1), 161–218.
De Pauw, D. & Vrijens, J. (1971). Untersuchungen bei Elite–Ruderern in Belgien. *Sportarzt und Sportmedizin* **8**, 176–9.
Deroanne, R., Pirney, F., Servais, C. & Petit, J. M. (1971). Contrôle physiologique du joueur de football en competition. *Sport* **14** (2), 66–71.
De Rose, H. (1973). O Exame Medico do Jogador de Futebol. *Med. do Esporte* **1**, 15–21.
Di Giovanna, V. (1943). The relationship of selected structural and functional measures to success in college athletes. *Res. Quart.* **14**, 199ff.
Dill, D. B. (1965). Marathoner De Mar: Physiological studies. *J. Nat. Cancer Inst.* **35**, 185–91.
Dill, D. B., Robinson, S., Balke, B. & Newton, J. L. (1964). Work tolerance: age and altitude. *J. Appl. Physiol.* **19**, 483–8.
Dill, D. B., Robinson, S. & Ross, J. C. (1967). A longitudinal study of 16 champion runners. *J. Sports Med.* **7**, 4–32.
Di Prampero, P. E. & Cerretelli, P. (1969). Maximal muscular power (aerobic and anaerobic) in African natives. *Ergonomics* **12**, 51–9.
Di Prampero, P. E., Limas, F. P. & Sassi, G. (1970). Maximal muscular power, aerobic and anaerobic, in 116 athletes performing at the XIXth Olympic games in Mexico. *Ergonomics* **13**, 665–74.
Di Prampero, P. E., Cortili, G., Celentano, F. & Cerretelli, P. (1971). Physiological aspects of rowing. *J. Appl. Physiol.* **31**, 853–7.

Dixon, R. W. & Faulkner, J. A. (1971). Cardiac outputs during maximum effort running and swimming. *J. Appl. Physiol.* **30**, 653–6.

Doll, E., Keul, J. & Maiwald, C. (1968). Oxygen tension and acid-base equilibria in venous blood of working muscle. *Amer. J. Physiol.* **215**, 23–9.

Donoso, H., Apud, E., Sañudo, M. C. & Santolaya, R. (1974*a*). Assessment of physical fitness in a population of Aymara speaking males native to high altitude. Personal communication to author.

Donoso, H., Arteaga, A., Valiente, S., Apud, E. & Rosales, E. (1974*b*). Physical working capacity and nutritic status in different groups of the Chilean population. Report to IBP Secretariat.

Dossetor, J. B., Haystead, J., Howson, W. T., Lockwood, B., McConnachie, P. R., Schaefer, O., Smith, L. & Wilson, J. (1971). *The HL-A Antigens in Eskimos. Annual Report 3, HA Project, Igloolik, NWT.* Toronto, Ontario: Dept. of Anthropology, University of Toronto.

Drake, V., Jones, G., Brown, J. R. & Shephard, R. J. (1968). Fitness performance tests and their relationship to maximum oxygen uptake. *Canad. Med. Ass. J.* **99**, 844–8.

Dreyer, G. (1920). *The Assessment of Physical Fitness.* London, UK: Cassell.

Driver, A. F. M. (1958). Climatic preferences in Hong Kong. *J. Appl. Physiol.* **13**, 430–4.

Drucker, P. (1955). *Indians of the Northeast Coast.* Garden City, New York: Natural History Press.

Dua, G. L., Ramaswamy, S. S. & Sen Gupta, J. (1966). Altitude effect on physical work performance. In: *Proceedings of Symposium on Human Adaptability to Environments and Physical Fitness.* Ed: Malhotra, M. S. Madras, India: Defence Institute of Physiology & Allied Sciences.

Dujardin, J., Deroanne, R., Pirnay, F. & Petit, J. M. (1967–8). Acide lactique et exercise musculaire. *Trav. Soc. Med. Belge d'Ed. Phys. et de Sports* **20**, 74–6.

Duncan, M. (1972). Aerobic work capacity in young untrained Asian men. *Quart. J. Exp. Physiol.* **57**, 247–56.

Düner, H. (1959). Oxygen uptake and working capacity during work on the bicycle ergometer with one or both legs. *Acta Physiol. Scand.* **46**, 55–61.

Durnin, J. V. G. A. (1966). Age, physical activity and energy expenditure. *Proc. Nutr. Soc.* **25**, 107–13.

Durnin, J. V. G. A. & Passmore, R. (1967). *Energy, Work and Leisure.* London, UK: Heinemann.

Durnin, J. V. G. A. & Ramahan, M. M. (1967). The assessment of the amount of fat in the human body from measurements of skinfold thickness. *Brit. J. Nutr.* **21**, 681–9.

Durnin, J. V. G. A. & Womersley, J. (1971). The relationship of total body fat, 'fat-free mass' and total body weight in male and female human populations of varying ages. *J. Physiol.* **213**, 33P.

Edgren, J., Bryngelson, C., Lewin, T., Fellman, J. & Skrobak-Kaczynski, J. (1976). Skeletal maturation of the hand and wrist in Finnish Lapps. In: *Circumpolar Health.* Ed: Shephard, R. J. & Itoh, S. Toronto, Ontario: University of Toronto Press.

Edholm, O. G. & Lewis, H. E. (1964). Terrestrial animals in cold: man in polar regions. In: *Handbook of Physiology, Section 4*, pp. 435–46. Ed: Dill, D. B. Washington, DC: American Physiological Society.

Edholm, O. G., Fox, R. H., Adam, J. M. & Goldsmith, R. (1963). Comparison of artificial and natural acclimatization. *Federation of Amer. Soc. for Experimental Biology. Fed. Proc.* **22**, 709–15.

Edholm, O. G., Humphrey, S., Lourie, J. A., Tredre, B. E. & Brotherhood, J.

(1973). Energy expenditure and climatic exposure of Yemenite and Kurdish Jews in Israel. *Phil. Trans. R. Soc. Lond. B* **266**, 127–40.
Edwards, R. H. T., Miller, G. J., Hearn, C. E. D. & Cotes, J. E. (1972). Pulmonary function and exercise responses in relation to body composition and ethnic origin in Trinidadian males. *Phil. Trans. R. Soc. Lond. B* **181**, 407–20.
Eichhorn, J., Brüner, H., Klein, K. E. & Wegmann, H. M. (1967). Fehleinschätzungen der maximalen Sauerstoffaufnahme bei ihrer Bestimmung mit indirekten Methoden. *Int. Z. angew. Physiol.* **24**, 275–83.
Ekblöm, B. (1969). Effect of physical training on adolescent boys. *J. Appl. Physiol.* **27**, 350–5.
Ekblöm, B. & Gjessing, E. (1968). Maximal oxygen uptake of the Easter Island population. *J. Appl. Physiol.* **25**, 124–9.
Eksmyr, R. (1971). Anthropometry in Ethiopian private school children. CNV Report No. 40. *Nutr. Metabol.* **13**, 7–20.
Elsner, R. W. (1963). Skinfold thickness in primitive people native to a cold climate. *Ann. N.Y. Acad. Sci.* **110**, 503–14.
Engström, L. M. (1972). *Idrott par Fritid. En Enkätstudie bland elever i Årskurs 8.* Stockholm, Sweden: Pedagogiska Institutionen Lärahogskolan.
Engström, I., Eriksson, B. O., Karlberg, P., Saltin, P. & Thoren, C. (1971). Preliminary report on the development of lung volumes in young girl swimmers. *Acta Paediatr. Scand. Suppl.* **217**, 73–6.
Erb, B. D. (Chairman) (1970). *Physicians' Handbook for Evaluation of Cardiovascular and Physical Fitness.* Nashville, Tennessee: Tennessee Heart Association Physical Exercise Committee.
Erikson, H. (1958). The respiratory response to acute exercise of Eskimos and whites. *Acta Physiol. Scand.* **41**, 1–11.
Eriksson, B. O. (1972). Physical training, oxygen supply and muscle metabolism in 11–13 year old boys. *Acta Physiol. Scand. Suppl.* **384**, 1–48.
Eriksson, B. O., Engström, I., Karlberg, P., Saltin, B. & Thoren, C. (1971). A physiological analysis of former girl swimmers. *Acta Paed. Scand. Suppl.* **217**, 68–72.
Eriksson, B. O., Gollnick, P. O. & Saltin, B. (1973*a*). Muscle metabolism and enzyme activities after training in boys 11–13 years old. *Acta Physiol. Scand.* **87**, 485–97.
Eriksson, B. O., Lundin, A. & Saltin, B. (1973*b*). Physical training of post-active girl swimmers. In: *Pediatric Work Physiology. Proceedings of the Fourth International Symposium*, pp. 217–18. Ed: Bar-Or, O. Natanya, Israel: Wingate Institute.
Espenschade, A. S. & Meleney, H. E. (1961). Motor performance of adolescent boys and girls of today in comparison with those of 24 years ago. *Res. Quart.* **32**, 186–9.
Evans, E. G. (1973). Basic fitness testing of rugby football players. *Brit. J. Sports Med.* **7**, 384–7.
Evans, J. G. & Prior, I. A. M. (1969). Indices of obesity derived from height and weight in two Polynesian populations. *Brit. J. Prev. Soc. Med.* **23**, 56–9.
Eyman, C. & Salter, E. (1976). Growth assessment through hand–wrist X-rays of some Fort Chimo males. In: *Circumpolar Health.* Ed: Shephard, R. J. & Itoh, S. Toronto, Ontario: University of Toronto Press.
Falls, H. B. & Humphrey, L. D. (1973). A comparison of methods for eliciting maximum oxygen uptake from college women during treadmill walking. *Med. Sci. Sports* **5**, 239–41.
Falls, H. B., Ismail, A. H. & MacLeod, D. F. (1966).Estimation of maximum

oxygen uptake in adults from AAHPER youth fitness test items. *Res. Quart.* **37**, 192–201.
Faulkner, J. F. (1973). Viewpoint – Exercise testing. Newsletter, American College of Sports Medicine, July, 1973, Madison, Wisconsin.
Faulkner, J. F. & Stoedefalke, K. (1975). *Guidelines for Graded Exercise Testing and Exercise Prescription and Behavioral Objectives for Physicians, Program Directors, Exercise Leaders and Exercise Technicians.* Philadelphia, Pennsylvania: Lea & Febiger.
Ferguson, R. J., Marcotte, G. G. & Montpetit, R. R. (1969). A maximal oxygen uptake test during ice-skating. *Med. Sci. Sports* **1**, 2078–11.
Ferro-Luzzi, A. & Ferro-Luzzi, G. (1962). Study on skinfold thickness of school children in some developing countries. I. Skinfold thickness of Libyan boys. II. Skinfold thickness of Moroccan boys. *Metabolism* **11**, 1064–72, 1072–6.
Flandrois, R., Puccinelli, R., LeFrancois, R. & Bouverot, P. (1962). Intérêt de la consommation maximale d'oxygène en physiologie appliquée: premiers resultats experimentaux sur une population français. *Presse Therm. Climat.* **99**, 179–84.
Foote, D. C. & Greer-Wootten, B. (1966). Man–environment interactions in an Eskimo hunting system. *Symposium on Man–Animal Linked Cultural Sub-Systems. 133rd Meeting of Amer. Assoc. Adv. Sci., Washington DC.*
Forbes, G. B. (1974). Stature and lean body mass. *Amer. J. Clin. Nutr.* **27**, 595–602.
Fox, E. L. (1973). A simple, accurate technique for predicting maximal aerobic power. *J. Appl. Physiol.* **35**, 914–16.
Fox, E. L. & Costill, D. L. (1972). Estimated cardiorespiratory responses during marathon running. *Arch. Env. Health* **24**, 316–24.
Fox, S. M. (1969). Exercise and stress testing workshop report. *J. S. Carol. Med. Ass. Suppl.* **1**, 77.
Fox, S. M. (1974). *Coronary Heart Disease. Prevention, Detection, Rehabilitation, with Emphasis on Exercise Testing.* Denver, Colorado: International Medical Corporation.
Freyschuss, V. & Strandell, T. (1968). Circulatory adaptation to one- and two-leg exercise in supine position. *J. Appl. Physiol.* **25**, 511–15.
Friis-Hansen, B. (1965). Hydrometry of growth and aging. In: *Human Body Composition: Approaches and Applications.* Ed: Brožek, J. New York: Pergamon Press.
Frisch, R. & Revelle, R. (1969). Variation in body weights and the age of the adolescent growth spurt among Latin American and Asian populations in relation to calorie supplies. *Human Biol.* **41**, 185–212.
Froelicher, V. F., Brammell, H., Davis, G., Noguera, I., Stewart, A. & Lancaster, M. C. (1974). A comparison of three maximal treadmill exercise protocols. *Chest* **65** (5), 512–17.
Fry, E. I. (1960). Subcutaneous tissue in Polynesian children from Rarotonga, Cook Islands. *Human Biol.* **32**, 239–48.
Fry, E. I., Chang, K. S. F., Lee, M. M. C. & Ng, C. K. (1965). The amount and distribution of subcutaneous tissue in Southern Chinese children from Hong Kong. *Amer. J. Phys. Anthropol.* **23**, 69–80.
Fugelli, P. (1976). Health of arctic fishermen as related to occupational capacity. In: *Circumpolar Health.* Ed: Shephard, R. J. & Itoh, S. Toronto, Ontario: University of Toronto Press.
Gandra, Y. R. & Bradfield, R. B. (1971). Energy expenditure and oxygen handling efficiency of anemic school children. *Amer. J. Clin. Nutr.* **24**, 1451–6.
Gardner, G. W. (1971). Physical fitness of primitive peoples – the Warao Indians

of Venezuela. In: *Physical Fitness*. Ed: Seliger, V. Prague, Czechoslovakia: Charles University Press.

Garn, S. M. (1965). The applicability of North American growth standards in developing countries. *Canad. Med. Ass. J.* **93**, 914–19.

Garn, S. M., Clark, A., Landkof, L. & Newell, L. (1960). Parental body build and developmental progress in the offspring. *Science* **132**, 1555–6.

Garn, S. M., Rohmann, C. G. & Silverman, F. N. (1967). Radiographic standards for post-natal ossification and tooth calcification. *Med. Radiograph. Photogr.* **43** (2), 45–66.

Gedda, L., Milani-Comparetti, M. & Brenci, G. (1968). *Rapporto Scientifico Sugli Atleti Della XVII Olimpiade, Roma 1960*. Rome, Italy: Istituto di Medicina dello Sport.

Getchell, L. H. (1968). Energy cost of playing golf. *Arch. Phys. Med. Rehab.* **49**, 31–5.

Ghesquière, J. L. A. (1971). Physical development and working capacity of Congolese. In: *Human Biology and Environmental Change*. Ed: Vorster, R. Malawi: IBP.

Ghesquière, J. & Andersen, K. L. (1971). Similarities and differences in physiological responses to muscular exercise among primitive African men of two genetically different population groups. In: *Physical Fitness*. Ed: Seliger, V. Prague, Czechoslovakia: Charles University Press.

Glagov, S., Rowley, D. A., Cramer, D. B. & Page, R. G. (1970). Heart rates during 24 hours of usual activity for 100 normal men. *J. Appl. Physiol.* **29**, 799–805.

Glaser, E. M. (1966). *The Physiological Basis of Habituation*. Oxford, UK: Oxford University Press.

Glaser, E. M. & Shephard, R. J. (1963). Simultaneous experimental acclimatization to heat and cold in man. *J. Physiol.* **169**, 592–603.

Glassford, R. C., Bayford, G. H. Y., Sedgwick, A. W. & MacNab, R. B. J. (1965). Comparison of maximal oxygen uptake values determined by predicted and actual methods. *J. Appl. Physiol.* **20**, 509–13.

Gleser, M. A. (1973). The effect of hypoxia and physical training on the hemodynamic adjustments to one-legged exercise. *J. Appl. Physiol.* **34**, 655–9.

Gleser, M., Horstman, D. H. & Mello, R. P. (1974). The effect on $\dot{V}_{O_2(max)}$ of adding arm work to maximal leg work. *Med. Sci. Sports* **6**, 104–7.

Glick, Z. & Shvartz, E. (1974). Physical working capacity of young men of different ethnic groups in Israel. *J. Appl. Physiol.* **37**, 22–6.

Godin, G. & Shephard, R. J. (1973). Activity patterns of the Canadian Eskimo. In: *Human Polar Biology*. Ed: Edholm, O. & Gunderson, E. K. Cambridge, UK: Heinemann.

Gold, A. J., Zornitzer, A. & Samueloff, S. (1969). Influence of season and heat on energy expenditure during rest and exercise. *J. Appl. Physiol.* **27**, 9–12.

Goldman, R. F., Haisman, M. F. & Pandolf, K. B. (1976). Metabolic energy cost and terrain co-efficients of walking on snow. In: *Circumpolar Health*. Ed: Shephard, R. J. & Itoh, S. Toronto, Ontario: University of Toronto Press.

Goldrick, R. B., Sinnett, P. F. & Whyte, H. M. (1970). An assessment of coronary heart disease and coronary risk factors in a New Guinea highlands population. In: *Atherosclerosis. Proceedings of the Second International Symposium*. Ed: Jones, R. J. New York: Springer Verlag.

Gollnick, P. & Hermansen, L. (1973). Biochemical adaptations to exercise. Anaerobic metabolism. *Exercise & Sports Science Rev.* **1**, 1–43.

Goodwin, A. B. & Cumming, G. R. (1966). Radio telemetry of the electrocardiogram, fitness tests, and oxygen uptake of water polo players. *Canad. Med. Ass. J.* **95**, 402–6.

Grande, F. (1961). In: *Techniques for Measuring Body Composition*. Ed: Brožek, J. M. & Henschel, A. Washington, DC: US National Academy of Sciences – National Research Council.

Green, H. (1967). Urban and rural differences in the work capacity of Alberta Secondary School Students as measured by the Åstrand predicted maximal oxygen intake test. University of Alberta, Edmonton: unpublished report, Fitness Research Unit.

Green, H. J., Houston, M. E. & Thomson, J. (1974). Cardiovascular and metabolic responses during maximal and submaximal ice-skating. *Proceedings, Canadian Association of Sports Sciences, Edmonton, Alberta.*

Green, H. J. & Houston, M. E. (1975). Effect of a season of ice-hockey on energy capacities and associated functions. *Med. Sci. Sports* **7**, 299–303.

Greulich, W. W. & Pyle, S. I. (1959). *Radiographic Atlas of Skeletal Development of the Hand and Wrist*. Oxford, UK: Oxford University Press.

Guberan, E., Williams, M. K., Walford, J. & Smith, M. M. (1969). Circadian variations of FEV in shift workers. *Brit. J. Industr. Med.* **26**, 121–5.

Hagerman, F. C. & Howie, J. A. (1971). Use of certain physiological variables in the selection of the 1967 New Zealand crew. *Res. Quart.* **42**, 264–73.

Hagerman, F. C. & Lee, W. D. (1971). Measurement of oxygen consumption, heart rate and work output during rowing. *Med. Sci. Sports* **3**, 155–60.

Hagerman, F. C., Addington, W. C. & Gaensler, E. A. (1972). A comparison of selected physiological variables among outstanding competitive oarsmen. *J. Sports Med. Phys. Fitness* **12**, 12–22.

Hamley, E. J. (1963). Respiratory changes in cyclists. *Bull. Brit. Assoc. Sports Med.* **1**, 5.

Hammel, H. T. (1964). Terrestrial animals in cold; recent studies of primitive man. In: *Handbook of Physiology, Section 4*, pp. 413–34. Ed: Dill, D. B., Adolph, E. F. & Wilber, C. G. Washington, DC: American Physiological Society.

Harrison, G. A., Weiner, J. S., Tanner, J. M. & Barnicot, N. A. (1964). *An Introduction to Human Evolution, Variation and Growth*. Oxford, UK: Clarendon Press.

Hart, J. S. (1967). Commentary. *Canad. Med. Ass. J.* **96**, 803–4.

Hartley, L. H. & Saltin, B. (1969). Blood gas tensions and pH in brachial artery, femoral vein and brachial vein during maximal exercise. *Med. Sci. Sport* **3**, 66–72.

Harvald, B. (1976). Current genetical trends in the Greenlandic population. In: *Circumpolar Health*. Ed: Shephard, R. J. & Itoh, S. Toronto, Ontario: University of Toronto Press.

Hasegawa, J., Ishiko, T., Matsuda, I., Meshizuka, T. & Noguchi, Y. (1966). Physical fitness status of Japanese Youth through sport test. *Proceedings of International Congress of Sport Sciences, 1964*. Tokyo: Japanese Union of Sport Sciences.

Hay, J. G. (1967). Rowing: an analysis of the New Zealand Olympic selection tests. *N.Z.J.H.P.E.R.* 83–90.

Heald, F. P., Hunt, E. E., Schwartz, R., Cook, C., Elliot, O. & Vajda, B. (1963). Measures of body fat and hydration in adolescent boys. *Pediatrics* **31**, 226–39.

Hebbelinck, M. & Borms, J. (1969). *Tests en normen schalen van Lichamelijke Prestatiegeschiktheid voor Jongens van 6 tot 13 Jaar iut het lager Onderwijs*. Ministerie Van Nederlandse Cultuur, Belgium.

Hellon, R. F., Jones, R. M., MacPherson, R. K. & Weiner, J. S. (1956). Natural and artificial acclimatization to hot environments. *J. Physiol.* **132**, 559–76.

Hermansen, L. & Oseid, S. (1971). Direct and indirect estimation of maximal oxygen uptake in pre-pubertal boys. *Acta Paed. Scand. Suppl.* **217**, 18–23.

Hermansen, L. & Rodahl, K. (1976). Comparison of values for maximal oxygen uptake obtained in cross-sectional and longitudinal studies. In: *Circumpolar Health.* Ed: Shephard, R. J. & Itoh, S. Toronto, Ontario: University of Toronto Press.

Hermansen, L. & Saltin, B. (1969). Oxygen uptake during maximal treadmill and bicycle exercise. *J. Appl. Physiol.* **26**, 31–7.

Heywood, P. F. & Latham, M. C. (1971). Use of the SAMI heart-rate integrator in children. *Amer. J. Clin. Nutr.* **24**, 1446–50.

Hicks, C. S. (1964). Terrestrial animals in cold: exploratory studies of primitive man. In: *Handbook of Physiology, Section 4*, pp. 405–12. Ed: Dill, D. B., Adolph, E. F. & Wilbur, C. G. Washington, DC: American Physiological Society.

Hiernaux, J. (1966). Peoples of Africa from 22° N to the Equator. In: *The Biology of Human Adaptability.* Ed: Baker, P. T. & Weiner, J. S. Oxford, UK: Clarendon Press.

Higgs, S. L. (1973). Telemetered heart rate during girls' competitive basketball as played under international rules. *First Canadian Congress of Sport & Physical Activity, Montreal.*

Hildes, J. A., Schaefer, O., Sayed, J. E., Fitzgerald, E. J. & Koch, E. A. (1976). Chronic pulmonary disease and associated cardiovascular disease in Igloo-ligmiuts. In: *Circumpolar Health.* Ed: Shephard, R. J. & Itoh, S. Toronto, Ontario: University of Toronto Press.

Hipsley, E. H. & Kirk, N. E. (1965). Studies of dietary intake and expenditure of New Guineans. *South Pacific Tech. Paper* **147**. Cited by Sinnett & Solomon (1968).

Hirata, K. I. (1966). Physique and age of Tokyo Olympic champions. *J. Sports Med. Fitness* **6**, 207–22.

Hirota, K., Asami, T., Toyoda, H. & Shimazu, D. (1969). Aerobic and anaerobic work capacity of the Ama. *Research J. Phys. Ed.* **13**, 260–5.

Hoes, M., Binkhorst, R. A., Smeekes-Kuyl, A. & Vissurs, A. C. (1968). Measurement of forces exerted on a pedal crank during work on the bicycle ergometer at different loads. *Int. Z. angew. Physiol.* **26**, 33–42.

Hofvander, Y. (1968). Hematological investigations in Ethiopia with special reference to a high iron intake. *Acta Med. Scand. Suppl.* **494**.

Hollmann, W. (1965). *Körperliches Training als Prävention von Herz-Kreislauf Krankheiten.* Stüttgart, West Germany: Hippokrates Verlag.

Hollmann, W. (1972). Lungenfunktion, Atmung und Stoffwechsel in Sport. In: *Zentrale Themem der Sportmedizin.* Ed: Hollmann, W. Berlin, Germany: Springer Verlag.

Hollmann, W., Venrath, H., Bonnekoh, A. & Nöcker, J. (1962). Untersuchungen zur Höchst- und Dauerleistungs-fähigkeit deutscher Fussball-Spitzenspieler. *Sportarzt* **6**, 172–83.

Hollmann, W., Scholtzmethner, R., Grunewald, B. & Werner, H. (1967). Untersuchungen zur Ausdauer-verbesserung 9 bis 11 jähriger Mädchen in Rahmen des Schulsonderturnens. *Die Liebererzeihung* **10**, Cologne, West Germany.

Hollmann, W., Schmücker, B., Heck, H., Stolte, A., Liesen, H., Fotescu, M. D. & Mathur, D. N. (1971). Über das Verhalten spiroergometrischer Messgrossen bei Radrenfahrern auf dem Laufband und auf dem Fahrradergometer. *Sportarzt und Sportmedizin* **7**, 153–7.

Holloszy, J. O. (1973). Biochemical adaptations to exercise: aerobic metabolism. *Exercise & Sports Science Rev.* **1**, 45–71.

Holmér, I. (1972). Oxygen uptake during swimming in man. *J. Appl. Physiol.* **33**, 502–9.

Holt (1948). Data on body weight of children in the U.K. Quoted by Sheldon, W. In: *Diseases of Infancy and Childhood.* London, UK: Churchill.
Holter, N. J. (1961). New method for heart studies. *Science* **134**, 1214–20.
Horvath, S. M. & Finney, B. R. (1969). Paddling experiments and the question of Polynesian voyaging. *Amer. Anthropologist* **71**, 271–6.
Horvath, S. M. & Michael, E. D. (1970). Responses of young women to gradually increasing and constant load maximal exercise. *Med. Sci. Sports* **2**, 128–31.
Howald, H. (1975). Ultrastructural adaptation of skeletal muscle to prolonged physical exercise. *Proceedings of the Second International Symposium on Biochemistry of Exercise, Magglingen.* Ed: Howald, H. & Poortmans, J. Basel: Birkhauser Verlag.
Howald, H., Hanselmann, E. & Jucker, P. (1974). Comparative study on the determination of aerobic power. *Standardization of Physical Fitness Tests, Twelfth Magglingen Symposium.* Basel, Switzerland: Birkhauser Verlag.
Howell, J. B. L. & Campbell, E. J. M. (1966). *Breathlessness.* Oxford, UK: Blackwell Scientific Publications.
Howell, M. L. & MacNab, R. B. J. (1968). *The Physical Work Capacity of Canadian Children aged 7 to 17.* Toronto, Ontario: Canadian Association for Health, Physical Education and Recreation.
Howell, M. L., Loiselle, D. S. & Lucas, W. G. (1965). Strength of Edmonton School Children. University of Alberta, Edmonton: unpublished report, Fitness Research Unit.
Howitt, J. S., Balkwill, J. S., Whiteside, T. C. D. & Whittingham, P. D. G. (1966). A preliminary study of flight deck work loads in civil air transport aircraft. *UK Ministry of Defence, F.P.R.C.* **1240**.
Hughes, A. L. & Goldman, R. F. (1970). Energy cost of hard work. *J. Appl. Physiol.* **29**, 570–2.
Hukuda, K. & Ishiko, T. (1966). Comparison between physical fitness of Japanese and European athletes. In: *Proceedings of International Congress of Sport Sciences, 1964.* Ed: Kato, K. Tokyo: Japanese Union of Sport Sciences.
Hyde, R. C. (1967). The Åstrand–Ryhming nomogram as a prediction of aerobic capacity for Secondary School Students. *Fitness Research Unit Report* **3**, University of Alberta, Edmonton.
Iampietro, P. F., Goldman, R. F., Buskirk, E. R. & Bass, D. E. (1959). Response of Negro and White males to cold. *J. Appl. Physiol.* **14**, 798–800.
Ikai, M. (1967). Comparative study on the maximum aerobic work capacity. *Jap. J. Physiol.* **29**, 517–22.
Ikai, M. & Katagawa, K. (1972). Maximum oxygen uptake of Japanese related to sex and age. *Med. Sci. Sports* **4**, 127–31.
Ikai, M., Shindo, M. & Miyamura, M. (1970). Aerobic work capacity of Japanese people. *Res. J. Phys. Ed.* **14**, 137–42.
Ikai, M., Ishii, K., Miyamura, M., Kusano, K., Bar-Or, O., Kollias, J. & Buskirk, E. R. (1971). Aerobic capacity of Ainu and other Japanese on Hokkaido. *Med. Sci. Sports* **3**, 6–11.
International Committee on the Standardization of Physical Fitness Tests (1969). Physical Fitness Measurements Standards. Tel Aviv, Israel: ICSPFT.
Ishiko, T. (1967). Aerobic capacity and external criteria of performance. *Canad. Med. Ass. J.* **96**, 746–9.
Ishiko, T. (1973). PWC_{170} of Japanese boys and girls. *Brit. J. Sports Med.* **7**, 215.
Ishiko, T. & Acki, L. (1973). Maximal oxygen uptake and athletic performance in long distance runners. In: *Sport in the Modern World – Chances and Problems.* Ed: Grupe, O., Kurz, D. & Teipel, J. M. Berlin, Germany: Springer Verlag.

Ishiko, T., Ikeda, N. & Enomoto, Y. (1968). Obese children in Japan. *Res. J. Phys. Educ.* **12**, 168–74.

Jamison, P. L. (1970). Growth of Wainwright Eskimos: Stature and weight. *Arctic Anthropol.* **7**, 86–9.

Jamison, P. L. & Zegura, S. L. (1970). An anthropometric study of the Eskimos of Wainwright, Alaska. *Arctic Anthropol.* **7**, 125–43.

Jansen, A. A. J. (1963). Skinfold measurements from early childhood to adulthood in Papuans from Western New Guinea. *Ann. N.Y. Acad. Sci.* **110**, 515–31.

Jéquier, J. C. (1974). Factors in the design of a longitudinal study. Paper presented to Canadian Pediatric Society, St. John's, Newfoundland.

Jeniček, M. & Demirjian, A. (1972). Triceps and subscapular skinfold thickness in French Canadian school-age children in Montreal. *Amer. J. Clin. Nutr.* **25**, 576–81.

Johnson, B. C., Epstein, F. H. & Kjelsberg, M. O. (1965). Distribution and familial studies of blood pressure and serum cholesterol levels in a total community, Tecumseh, Michigan. *J. Chr. Dis.* **18**, 147–60.

Johnson, La Von E. (1970). Effects of 5-day a week versus 2- and 3-day a week physical education class on fitness skill, adipose tissue and growth. *Res. Quart.* **40**, 93–8.

Johnston, F. E., Hamill, P. V. V. & Lemeshow, S. (1972). *Skinfold Thickness of Children 6–11 Years, United States.* Washington, DC: National Center for Health Statistics, Series 11, **120**.

Jokl, E. & Jokl, P. (1968). *The Physiological Basis of Athletic Records.* Springfield, Illinois: C. C. Thomas.

Jones, H. E. (1947). Sex differences in physical abilities. *Human Biol.* **19**, 12–25.

Jones, H. E. (1949). *Motor Performance and Growth.* Berkeley, California: University of California Press.

Jones, N. L. & Haddon, R. W. T. (1973). Effect of a meal on cardiopulmonary and metabolic changes during exercise. *Can. J. Physiol. Pharm.* **51**, 445–50.

Jones, P. R. M. & Pearson, J. (1969). Anthropometric determination of leg fat and muscle plus bone volumes in young male and female adults. *J. Physiol.* **204**, 63P.

Joseph, N. T., Srinivasulu, N., Sampath Kumar, T. & Sen Gupta, J. (1973). Indirect estimation of aerobic capacity in Indians. *Indian J. Med. Res.* **61**, 252–9.

Joshi, R. C., Madan, R. N. & Eggleston, F. C. (1973). Clinical spirometry in normal North Indian males. *Respiration* **30**, 39–47.

Kaijser, L. (1970). Limiting factors for aerobic muscle performance. *Acta Physiol. Scand. Suppl.* **346**, 1–96.

Kamon, E. & Pandolf, K. B. (1972). Maximal aerobic power during laddermill climbing, uphill running and cycling. *J. Appl. Physiol.* **32**, 467–73.

Karlsson, J. (1970). Maximum oxygen uptake in Skolt Lapps. *Arctic Anthropol.* **7**, 19–20.

Kasch, F. W., Phillips, W. H., Ross, W. D. & Carter, J. E. L. (1965). A step test for inducing maximal work. *J. Assoc. Phys. Ment. Rehab.* **19**, 84–6.

Kasch, F. W., Phillips, W. H., Ross, W. D., Carter, J. E. L. & Boyer, J. L. (1966). A comparison of maximal oxygen uptake by treadmill and step-test procedures. *J. Appl. Physiol.* **21**, 1387–8.

Katch, F. I., Girandola, F. N. & Katch, V. L. (1971). The relationship of body weight on maximum oxygen uptake and heavy work endurance capacity on the bicycle ergometer. *Med. Sci. Sports* **3**, 101–6.

Katch, F. I., McArdle, W. D. & Pechar, G. S. (1974). Relationship of maximal leg force and leg composition to treadmill and bicycle ergometer maximum oxygen uptake. *Med. Sci. Sports* **6**, 38–43.

Kattus, A. A. (Chairman) (1972). Exercise Testing and Training of Apparently Healthy Individuals: A Handbook for Physicians. New York: American Heart Association.

Kay, C. & Shephard, R. J. (1969). On muscle strength and the threshold of anaerobic work. *Int. Z. angew. Physiol.* **27**, 311–28.

Kay, J. D. S., Petersen, E. S. & Vejby-Christensen, H. (1974). Breathing patterns in man during bicycle exercise at differing pedalling frequencies. *J. Physiol.* **241**, 123P.

Kemp, W. H. (1971). The flow of energy in a hunting society. *Scientific American* **225**, 105–15.

Kemper, H. C. G. (1971). The influence of extra lessons of physical education on physical and mental development of 12–13 year old boys. In: *Physical Fitness*. Ed: Seliger, V. Prague, Czechoslovakia: Charles University Press.

Kemper, H. C. G. (1973). Heart rate during bicycle ergometer exercise in watts per kilogram body weight of 12 and 13 year old boys. In: *Pediatric Work Physiology. Proceedings of the Fourth International Symposium*. Ed: Bar-Or, O. Natanya, Israel: Wingate Institute.

Kemper, H. C. G., Ras, J. G. A., Snel, J., Splinter, P. G., Tavecchio, L. W. C. & Verschuur, R. (1974). Investigation to the effects of two extra lessons in physical education a week during one school-year upon the physical development of 12 and 13 year old boys. *Proceedings of the Sixth International Symposium on Pediatric Work Physiology, Seč, Czechoslovakia.*

Kemsley, W. F. F., Billewicz, W. Z. & Thomson, A. M. (1962). A new weight-for-height standard based on British anthropometric data. *Brit. J. Prev. Soc. Med.* **16**, 189–95.

Keul, J. (1973). *Limiting Factors of Physical Performance*. Stüttgart, West Germany: Thieme.

Keys, A., Anderson, J. J. & Grande, F. (1959). Serum cholesterol response in man to oral ingestion of arachidonic acid. *Amer. J. Clin. Nutr.* **7**, 444–50.

Khosla, T. & Lowe, C. R. (1967). Indices of obesity derived from body weight and height. *Brit. J. Prev. Soc. Med.* **21**, 122–8.

Kilbom, A. & Åstrand, I. (1971). Physical training with sub-maximal intensities in women. II. Effect on cardiac output. *Scand. J. Clin. Lab. Invest.* **28**, 163–75.

Kiss, M. A., Rocha-Ferreira, M. B., Souza, P., Vasconcelos, R. M., Santos, F. B., Anzai, K., Pagan, V., André, J., Rodrigues, R., Bosco, J., Baccala, L. T. & Pini, M. C. (1973). Potencia Maxima Aerobica em Atletas de Seleçoes Paulistas e Brasileiras. *Med. do Esporte* **1**, 23–30.

Klavora, P. (1973). Effect of training on working capacity of rowers and their aerobic responses to training. *First Canadian Congress of Sport & Physical Activity, Montreal.*

Klein, K. E., Wegmann, H. M. & Bruner, H. (1968). Circadian rhythms in indices of human performance, physical fitness and stress resistance. *Aerospace Med.* **39**, 512–18.

Kleitman, N. & Kleitman, E. (1953). Effect of non-24 hour routines of living on oral temperature and heart rate. *J. Appl. Physiol.* **6**, 283–91.

Klimt, F. (1966). Telemotorische Herzschlagfrequenz registrierungen bei Kleinkindern wahrend einer körperlichen Tätigkeit. *Deutsches Gesundheitwesen* **21**, 599.

Klimt, F. & Voigt, G. B. (1971). Investigations on the standardization of ergometry in children. *Acta Paed. Scand. Suppl.* **217**, 35–6.

Klimt, F., Pannier, R. & Paufler, D. (1974). Ausdauerbelastungen bei Vorschulkindern. *Schweiz. Zsch. Sportmedizin* **201**, 7–23.

Klissouras, V. (1971). Heritability of adaptive variation. *J. Appl. Physiol.* **31**, 338–44.

Klissouras, V. (1972). Genetic limit of functional adaptability. *Int. Z. angew. Physiol.* **30**, 85–94.

Klissouras, V. & Weber, G. (1973). Training: growth and heredity. In: *Pediatric Work Physiology. Proceedings of the Fourth International Symposium.* Ed: Bar-Or, O. Natanya, Israel: Wingate Institute.

Knittle, J. L. & Hirsch, J. (1968). Effect of early nutrition on the development of epididymal fat fads: cellularity and metabolism. *J. Clin. Invest.* **47**, 2091–8.

Knuttgen, H. G. (1961). Comparison of fitness of Danish and American school children. *Res. Quart.* **32**, 190–6.

Knuttgen, H. G. & Steendahl, K. (1963). Fitness of Danish school children during the course of one academic year. *Res. Quart.* **34**, 34–40.

Kofranyi, E. & Michaelis, H. F. (1949). Ein Tragbarer Apparat zur Bestimmung des Gasstoffwechsels. *Arbeitsphysiologie* **11**, 148–50.

Kohlraush, W. (1929). Zusammenhange von Körperform und Leistung. Ergebnisse der anthropometrischen Messungen an den Athletes der Amsterdamer Olympiade. *Arbeitsphysiologie* **2**, 187–204.

Kollias, J., Moody, D. L. & Buskirk, E. R. (1967). Cross-country running: treadmill simulation and suggested effectiveness of supplemental treadmill training. *J. Sports Med. Phys. Fitness* **7**, 148–54.

Kollias, J., Buskirk, E. R., Howley, E. T. & Loomis, S. L. (1972). Cardiorespiratory and body composition measurements of high school football players. *Res. Quart.* **43**, 472–8.

Komi, P. V., Klissouras, V. & Karvinen, E. (1973). Genetic variation in neuromuscular performance. *Int. Z. angew. Physiol.* **31**, 289–304.

Konig, G., Reindell, H., Keul, J. & Roskamm, H. (1961). Untersuchungen über das verhalten von Atmung und Kreislauf im Belastungs versuch bei Kindern und Jugendlichen im alter von 10–19 Jähren. *Int. Z. angew. Physiol.* **18**, 393–434.

Kozlowski, S., Nowakowska, A., Kirschner, H. & Obuchowicz-Lozynska, I. (1968). Determination of aerobic working capacity, as related to direct measurement of maximum oxygen uptake. *Wych. Fizycne.* (Authors' translation from original Polish in *Sport* **12** (3).)

Kozlowski, S., Kirschner, H., Kaminski, A. & Starnowski, R. (1969). The relationship between the predicted maximum oxygen uptake and the age of workers employed in various professions. *Polish Med. J.* **8**, 1303–11.

Kramer, J. D. & Lurie, P. R. (1964). Maximal exercise tests in children. *Amer. J. Dis. Childh.* **108**, 283–97.

Ladell, W. S. S. (1947). Effects on man of high temperatures. Effects on man of restricted water supply. *Brit. Med. Bull.* **5**, 5–8, 9–12.

Ladell, W. S. S. (1951). Assessment of group acclimatization to heat and humidity. *J. Physiol.* **115**, 296–312.

Ladell, W. S. S. (1955*a*). Effects of water and salt intake upon the performance of men working in hot and humid environments. *J. Physiol.* **127**, 11–46.

Ladell, W. S. S. (1955*b*). Physiological observations on men working in supposedly limiting environments in a West African gold-mine. *Brit. J. Indust. Med.* **12**, 111–25.

Ladell, W. S. S. (1957). Disorders due to heat. *Trans. Roy. Soc. Trop. Med. Hyg.* **51**, 189–207.

Ladell, W. S. S. (1964). Terrestrial animals in humid heat: man. In: *Handbook of*

Physiology, Section 4. Ed: Dill, D. B. Washington, DC: American Physiological Society.

Ladell, W. S. S. & Kenney, R. A. (1955). Some laboratory and field observations on the Harvard pack test. *Quart. J. Exp. Physiol.* **40**, 283–96.

Ladell, W. S. S., Waterlow, J. C. & Hudson, M. F. (1944). Desert climate. Physiological and clinical observations. *Lancet* **247**, 491–7, 527–31.

Lammert, O. (1972). Maximal aerobic power and energy expenditure of Eskimo hunters in Greenland. *J. Appl. Physiol.* **33**, 184–8.

Lapiccirella, V., Lapiccirella, R., Abboni, F. & Liotta, S. (1962). Enquête clinique, biologique et cardiographique parmi les tribes nomades de la Somalie qui se nourissent seulement du lait. *Bull. Wld Health Org.* **27**, 681–97.

Laughlin, W. S. (1966). Genetic and anthropological characteristics of arctic populations. In: *The Biology of Human Adaptability*. Ed: Baker, P. T. & Weiner, J. S. Oxford, UK: Clarendon Press.

Lavallée, H., Larivière, G. & Shephard, R. J. (1974). Correlations between field tests of performance and laboratory measurements of fitness. Results in the ten year old school child. *Acta Paed. Belg.* **28** (Suppl.), 29–39.

Leary, W. P. & Wyndham, C. H. (1965). The capacity for maximum physical efforts of Caucasian and Bantu athletes of international class. *S. Afr. Med. J.* **39**, 651–5.

Lee, D. H. K. (1964). Terrestrial animals in dry heat: man in the desert. In: *Handbook of Physiology. Section 4*. Ed: Dill, D. B. Washington, DC: American Physiological Society.

Lee, R. B. (1969). !Kung Bushmen subsistence: An input–output analysis. In: *Environment and Cultural Behaviour*. Ed: Vayda, A. P. New York: Natural History Press.

Lee, R. B. (1972*a*). Population growth and the beginning of sedentary life among the !Kung Bushmen. In: *Population Growth: Anthropological Implications*. Ed: Spooner, B. Cambridge, Massachusetts: MIT Press.

Lee, R. B. (1972*b*). !Kung spatial organization: an ecological and historical perspective. *Human Ecology* **1**, 125–47.

Lewin, T., Jurgens, H. W. & Louekari, L. (1970). Secular trend in the adult height of Skolt Lapps. *Arctic Anthropol.* **7**, 53–62.

Lewis, H. E.,. Masterton, J. P. & Rosenbaum, S. (1960). Body weight and skinfold thickness of men on a polar expedition. *Clin. Sci.* **19**, 551–61.

Lewis, P. R. & Lobban, M. C. (1957). Dissociation of diurnal rhythms in human subjects living on abnormal time routines. *Quart J. Exp. Physiol.* **42**, 371–86.

Li, C. C. (1961). *Human Genetics, Principles and Methods*. New York: McGraw-Hill.

Lobban, M. C. (1965). Time, light and diurnal rhythms. In: *The Physiology of Human Survival*. Ed: Edholm, O. G. & Bacharach, A. L. London, UK: Academic Press.

Lobban, M. C. (1967). Daily rhythms of renal excretion in arctic-dwelling Indians and Eskimos. *Quart. J. Exp. Physiol.* **52**, 401–10.

Lobban, M. C. (1969). Human renal diurnal rhythms at the equator. *J. Physiol.* **204**, 133–4P.

Lobban, M. C. (1976). Seasonal variations in daily patterns of urinary secretion in Eskimo subjects. In: *Circumpolar Health*. Ed: Shephard, R. J. & Itoh, S. Toronto, Ontario: University of Toronto Press.

Lowenstein, F. W. & Connell, D. E. (1974). Selected body measurements in boys ages 6–11 years from six villages in Southern Tunisia: An international comparison. *Human Biol.* **46**, 471–82.

Lumholtz, C. (1902). *Unknown Mexico*. New York: Scribner.

Máček, M., Vávra, J. & Zika, K. (1970). The comparison of the W_{170} values during growth. In: *Physical Fitness*. Ed: Seliger, V. Prague, Czechoslovakia: Charles University Press.

MacKinnon, M. J. (1923). Education in its relation to the physical and mental development of European children of school age in Kenya. *J. Trop. Med. Hyg.* **26**, 136–40.

MacMillan, M. G., Reid, C, M., Shirling, D. & Passmore, R. (1965). Body composition, resting oxygen consumption, and urinary creatinine in Edinburgh students. *Lancet* (i) 728–9.

MacNab, R. B. J. & Conger, P. (1966). Observations on the use of the Åstrand–Rhyming nomogram in University women. Paper presented to American College of Sports Medicine, Madison, Wisconsin.

MacPherson, R. K. (1966). Physiological adaptation, fitness, and nutrition in the peoples of the Australian and New Guinea regions. In: *The Biology of Human Adaptability*. Ed: Baker, P. T. & Weiner, J. S. Oxford, UK: Clarendon Press.

Magel, J. R. (1971). Comparison of the physiological response to varying intensities of submaximal work in tethered swimming and treadmill running. *J. Sports Med. Fitness* **11**, 203–12.

Magel, J. R. & Faulkner, J. F. (1967). Maximum oxygen uptake of college swimmers. *J. Appl. Physiol.* **22**, 929–38.

Magel, J. R., McArdle, W. D. & Glaser, R. M. (1969). Telemetered heart rate response to selected competitive swimming events. *J. Appl. Physiol.* **26**, 764–70.

Malhotra, M. S. (1966). People of India, including primitive tribes – a survey on physiological adaptation, physical fitness and nutrition. In: *The Biology of Human Adaptability*. Ed: Baker, P. T. & Weiner, J. S. Oxford, UK: Clarendon Press.

Malhotra, M. S., Ramaswamy, S. S., Joseph, N. T. & Gupta, J. S. (1972*a*). Functional capacity and body composition of different classes of Indian athletes. *Ind. J. Physiol. Pharm.* **16**, 301–8.

Malhotra, M. S., Ramaswamy, S. S., Joseph, N. T. & Gupta, J. S. (1972*b*). Physiological assessment of Indian athletes. *Ind. J. Physiol. Pharm.* **16**, 55–62.

Maritz, J. S., Morrison, J. F., Peter, J., Strydom, N. B. & Wyndham, C. H. (1961). A practical method of estimating an individual's maximum oxygen intake. *Ergonomics* **4**, 97–122.

Masironi, R. (1971). Development of a multi-dial wrist-watch pulse counter. Personal communication to author.

Masironi, R. (1974). Determination of habitual physical activity in WHO study. *ISC Symposium on Daily Energy Expenditure, Bad Kissingen, West Germany*.

Massie, J. & Shephard, R. J. (1971). Physiological and psychological effects of training. *Med. Sci. Sports* **3**, 110–17.

Matsui, H. (1968). Physical fitness of city school children. *JIBP-HA* (*Work Capacity*) *Report* **2**, 26–34, Tokyo.

Matsui, H., Miyashita, M., Miura, M., Amano, K., Mizutani, S., Hoshikawa, T., Toyoshima, S. & Kamei, S. (1971). Aerobic work capacity of Japanese adolescents. *J. Sports Med. Phys. Fitness* **11**, 28–35.

Mayer, J. (1972). *Human Nutrition. Its Physiological, Medical and Social Aspects. A Series of Eighty-two Essays*. Springfield, Illinois: C. C. Thomas.

Maynard, J. (1976). Coronary heart disease risk factors in relation to urbanization in Alaskan Eskimo men. In: *Circumpolar Health*. Ed: Shephard, R. J. & Itoh, S. Toronto, Ontario: University of Toronto Press.

Mazess, R. B. (1969). Exercise performance at high altitude in Peru. *Fed. Proc.* **28**, 1301–6.
McArdle, W. D., Zwiren, L. & Magel, J. R. (1969). Validity of the post-exercise heart rate as a means of estimating heart rate during work of varying intensities. *Res. Quart.* **40**, 523–8.
McArdle, W. D., Glaser, R. M. & Magel, J. R. (1971). Metabolic and cardiorespiratory responses during free swimming and treadmill walking. *J. Appl. Physiol.* **30**, 733–8.
McArdle, W. D., Katch, F. I. & Pechar, G. S. (1973). Comparison of continuous and discontinuous treadmill and bicycle tests for max $\dot{V}_{O_2}$. *Med. Sci. Sports* **5**, 156–60.
McDonough, J. R. & Bruce, R. A. (1969). Maximal exercise testing in assessing cardiovascular function. In: Proceedings of the National Conference on Exercise in the Prevention, in the Evaluation and in the Treatment of Heart disease. *J. S. Carol. Med. Ass.* **65** (Suppl. **1**), 26–33.
McDowell, A. J., Tasker, A. D. & Sarhan, A. E. (1970). *Height and Weight of Children in the United States, India and the United Arab Republic.* Washington, DC: US National Center for Health Statistics, Series 3, **14**, 1–51.
McGregor, I. A., Billewicz, W. Z. & Thompson, A. M. (1961). Growth and mortality in children in an African village. *Brit. Med. J.* (ii), 1661–6.
Medalie, J. H. (1970). Current developments in the epidemiology of atherosclerosis in Israel. In: *Atherosclerosis. Proceedings of the Second International Symposium.* Ed: Jones, R. J. New York: Springer Verlag.
Medical Research Council (1960). *Physiological Responses to Hot Environments. MRC Special Report* **298**. London, UK: HMSO.
Medved, R. (1966). Body height and predisposition for certain sports. *J. Sports Med. Fitness* **6**, 89–91.
Mellerowicz, H. (1962). *Ergometrie.* Munich, West Germany: Von Urban & Schwarzenburg.
Mellerowicz, H. (1966). Bericht über die Sitzung der Internationales Ergometrie – Standardisierungskommission. *Sportarzt und Sportmedizin* **8**, 405–8.
Mellerowicz, H. & Hansen, G. (1965). Sauerstoffkapazität und andere Spiroergometrische Maximalwerte der Ruder-Olympiasieger im Vierer mit Steuermann vom Berliner Ruderclub. *Sportarzt und Sport Medizin* **16**, 188–91.
Menier, D. R. & Pugh, L. G. C. E. (1968). The relation of oxygen intake and velocity of walking and running in competition walkers. *J. Physiol.* **197**, 717–21.
Meshizuka, T. & Nakanashi, M. (1972). A report on the results of the ICSPFT performance test applied to the people in Asian countries. In: *Proceedings of the ACSPFT and the ICSPFT – 1972.* Ed: Simri, V. Natanya, Israel: Wingate Institute.
Metayer, M. (1976). Eskimo personality and society, yesterday and today. In: *Circumpolar Health.* Ed: Shephard, R. J. & Itoh, S. Toronto, Ontario: University of Toronto Press.
Metheny, E. (1939). Some differences in bodily proportions between American Negro and White male college students as related to athletic performance. *Res. Quart.* **10**, 41–53.
Metivier, G. & Orban, W. A. R. (1971). *The Physical Fitness Performance and Work Capacity of Canadian Adults, Aged 18 to 44 Years.* Ottawa, Ontario: Canadian Association for Physical Health Education and Recreation.
Miall, W. E., Kass, E. H., Ling, J. & Stuart, K. L. (1962). Factors influencing

arterial blood pressure in the general population of Jamaica. *Brit. Med. J.* (ii), 497–500.

Miall, W. E. & Oldham, P. D. (1963). The hereditary factor in arterial blood pressure. *Brit. Med. J.* (i), 75–80.

Michaunt, E., Niang, I. & Dan, V. (1972). La maturation osseuse pendant la période pubertaire. A propos de l'étude de 227 adolescents dakarois. *Ann. Radiol.* **15**, 767–79.

Milan, F. A. (1960). A study of the maintenance of thermal balance in the Eskimo. *USAF Arctic Aeromed. Lab. Tech. Rept. 60-40.* Ladd Air Force Base, Alaska.

Milan, F. A. & Rodahl, K. (1961). Calorie requirements of man in the Antarctic. *J. Nutr.* **75**, 152–6.

Milic Emili, G., Cerretelli, P., Petit, J. M. & Falconi, G. (1959). La consommation d'oxygène en fonction de l'intensité de l'exercice musculaire. *Arch. Int. Physiol. Biochim.* **67**, 10–14.

Millahn, H. P. & Helke, H. (1968). Über Beziehungen zwischen der Herzfrequenz während Arbeitsleistung und in der Erholungsphase in Abhängigkeit von der Leistung und der Erholungsdauer. *Int. Z. angew. Physiol.* **26**, 245–57.

Miller, G. J., Ashcroft, M. T., Swan, A. V. & Beadnell, H. M. S. G. (1970). Ethnic variation in forced expiratory volume and forced vital capacity of African and Indian adults in Guyana. *Amer. Rev. Resp. Dis.* **102**, 979–81.

Miller, G. J., Cotes, J. E., Hall, A. M., Salvosa, C. B. & Ashworth, A. (1972). Lung function and exercise performance of healthy Caribbean men and women of African ethnic origin. *Quart. J. Exp. Physiol.* **57**, 325–41.

Mills, J. N. (1973). *Biological Aspects of Circadian Rhythms.* London, UK: Plenum Press.

Minard, D., Belding, H. S. & Kingston, J. R. (1957). Prevention of heat casualties. *J. Amer. Med. Ass.* **165**, 1813–18.

Ministry of Education, Japan (1959). *Official Report of Motor Performance Test on School Children in 1959.*

Mitchell, J. H., Sproule, B. J. & Chapman, C. B. (1958). The physiological meaning of the maximum oxygen intake test. *J. Clin. Invest.* **37**, 538–47.

Mitrevski, P. J. (1969). Incomplete right bundle branch block (Lead V_1) in athletes. *Med. Sci. Sports* **1**, 152–5.

Miyashita, M., Hayashi, Y. & Furuhashi, H. (1970). Maximum oxygen intake of Japanese top swimmers. *J. Sports Med. Fitness* **10**, 211–16.

Mocellin, R. & Wasmund, U. (1973). Investigations on the influence of a running-training programme on the cardiovascular and motor performance capacity in 53 boys and girls of a second and third primary school class. In: *Pediatric Work Physiology. Proceedings of the Fourth International Symposium.* Ed: Bar-Or, O. Natanya, Israel: Wingate Institute.

Mocellin, R., Lindemann, H., Rutenfranz, J. & Sbresny, W. (1971*a*). Determination of W_{170} and maximal oxygen uptake in children by different methods. *Acta Paediatr. Scand. Suppl.* **217**, 13–16.

Mocellin, R., Rutenfranz, J. & Singer, R. (1971*b*). Zur Frage von Normwerten der Körperlichen Leistungsfähigkeit (W_{170}) in Kindes und Jugendalter. *Z. Kinderheilkd.* **110**, 140–65.

Montoye, H. J. (1971). Estimation of habitual activity by questionnaire and interview. *Amer. J. Clin. Nutr.* **24**, 1113–18.

Montoye, H. J., Van Huss, W. D., Olson, H. W., Pierson, W. O. & Hudec, A. J. (1957). *The Longevity and Morbidity of College Athletes.* Michigan State University: Phi Epsilon Kappa Fraternity.

Montoye, H. J., Epstein, F. H. & Kyelsberg, M. O. (1964). The measurement of body fatness: a study in a total community. *Amer. J. Clin. Nutr.* **16**, 417–27.

Moody, D. L., Kollias, J. & Buskirk, E. R. (1969). Evaluation of aerobic capacity in lean and obese women with four test procedures. *J. Sports Med. Fitness* **9**, 1–9.

Morris, J. N. (1951). Recent history of coronary disease. *Lancet* (i), 1–7.

Morse, M., Schultz, F. W. & Cassels, D. E. (1949). Relation of age to physiological responses of the older boy (10–17 years) to exercise. *J. Appl. Physiol.* **1**, 683–709.

Motulsky, A. G. (1960). Metabolic polymorphisms and the role of infectious diseases in human evolution. *Human Biol.* **32**, 28–62.

Mrzena, B. & Máček, M. (1974).; Methodological study of working capacity in preschool children. *Proceedings of the Sixth International Symposium on Pediatric Work Physiology, Seč, Czechoslovakia.*

Müller, E. A. (1950). Ein Leistungs – Pulsindex als Mass der Leistungsfähigkeit. *Arbeitsphysiologie* **14**, 271–84.

Munro, A. F. (1950). Basal metabolic rates and physical fitness scores of British and Indian males in the tropics. *J. Physiol.* **110**, 356–66.

Myhre, L. G. & Kessler, W. V. (1966). Body density and potassium 40 measurements as related to age. *J. Appl. Physiol.* **21**, 1251–5.

Nagle, F. J., Balke, B. & Naughton, J. P. (1965). Gradational step tests for assessing work capacity. *J. Appl. Physiol.* **20**, 745–8.

Nakagawa, A. & Ishiko, T. (1970). Assessment of aerobic capacity with special reference to sex and age of junior and senior high school students in Japan. *Jap. J. Physiol.* **20**, 118–29.

Nakanashi, M., Tsukagoshi, K., Yorikane, Y. & Aoki, J. (1966). Study on the measurement of maximum O_2 intake and maximum O_2 debt. In: *Proceedings of International Congress of Sports Sciences, 1964.* Ed: Kato, K. Tokyo: Japanese Union of Sport Sciences.

Nansen, F. (1892). *The First Crossing of Greenland.* London, UK: Longmans Green. Cited by Edholm, O. G. & Lewis, H. E. (1964).

Nansen, F. (1898). *Farthest North.* Vol. II, p. 7. London, UK: George Newnes. Cited by Edholm, O. G. & Lewis, H. E. (1964).

Ness, G. W., Cunningham, D. A., Eynon, R. B. & Shaw, D. B. (1974). Cardiopulmonary function in prospective competitive swimmers and their parents. *J. Appl. Physiol.* **37**, 27–31.

Newman, M. T. (1960). Adaptations in the physique of American Aborigines to nutritional factors. *Human Biol.* **32**, 288–313.

Newman, M. T. (1961). Biological adaptation of man to his environment: Heat, cold, altitude and nutrition. *Ann. N.Y. Acad. Sci.* **91**, 617–33.

Niinimaa, V. M. J., Wright, G., Clarke, A. J., Clapp, J. & Shephard, R. J. (1974). Physiological profile of competitive dinghy sailors. *Proceedings, Canadian Association of Sports Sciences, Edmonton, Alberta.*

Niu, H., Ito, K., Takagi, K. & Ito, M. (1968). A study of the development of cardio-respiratory function of oarsmen. In: *Proceedings of International Congress of Sports Sciences, 1964.* Ed: Kato, K. Tokyo: Japanese Union of Sport Sciences.

Norman, J. N. (1960). Man in the Antarctic. M.D. Thesis, Glasgow University.

Norman, J. N. (1962). Micro-climate of man in Antarctica. *J. Physiol.* **160**, 27–8P.

Norris, A. H., Lundy, T. & Shock, N. W. (1963). Trends in selected indices of body composition in men between the ages 30 and 80 years. *Ann. N.Y. Acad. Sci.* **110**, 623–39.

Novak, L. P. (1963). Age and sex differences in body density and creatinine excretion of high-school children. *Ann. N.Y. Acad. Sci.* **110**, 545–77.

Novak, L. P. (1970). Comparative study of body composition of American and Filipino women. *Human Biol.* **42**, 206–16.

Novak, L. P., Hyatt, R. E. & Alexander, J. F. (1968). Body composition and physiologic function of athletes. *J. Amer. Med. Assoc.* **205**, 764–70.

Nowacki, P., Krause, R. & Adam, K. (1967). Maximal oxygen uptake by the rowing crew winning the Olympic Gold Medal 1968. *Pflüg. Arch.* **312**, 66–7.

Oja, P., Partanen, T. & Teräslinna, P. (1970). The validity of three indirect methods of measuring oxygen uptake and physical fitness. *J. Sports Med. Fitness* **10**, 67–71.

Olavi, E., Hirvonen, L., Peltonen, T. & Välimäki, I. (1965). Physical working capacity of normal and diabetic children. *Ann. Pediat. Fenn.* **11**, 25–31.

Oscherwitz, M., Edlavitch, S. A., Baker, T. R. & Jarboe, T. (1971). Differences in pulmonary functions in various racial groups. *Amer. J. Epidemiol.* **96**, 319–27.

Oseid, S. & Hermansen, L. (1973). Evaluation of physical work capacity in Norwegian female gymnasts. *Brit. J. Sports Med.* **7**, 38.

Oseid, S., Hövde, R., Osnes, J. B. & Hermansen, L. (1969). Circulatory responses to prolonged exercise in pre-pubertal boys. *Acta Physiol. Scand. Suppl.* **330**.

Pales, L. (1950). *Les Sels Alimentaires: Sel Mineraux.* Dakar, Senegal: Gouvern. Gen. de l'A.O.F. Cited by Ladell (1964).

Pařízková, J. (1961). Total body fat and skinfold thicknesses in children. *Metabolism* **10**, 794–807.

Pařízková, J. (1963). The impact of age, diet, and exercise on man's body composition. *Ann. N.Y. Acad. Sci.* **110**, 661–74.

Pařízková, J. (1974). Physical anthropology and nutritional status. In: *Nutrition and Malnutrition. Proceedings of the Burg Wartenstein Symposium 60.* Ed: Roche, A. & Faulkner, F. New York: Plenum Press.

Pařízková, J., Merhautová, J. & Prokopec, M. (1972). Comparaison entre la croissance des jeunes Tunisiens et celle des jeunes Tcheques agés de 11 et 12 ans. *Biometre Hum. (Paris)* **7**, 1–10.

Park, C. B. (1972). Studies on the physical fitness of Korean boys and girls. In: *Proceedings of the ACSPFT and the ICSPFT – 1972.* Ed: Simri, U. Natanya, Israel: Wingate Institute.

Pascale, L. R., Grossman, M. I., Sloane, H. S. & Frankel, T. (1956). Correlations between thickness of skinfolds and body density in 88 soldiers. *Human Biol.* **28**, 165–76.

Patrick, J. M. & Cotes, J. E. (1971). Cardiac determinants of aerobic capacity in New Guineans. In: *Physical Fitness.* Ed: Seliger, V. Prague, Czechoslovakia: Charles University Press.

Paul, R., Fletcher, G. H. & Addison, G. (1960). A comparative study between Europeans and Africans in the mining industry of Northern Rhodesia. *Med. Proc.* **6**, 69ff.

Peart, A. F. W. & Nagler; F. P. (1954). Measles in the Canadian Arctic, 1952. *Can. J. Publ. Health* **45**, 146–56.

Phillips, P. G. (1954). Metabolic cost of common West African agricultural activities. *J. Trop. Med. Hyg.* **57**, 12–20.

Phillips, W. H. & Ross, W. D. (1967). Timing error in determining maximal oxygen uptake. *Research Quart.* **38**, 315–16.

Pinto, I. J., Thomas, P., Colaco, F. & Datey, K. K. (1970). Current developments in India. In: *Atherosclerosis. Proceedings of the Second International Symposium.* Ed: Jones, R. J. New York: Springer Verlag.

Pirnay, F., Petit, J. M., Bottin, R., Deroanne, R., Juchmes, J. & Belge, G. (1966). Comparaison de deux méthodes de mesure de la consommation maximum d'oxygène. *Int. Z. angew. Physiol.* **23**, 203–11.

Pirnay, F., Petit, J. M. & Deroanne, R. (1969). Consommation maximum d'oxygène et température corporelle. *J. Physiol. (Paris) Suppl.* **2**, 376.

Piwonka, R. W., Robinson, S., Gay, V. L. & Manalis, R. S. (1965). Preacclimatization of men to heat by training. *J. Appl. Physiol.* **20**, 379–84.

Placheta, Z., Dražil, V., Rouš, J., Kočner, K. & Matějková, I. (1973). Basic values of the physical fitness of top cyclists. In: *Cycling and Health.* Ed: Rouš, J. Prague: Czechoslovak Sports Organization.

Ponthieux, N. A. & Barker, D. G. (1965). Relationships between race and physical fitness. *Res. Quart.* **36**, 468–72.

President's Council on Physical Fitness and Sports (1973). National adult physical fitness survey. Newsletter – Special Edition, May 1973, pp. 1–27.

Prior, I. A. M. & Evans, J. G. (1970). Current developments in the Pacific. In: *Atherosclerosis. Proceedings of the Second International Symposium.* Ed: Jones, R. J. New York: Springer Verlag.

Provis, H. S. & Ellis, R. W. B. (1955). An anthropometric study of Edinburgh school children. *Arch. Dis. Childh.* **30**, 328–37.

Pugh, L. G. C. E. (1971). Deaths from exposure on Four Inns Walking Competition, March 14–15, 1964. In: *Exercise and Cardiac Death*, pp. 112–20. Ed: Jokl, E. & McClellan, J. T. Baltimore, Maryland: University Park Press.

Pugh, L. G. C. E., Edholm, O. G., Fox, R. H., Wolff, H. S., Harvey, G. R., Hammond, W. H., Tanner, J. M. & Whitehouse, R. H. (1960). A physiological study of channel swimming. *Clin. Sci.* **19**, 257–73.

Rabinowitch, I. M. (1936). Clinical and other observations on Canadian Eskimos in the Eastern Arctic. *Canad. Med. Ass. J.* **34**, 487–501.

Ramaswamy, S. S., Malhotra, M. S., Sen Gupta, J. & Dua, G. L. (1966). Comparison of physical work performance between natives and newcomers at high altitude. In: *Proceedings of Symposium on Human Adaptability to Environments and Physical Fitness.* Ed: Malhotra, M. S. Madras, India: Defence Institute of Physiology and Allied Sciences.

Rathbun, E. N. & Pace, N. (1945). Studies on body composition. I. Determination of body fat by means of the body specific gravity. *J. Biol. Chem.* **158**, 667–76.

Reed, R. B. & Stuart, H. C. (1959). Patterns of growth in height and weight from birth to eighteen years of age. *Pediatrics* **24**, 904–21.

Renburn, E. T. (1972). *Materials and Clothing in Health and Disease.* London, UK: H. L. Lewis.

Rennie, D. W. (1963). Comparison of non-acclimatized Americans and Alaskan Eskimos. *Fed. Proc.* **22**, 828–30.

Rennie, D. W. & Adams, T. (1957). Comparative thermoregulatory responses of Negroes and White persons to acute cold stress. *J. Appl. Physiol.* **11**, 201–4.

Rennie, D. W., Covino, B. G., Blair, M. R. & Rodahl, K. (1962). Physical regulation of temperature in Eskimos. *J. Appl. Physiol.* **17**, 326–32.

Rennie, D. W., Di Prampero, P., Fitts, R. W. & Sinclair, L. (1970). Physical fitness and respiratory function of Eskimos of Wainwright, Alaska. *Arctic Anthropol.* **7**, 73–82.

Rennie, D. W., Prendergast, D. R., Di Prampero, P., Wilson, D., Lanphier, E. H. & Myers, C. R. (1973). Energetics of the overarm crawl. *Med. Sci. Sports* **5**, 65.

Ribisl, P. M. & Herbert, W. G. (1970). Effects of rapid weight reduction and subsequent rehydration upon the physical working capacity of wrestlers. *Res. Quart.* **41**, 536–41.

Roberts, D. F. (1953). Body weight, races and climate. *Amer. J. Phys. Anthrop.* **11**, 533–58.
Roberts, D. F. (1970). Genetic problems of hot desert populations of simple technology. *Human Biol.* **42**, 469–85.
Robinson, S. (1938). Experimental studies in physical fitness in relation to age. *Arbeitsphysiologie* **10**, 251–323.
Robinson, S. (1952). Physiological effects of heat and cold. *Ann. Rev. Physiol.* **14**, 73–96.
Robinson, S., Edwards, H. T. & Dill, D. B. (1937). New records in human power. *Science* **85**, 409.
Robinson, S., Dill, D. B., Harmon, P. M., Hall, F. G. & Wilson, J. W. (1941). Adaptation to exercise of Negro and White share-croppers in comparison with Northern Whites. *Human Biol.* **13**, 139–58.
Robson, J. R. K. (1964). Skinfold thickness in apparently normal African adolescents. *J. Trop. Med. Hyg.* **67**, 209–10.
Robson, J. R. K., Bazin, M. & Soderstrom, R. (1971). Ethnic differences in skinfold thickness. *Amer. J. Clin. Nutr.* **24**, 864–8.
Roche, A. F. & Cahn, A. (1962). Subcutaneous fat thickness and caloric intake in Melbourne children. *Med. J. Australia* (i), 595–7.
Rodahl, K. (1958). Physical fitness. *J. Amer. Geriatr. Soc.* **6**, 205–9.
Rodahl, K., Åstrand, P. O., Birkhead, N. C., Hettinger, T., Issekutz, B., Jones, D. M. & Weaver, R. (1961). Physical work capacity. A study of some children and young adults in the United States. *AMA Arch. Env. Health.* **2**, 499–510.
Rode, A. & Shephard, R. J. (1971). Cardio-respiratory fitness of an Arctic community. *J. Appl. Physiol.* **31**, 519–26.
Rode, A. & Shephard, R. J. (1973*a*). Pulmonary function of Canadian Eskimos. *Scand. J. Resp. Dis.* **54**, 191–205.
Rode, A. & Shephard, R. J. (1973*b*). Fitness of the Canadian Eskimo – the influence of season. *Med. Sci. Sports* **5**, 170–3.
Rode, A. & Shephard, R. J. (1973*c*). On the mode of exercise appropriate to an Arctic community. *Int. Z. angew. Physiol.* **31**, 187–96.
Rode, A. & Shephard, R. J. (1973*d*). Growth, development and fitness of the Canadian Eskimo. *Med. Sci. Sports* **5**, 161–9.
Rose, G. A. & Blackburn, H. (1968). *Cardiovascular Survey Methods.* Geneva, Switzerland: World Health Organization.
Roskamm, H. (1973). Limits and age dependency in the adaptation of the heart to physical stress. In: *Sport in the Modern World – Chances and Problems.* Ed: Grupe, O., Kurz, D. & Teipel, J. M. Berlin, Germany: Springer Verlag.
Rossiter, C. E. & Weill, H. (1974). Ethnic differences in lung function: evidence for proportional differences. *Int. J. Epidemiol.* **3**, 55–61.
Rouš, J. (1973). *Cycling and Health.* Prague: Czechoslovakia Sports Organization.
Rowell, L. B. (1974). Human cardiovascular adjustments to exercise and thermal stress. *Physiol. Rev.* **54**, 75–159.
Rowell, L. B., Taylor, H. L. & Wang, Y. (1964). Limitations to prediction of maximal oxygen intake. *J. Appl. Physiol.* **19**, 919–27.
Royce, J. (1959). Isometric fatigue curves in human muscle with normal and occluded circulation. *Res. Quart.* **29**, 204–12.
Rutenfranz, J. (1964). *Entwicklung und Beurteilung der Körperlichen Leistungs fähigkeit bei Kindern und Jugendlichen.* Basel, Switzerland: Karger.
Rutenfranz, J. & Mocellin, R. (1967). Investigations on children and youths regarding the relationships between various parameters of the physical

development and the working capacity. *Second International Seminar for Ergometry, Berlin.*

Ryhming, I. (1954). A modified Harvard step test for the evaluation of physical fitness. *Arbeitsphysiologie* **15**, 235–50.

Sagild, U. (1972). Personal communication, cited by Bang & Dyerberg (1972).

Saltin, B. (1964). Circulatory response to submaximal and maximal exercise after thermal dehydration. *J. Appl. Physiol.* **19**, 1125–32.

Saltin, B. (1973). Oxygen transport by the circulatory system during exercise. In: *Limiting Factors of Physical Performance.* Ed: Keul, J. Stuttgart, West Germany: Thieme.

Saltin, B. & Åstrand, P. O. (1967). Maximal oxygen uptake in athletes. *J. Appl. Physiol.* **23**, 353–8.

Saltin, B. & Grimby, G. (1968). Physiological analysis of middle-aged and old former athletes: comparisons with still active athletes of the same ages. *Circulation* **38**, 1104–15.

Saltin, B. & Karlsson, J. (1973). Muscle glycogen utilization during work of different intensities. In: *Muscle Metabolism During Exercise.* Ed: Pernow, B. & Saltin, B. New York: Plenum Press.

Samueloff, S., Davies, C. T. M. & Schvartz, E. (1973). The physical working capacity of Yemenite and Kurdish Jews in Israel. *Phil. Trans. Roy. Soc. Lond. B* **266**, 141–7.

Samuelsson, G. (1971). An epidemiological study of child health and nutrition in a Northern Swedish county. I. Food consumption survey. *Acta Paediatr. Scand. Suppl.* **214**.

Šaric, M. & Palaić, S. (1971). The prevalence of respiratory symptoms in a group of miners and the relationship between symptoms and some functional parameters, pp. 863–71. In: *Inhaled Particles.* Ed: Walton, W. H. London, UK: Unwin.

Sauberlich, H. E., Goad, W., Herman, Y. F., Milan, F. & Jamison, P. (1970). Preliminary report on the nutrition survey conducted among the Eskimos of Wainwright, Alaska, 21–27 January, 1969. *Arctic Anthropol.* **7**, 122–4.

Sayed, J., Schaefer, O., Hildes, J. A. & Lobban, M. A. (1976*a*). Biochemical indices of nutrient intake by Eskimos of Northern Foxe Basin, N.W.T. In: *Circumpolar Health.* Ed: Shephard, R. J. & Itoh, S. Toronto, Ontario: University of Toronto Press.

Sayed, J. E., Hildes, J. A. & Schaefer, O. (1976*b*). Feeding practices and growth of Igloolik infants. In: *Circumpolar Health.* Ed: Shephard, R. J. & Itoh, S. Toronto, Ontario: University of Toronto Press.

Scano, A. & Venerando, A. (1968). *Studi sulla Acclimatazione degli Atleti Italiani a Citta del Messico.* Rome, Italy: Institutio di Medicina dello Sport.

Schaefer, O., Hildes, J. A., Greidanus, P. & Leung, D. (1976). Regional sweating in Eskimos compared to Caucasians. In: *Circumpolar Health.* Ed: Shephard, R. J. & Itoh, S. Toronto, Ontario: University of Toronto Press.

Schmidt, C. F. & Comroe, J. H. (1944). Dyspnea. *Mod. Concepts Cardiovasc. Dis.* **13** (3).

Schreider, E. (1957). Ecological rules and body-heat regulation in man. *Nature* **179**, 915–16.

Schwartz, R. A. (1973). Oxygen consumption during rowing: implications for training. *Brit. J. Sports Med.* **7**, 188.

Scott, R. F., Likimani, J. C., Morrison, E. S., Thuku, J. J. & Thomas, W. A. (1963). Esterified serum fatty acids in subjects eating high and low cholesterol diets. A comparative study of serum lipid metabolism in New Yorkers,

indigenous poor East Africans and upper class East Africans. *Amer. J. Clin. Nutr.* **13**, 82–91.
Seliger, V. (1966). Circulatory responses to sports activities. In: *Physical Activity in Health and Disease.* Ed: Evang, K. & Andersen, K. L. Baltimore, Maryland: Williams & Wilkins.
Seliger, V. (1970). Physical fitness of Czechoslovak children at 12 and 15 years of age. International Biological Programme Results of Investigations 1968–1969. *Acta Univ. Carol. Gymnica* **5**, 6–169.
Seliger, V. & Pachlopniková, I. (1967). I. Vergleichende ergometrische Untersuchungen bei den Gymnastikturnern. In: *Second International Seminar for Ergometry, Berlin.* Ed: Mellerowicz, H. & Hansen, G. Berlin, Germany: Institüt für Leistungs-medizin.
Seliger, V., Pachlopinková, I., Mann, M., Selecká, R. & Treml, J. Energy expenditure during paddling. *Physiol. Bohem.* **18**, 49–55.
Seliger, V., Navara, M. & Pachlopniková, I. (1970). Der energetische Metabolismus in Verlauf des Fussballspiels. *Sportarzt und Sportmedizin* **54**, 114–18.
Seliger, V., Kostka, V., Grusova, D., Kovac, J., Machovcová, J., Pauer, M., Pribylová, A. & Urbánková, R. (1972). Energy expenditure and physical fitness of ice-hockey players. *Int. Z. angew. Physiol.* **30**, 283–91.
Seliger, V., Bartuněk, Zd. & Trefný, Zd. (1974). Comparison of the habitual activity in two groups of boys. *Proceedings of the Sixth International Symposium on Pediatric Work Physiology, Seč, Czechoslovakia.*
Sellers, F. J., Wood, W. J. & Hildes, J. A. (1959). The incidence of anemia in infancy and early childhood among Central Arctic Eskimos. *Canad. Med. Ass. J.* **81**, 656–7.
Sempé, M. (1964). Surveillance de la croissance de l'enfant. *Conc. Méd.* **43** (Suppl.), 52.
Serfass, R. C. (1971). Changes in cardio-respiratory fitness and body composition of participants in selected physical education classes. Ph.D. Thesis, University of Minnesota. Cited by Kollias *et al.* (1972).
Shaper, A. G., Jones, K. W., Jones, M. & Kyobe, J. (1963). Serum lipids in three nomadic tribes of Northern Kenya. *Amer. J. Clin. Nutr.* **13**, 135–46.
Shephard, R. J. (1966*a*). The relative merits of the step test, bicycle ergometer, and treadmill in the assessment of cardio-respiratory fitness. *Int. Z. angew. Physiol.* **23**, 219–30.
Shephard, R. J. (1966*b*). Initial 'fitness' and personality as determinants of the response to a training regime. *Ergonomics* **9**, 3–16.
Shephard, R. J. (1966*c*). On the timing of post-exercise pulse readings. *J. Sports Med. Fitness* **6**, 23–7.
Shephard, R. J. (1966*d*). World standards of cardio-respiratory performance. *AMA Arch. Env. Health* **13**, 664–72.
Shephard, R. J. (1967*a*). The prediction of 'maximal' oxygen consumption using a new progressive step test. *Ergonomics* **10**, 1–15.
Shephard, R. J. (1967*b*). Normal levels of activity in Canadian city dwellers. *Canad. Med. Ass. J.* **96**, 912–14.
Shephard, R. J. (1967*c*). Ethical considerations in human experimentation. *J. Canad. Assoc. Health, Phys. Ed., Rec.* **33**, 13–16.
Shephard, R. J. (1967*d*). The prediction of maximum oxygen intake from post-exercise pulse readings. *Int. Z. angew. Physiol.* **24**, 31–8.
Shephard, R. J. (1969*a*). *Endurance Fitness.* Toronto, Ontario: University of Toronto Press. (Second edition, 1977.)

Shephard, R. J. (1969*b*). Learning, habituation, and training. *Int. Z. angew. Physiol.* **28**, 38–48.
Shephard, R. J. (1969*c*). A nomogram to calculate the oxygen cost of running at slow speeds. *J. Sports Med. Fitness* **9**, 10–16.
Shephard, R. J. (1970). For exercise testing, please. A review of procedures available to the clinician. *Bull. Physio. Path. Resp.* **6**, 425–74.
Shephard, R. J. (1971*a*). The oxygen conductance equation. In: *Frontiers of Fitness*. Ed: Shephard, R. J. Springfield, Illinois: C. C. Thomas.
Shephard, R. J. (1971*b*). Standard tests of aerobic power. In: *Frontiers of Fitness*. Ed: Shephard, R. J. Springfield, Illinois: C. C. Thomas.
Shephard, R. J. (1971*c*). The working capacity of school-children. In: *Frontiers of Fitness*. Ed: Shephard, R. J. Springfield, Illinois: C. C. Thomas.
Shephard, R. J. (1971*d*). Prediction formulas and normal values for lung volumes: Man. In: *Biological Handbooks – Respiration and Circulation*. Ed: Altman, P. & Dittmer, D. S. Bethesda, Maryland: Federation of American Societies for Experimental Biology.
Shephard, R. J. (1972*a*). Exercise and the lungs. In: *Fitness and Exercise*. Ed: Alexander, J. F., Serfass, R. C. & Tipton, C. M., Chicago, Illinois: Athletic Institute.
Shephard, R. J. (1972*b*). *Alive, Man – The Physiology of Physical Activity*. Springfield, Illinois: C. C. Thomas.
Shephard, R. J. (1973*a*). Exercise prescription and the risk of sudden death. *Can. Fam. Physician*, August, 57–60.
Shephard, R. J. (1973*b*). The age factor in sport. In: *Sport in the Modern World – Chances and Problems*. Ed: Grupe, O., Kurz, D. & Teipel, J. M. Berlin, Germany: Springer Verlag.
Shephard, R. J. (1974*a*). A new look at aerobic power. *International Committee on the Standardization of Physical Fitness Tests, Jerusalem.*
Shephard, R. J. (1974*b*). What causes 'second wind'? *The Physician in Sports Medicine* **2** (11), 36–42.
Shephard, R. J. (1974*c*). Work physiology and activity patterns of circumpolar Eskimos and Ainu. A synthesis of IBP data. *Human Biol.* **46**, 263–94.
Shephard, R. J. (1975*a*). Work physiology and activity patterns. In: IBP Synthesis Volume: Circumpolar Peoples. Ed: Milan, F. London, UK: Cambridge University Press.
Shephard, R. J. (1975*b*). Specificity versus generality: the training of the swimmer. *Med. del Sport* **28**, 89–98.
Shephard, R. J. (1975*c*). On the prediction of athletic performance. *Proceedings, Pan American Congress of Sports Medicine, Sao Paulo.*
Shephard, R. J. (1976). Environment. In: *Sports Medicine* (2nd edition), pp. 76–84. Ed: Williams, J. G. P. & Sperryn, P. N. London, UK: Arnold
Shephard, R. J. & Andersen, K. L. (1971). IBP Workshop on fitness of traditional communities. In: *Physical Fitness*. Ed: Seliger, V. Prague, Czechoslovakia: Charles University Press.
Shephard, R. J. & Callaway, S. (1966). Principal component analysis of the responses to standard exercise training. *Ergonomics* **9**, 141–54.
Shephard, R. J. & Godin, G. (1976). Energy balance of an Eskimo community. In: *Circumpolar Health*. Ed: Shephard, R. J. & Itoh, S. Toronto, Ontario: University of Toronto Press.
Shephard, R. J. & Itoh, S. (1976). Circumpolar Health. Toronto, Ontario: University of Toronto Press.

Shephard, R. J. & Olbrecht, A. J. (1970). Body weight and the estimation of working capacity. *S. Afr. Med. J.* **44**, 296–8.

Shephard, R. J. & Rode, A. (1973). Fitness for arctic life: the cardio-respiratory status of the Canadian Eskimo. In: *Polar Human Biology*. Ed: Edholm, O. G. & Gunderson, E. K. E. London, UK: Heinemann.

Shephard, R. J. & Rode, A. (1975). *Growth and Development in the Eskimo. Proceedings of Second Canadian Workshop on Child Growth and Development, Saskatoon. Na'pão* **5**, 20–6. (University of Saskatoon Dept. of Anthropology.)

Shephard, R. J., Allen, C., Benade, A. J. S., Davies, C. T. M., Di Prampero, P. E., Hedman, R., Merriman, J. E., Myhre, K. & Simmons, R. (1968*a* & *b*). The maximum oxygen intake – an international reference standard of cardio-respiratory fitness. *Bull. Wld Health Org.* **38**, 757–64. Standardization of sub-maximal exercise tests. *Bull. Wld Health Org.* **38**, 765–76.

Shephard, R. J., Allen, C., Bar-Or, O., Davies, C. T. M., Degré, S., Hedman, R., Ishii, K., Kaneko, M., La Cour, J. R., Di Prampero, P. E. & Seliger, V. (1968*c*). The working capacity of Toronto school-children. *Canad. Med. Ass. J.* **100**, 560–6, 705–14.

Shephard, R. J., Weese, C. H. & Merriman, J. E. (1971). Prediction of maximal oxygen intake from anthropometric data – some observations on pre-adolescent school children. *Int. Z. angew. Physiol.* **29**, 119–30.

Shephard, R. J., Godin, G. & Campbell, R. (1973*a*). Characteristics of sprint, medium and middle-distance swimmers. *Int. Z. angew. Physiol.* **32**, 1–19.

Shephard, R. J., Hatcher, J. & Rode, A. (1973*b*). On the body composition of the Eskimo. *Europ. J. Appl. Physiol.* **30**, 1–13.

Shephard, R. J., Lavallée, H., Larivière, G., Rajic, M., Brisson, G., Beaucage, C., Jéquier, J. C. & LaBarre, R. (1974*a*). La capacité physique des enfants canadiens: une comparaison entre les enfants canadiens–français, canadiens–anglais et esquimaux. I. Consommation maximale d'oxygène et débit cardiaque. *Union Méd.* **103**, 1767–77.

Shephard, R. J., Lavallée, H., Rajic, K. M., Jéquier, J. C., Brisson, G. & Beaucage, C. (1974*b*). Radiographic age in the interpretation of physiological and anthropological data. *Proceedings of the Sixth International Symposium on Pediatric Work Physiology, Seč, Czechoslovakia.*

Shephard, R. J., Lavallée, H., Larivière, G., Rajic, M., Brisson, G., Beaucage, C., Jéquier, J. C. & LaBarre, R. (1975*a*). La capacité physique des enfants canadiens: une comparaison entre les enfants canadiens français, canadien–anglais et esquimaux. II. Anthropométrie et volumes pulmonaire. *Union Méd.* **104**, 259–69.

Shephard, R. J., Lavallée, H., Rajic, M., Brisson, G., Beaucage, C., Perusse, M., Jéquier, J. C. & LaBarre, R. (1975*b*). La Capacité physique des enfants canadiens: une comparaison entre les enfants canadiens–français, canadiens–anglais et esquimaux. III. Facteurs psychologique et sociologique. *Union Méd.* **104**, 1131–6.

Sidney, K. (1973). Effects of frequency of training on the fitness of elderly men and woman. *Med. Sci. Sports* **5**, 63.

Sidney, K. H. & Shephard, R. J. (1973). Physiological characteristics and performance of the whitewater paddler. *Int. Z. angew. Physiol.* **32**, 55–70.

Sidney, K. H. & Shephard, R. J. (1977). Activity patterns of elderly men and women. *J. Gerontol.* **32**, 25–32.

Sidor, R. & Peters, J. M. (1973). Differences in ventilatory capacity of Irish and Italian fire-fighters. *Amer. Rev. Resp. Dis.* **108**, 669–71.

Sigurjonsson, J. (1969). Urban–rural differences in mortality from ischemic heart disease. *Amer. J. Med. Sci.* **257**, 253–8.

Simmons, R. & Shephard, R. J. (1971*a*). Measurement of cardiac output in maximum exercise. Application of an acetylene rebreathing method to arm and leg exercise. *Int. Z. angew. Physiol.* **29**, 159–72.

Simmons, R. & Shephard, R. J. (1971*b*). Effects of physical conditioning upon the central and peripheral circulatory responses to arm work. *Int. Z. angew. Physiol.* **30**, 73–84.

Simons, J., Beunen, G. & Renson, R. (1974). The Louvain boy's growth study. Preliminary report. University of Louvain.

Simonson, E. (1971). *The Physiology of Work Capacity and Fatigue.* Springfield, Illinois: C. C. Thomas.

Simpson, H. W. & Bohlen, J. G. (1973). Latitude and the human circadian system. In: *Biological Aspects of Circadian Rhythms.* Ed: Mills, J. N. London, U.K.: Plenum Press.

Simpson, N. E. & McAlpine, P. J. (1976). A comparison of genetic markers in the blood of circumpolar populations. In: *Circumpolar Health.* Ed: Shephard, R. J. & Itoh, S. Toronto, Ontario: University of Toronto Press.

Singh, H. D., Abraham, D. L. & Antony, N. J. (1970). Expiratory flow rates and timed expiratory capacities in South Indian men. *J. Ind. Med. Ass.* **54**, 412–15.

Singh, M., Howell, M. L. & MacNab, R. B. J. (1968). The strength and physical work capacity of the native Indian children of Alberta. *Proceedings, Seventeenth World Congress of Sports Medicine, Mexico City.*

Sinnett, P. (1972). Nutrition in a New Guinea highland community. *Human Biology of Oceania* **1**, 299–305.

Sinnett, P. F. & Solomon, A. (1968). Physical fitness in a New Guinea highland population. *Papua & New Guinea Med. J.* **11**, 56–9.

Sinnett, P. F. & Whyte, H. M. (1973*a*). Epidemiological studies in a total highland population, Tukisenta, New Guinea. Cardiovascular disease and relevant clinical, electrocardiographic, radiological and biochemical findings. *J. Chron. Dis.* **26**, 265–90.

Sinnett, P. F. & Whyte, H. M. (1973*b*). Epidemiological studies in a highland population of New Guinea: Environment, culture and health status. *Human Ecol.* **1**, 245–77.

Skinner, J. S., Hutsler, R., Bergsteinova, V. & Buskirk, E. R. (1973). The validity and reliability of a rating scale of perceived exertion. *Med. Sci. Sports* **5**, 94–6.

Škranc, O., Havel, V. & Barták, K. (1970). A comparison of work capacity measured by graded step-test and on a bicycle ergometer. *Ergonomics* **13**, 675–83.

Skrobak-Kaczynski, J. & Lewin, T. (1976). Secular changes in Lapps of Northern Finland. In: *Circumpolar Health.* Ed: Shephard, R. J. & Itoh, S. Toronto, Ontario: University of Toronto Press.

Sloan, A. W. (1966*a*). Physical fitness tests of Cape Town high-school children. *S. Afr. Med. J.* **40**, 682–7.

Sloan, A. W. (1966*b*). Physical fitness of South African compared with British and American high-school children. *S. Afr. Med. J.* **40**, 688–90.

Sloan, A. W. (1969). Physical fitness and body build of young men and women. *Ergonomics* **12**, 25–32.

Sloan, A. W. & Hansen, J. D. L. (1969). Nutrition and physical fitness of white, coloured, and Bantu high-school children. *S. Afr. Med. J.* **43**, 508–11.

Sloan, A. W., Burt, J. J. & Blyth, C. S. (1962). Estimation of body fat in young women. *J. Appl. Physiol.* **17**, 967–70.

Smit, C. M. (1961). *J. Soc. Res.* **12**, 1. Quoted by Sloan (1966*a*).

Smit, P. J. (1973). Physiological responses to exercise in white and coloured children from South Africa. In: *Pediatric Work Physiology. Proceedings of the Fourth International Symposium.* Ed: Bar-Or, O. Natanya, Israel: Wingate Institute.

Smodlaka, V. N. (1947). Physical development of soccer players in relation to their station in the team. *Fiz. Kult.* **398**.

Sobolova, V., Seliger, V., Grussova, D., Machovcova, J. & Zelenka, V. (1971). The influence of age and sports training in swimming on physical fitness. *Acta Paed. Scand. Suppl.* **217**, 63–7.

Society of Actuaries (1959). *Build and Blood Pressure Study.* Chicago, Illinois: Society of Actuaries.

Song, S. H., Kang, D. H., Kang, B. S. & Hong, S. K. (1963). Lung volumes and ventilatory responses to high CO_2 and low O_2 in the Ama. *J. Appl. Physiol.* **18**, 466–70.

Šprynarova, S. (1966). Development of the relationship between aerobic capacity and the circulatory and respiratory reaction to moderate activity in boys 11–13 years old. *Physiol. Bohem.* **15**, 253–64.

Šprynarova, S. & Pařízková, J. (1969). Comparison of the functional, circulatory and respiratory capacity in girl gymnasts and swimmers. *J. Sports Med. Phys. Fitness* **9**, 165–72.

Šprynarova, S. & Pařízková, J. (1971). Functional capacity and body composition in top weight-lifters, swimmers, runners, and skiers. *Int. Z. angew. Physiol.* **29**, 184–94.

Stanesçu, N. (1970). *Medical Researches in Football Game.* Bucharest, Rumania: Ministry for Health – Sports Medicine Centre.

Stefanik, P. A., Heald, F. P. & Mayer, J. (1959). Caloric intake in relation to energy output of obese and non-obese adolescent boys. *Amer. J. Clin. Nutr.* **7**, 55–62.

Stein, E. S., Rothstein, M. S. & Clements, C. J. (1967). *Calibration of Two Bicycle Ergometers used by the Health Examination Survey.* Washington, DC: National Center for Health Statistics **2**, 21.

Stenberg, J., Åstrand, P. O., Ekblöm, B., Royce, J. & Saltin, B. (1967). Hemodynamic response to work with different muscle groups, sitting and supine. *J. Appl. Physiol.* **22**, 61–70.

Steplock, D. A., Veicsteinas, A. & Mariani, M. (1971). Maximal aerobic and anaerobic power and stroke volume of the heart in a sub-alpine population. *Int. Z. angew. Physiol.* **29**, 203–14.

Stigler, R. (1952). Rassenphysiologische Ergebnisse meiner Forschungsreise in Uganda 1911–1912. *Oesterr. Akad. Wiss. Math-Natur. Kl. Sitzber,* **109** (section 1), 1–44. Cited by Ladell (1964).

Strydom, N. B., Wyndham, C. H. & Greyson, J. S. (1967). A scientific approach to the selection and training of oarsmen. *S. Afr. Med. J.* **41**, 1100–2.

Stuart, H. C. & Meredith, H. V. (1946). Use of body measurements in the school health program. *Amer. J. Publ. Health* **36**, 1365–73.

Suchý, J. (1971). Trend tělesného vývoje české mládeže ve 20. *Stol. Cas. Lek Ces.* **110**, 935ff. Cited by Pařízková, J. (1977) in: *Body Fat and Physical Fitness.* The Hague, Netherlands: Nijhoff.

Suzuki, S. (1970). Experimental studies on factors in growth. *Soc. Res. Child Developm., Washington, Monograph* **35**, 6–11.

Taguchi, S., Raven, P. B. & Horvath, S. M. (1971). Comparisons between bicycle ergometry and treadmill walking maximum capacity tests. *Jap. J. Physiol.* **21**, 681–90.

Tanner, J. M. (1962). *Growth at Adolescence.* Oxford, UK: Blackwell.

Tanner, J. M. (1964). *The Physique of the Olympic Athlete.* London, UK: Unwin.

Tanner,. J. M. & Whitehouse, R. H. (1962). Standards for subcutaneous fat in British children. Percentiles for thickness of skinfolds over triceps and below scapula. *Brit. Med. J.* (i), 446–50.

Tanner, J. M., Whitehouse, R. H. & Takaishi, M. (1966). Standards from birth to maturity for height, weight, height velocity and weight velocity, British children, 1965, Part II. *Arch. Dis. Child.* **41**, 613–35.

Tanner, J. M., Whitehouse, R. H., Marshall, W. A., Healy, M. J. R. & Goldstein, H. (1975). *Assessment of Skeletal Maturity and Prediction of Adult Height (TW2 Method).* London, UK: Academic Press.

Taylor, H. L., Buskirk, E. & Henschel, A. (1955). Maximal oxygen intake as an objective measure of cardio-respiratory performance. *J. Appl. Physiol.* **8**, 73–80.

Taylor, H. L., Haskell, W. L., Fox, S. M. & Blackburn, H. (1969). Exercise tests: a summary of procedures and concepts of stress testing for cardiovascular diagnosis and function evaluation. In: *Measurement in Exercise Electrocardiography.* Ed: Blackburn, H. Springfield, Illinois: C. C. Thomas.

Teräslinna, P., Ismail, A. H. & MacLeod, D. F. (1966). Nomogram by Åstrand and Ryhming as a prediction of maximum oxygen intake. *J. Appl. Physiol.* **21**, 513–15.

Thinkaran, T., Chan, O. L., Duncan, M. T. & Klissouras, V. (1974). Absence of the influence of ethnic origin on the maximal aerobic power of Malaysians. *World Congress of Sports Medicine, Melbourne, Australia.*

Thomson, M. L. (1954). Comparison between number and distribution of functioning sweat glands in Europeans and Africans. *J. Physiol.* **123**, 225–33.

Tobias, P. V. (1966). The peoples of Africa south of the Sahara. In: *The Biology of Human Adaptability.* Ed: Baker, P. T. & Weiner, J. S. Oxford, UK: Clarendon Press.

Topp, S. G., Cook, J., Holland, W. W. & Elliott, A. (1970). Influence of environmental factors on height and weight of school children. *Brit. J. Prev. Soc. Med.* **24**, 154–62.

US Department of Health, Education and Welfare (1964–8). *Current Estimates from the Health Interview Survey.* National Center for Health Statistics, Series 10, nos. 5, 13, 25, 37, 43. Washington, DC: Superintendent of Documents, US Government Printing Office.

van Graan, C. H. & Greyson, J. S. (1970). A comparison between the bicycle ergometer and the step-test for determining maximum oxygen intake on Kalahari bushmen. *Int. Z. angew. Physiol.* **28**, 344–8.

van Huss, W. D. & Cureton, T. K. (1955). Relationship of selected tests with energy metabolism and swimming performance. *Res. Quart.* **26**, 205–21

van Uytvanck, P. & Vrijens, J. (1966). Investigations about some body circumference measurements for the appreciation of physical fitness in adolescence. *J. Sports Med. Phys. Fitness* **6**, 176–82.

Vank, L. (1973). Somatische Charakteristik der Radrennfahrer in Vorbeitung an die Weltmeisterschafter 1969. In: *Cycling and Health.* Ed: Rouš, J. Prague: Czechoslovak Sports Organization.

Viteri, F. E., Torun, B., Galicia, J. C. & Herrera, E. (1971). Determining energy costs of agricultural activities by respirometer and energy balance techniques. *Amer. J. Clin. Nutr.* **24**, 1418–30.

Vines, A. P. (1967). Malaria control and haemoglobin. *Papua & New Guinea Med. J.* **10**, 47–50.

Voigt, E. D., Engel, P. & Klein, H. (1967). Tages rhythmische Schwankungen des Leistungs-pulsindex. (Daily fluctuations of the performance-pulse index.) *German Medical Monthly* **12**, 394–5.

von Döbeln, W. (1954). A simple bicycle ergometer. *J. Appl. Physiol* **7**, 222–4.

von Döbeln, W. (1966). Kroppsstorlek, Energiomsättning och Kondition. In: *Handbok i Ergonomi.* Ed: Luthman, G., Åberg, U. & Lundgren, N. Stockholm, Sweden: Almqvist & Wiksell.

von Döbeln, W. & Eriksson, B. O. (1973). Physical training, growth and maximal oxygen uptake of boys aged 11–13 years. In: *Pediatric Work Physiology. Proceedings of the Fourth International Symposium.* Ed: Bar-Or, O. Natanya, Israel: Wingate Institute.

von Döbeln, W., Åstrand, I. & Bergström, A. (1967). An analysis of age and other factors related to maximal oxygen uptake. *J. Appl. Physiol.* **22**, 934–8.

Vuori, I. (1974). Studies on the feasability of long-distance (20–90 km) ski-hikes as a mass sport. *Twentieth World Congress of Sports Medicine, Melbourne, Australia.*

Wadsworth, G. R. & Lee, T. S. (1960). The height, weight, and skinfold thickness of Muar school children. *J. Trop. Pediatr.* **6** (2), 48.

Wahlund, H. (1948). Determination of physical working capacity. *Acta Med. Scand.* **215**, Suppl. **9**, 1–127.

Webb, C. G. (1959). An analysis of some observations of thermal comfort in an equatorial climate. *Brit. J. Industr. Med.* **16**, 297–310.

Weidermann, H., Görmudt, L., Roskamm, H., Samek, L. & Reindell, H. (1968). Herzfrequenzmessungen mit EKG-Speichergeraten beim Weltmeisterschafts-training von Hochleistungsruderern. *Med. u. Sport* **8**, 85–9.

Weiner, J. S. (1950). Observations on the working ability of Bantu mine workers with reference to acclimatization to hot humid conditions. *Brit. J. Industr. Med.* **7**, 17–26.

Weiner, J. S. (1964). *Proposals for International Research. Human Adaptability Project – Document 5.* London, UK: Royal Anthropological Institute.

Weiner, J. S. & Lourie, J. A. (1969). *Human Biology: A Guide to Field Methods.* Oxford, UK: Blackwell.

Wessel, J. A., Ufer, A., van Huss, D. & Cederquist, D. (1963). Age trends of various components of body composition and functional characteristics in women aged 20–69 years. *Ann. N.Y. Acad. Sci.* **110**, 608–22.

Whitt, F. R. (1971). Estimation of the energy expenditure of sporting cyclists. *Ergonomics* **14**, 419–24.

Whittow, G. C. (1961). Heart rates and the sub-lingual temperatures of heat-acclimatized people. *Nature* **192**, 126–33.

Williams, C., Reid, R. M. & Coutts, R. (1973). Observations on the aerobic power of university rugby players and professional soccer players. *Brit. J. Sports Med.* **7**, 390–1.

Wilmore, J. H. & Brown, C. H. (1974). Physiological profiles of women distance runners. *Med. Sci. Sports* **6**, 178–81.

Wilmore, J. H. & McNamara, J. J. (1974). Prevalence of coronary heart disease risk factors in boys, 8 to 12 years of age. *J. Pediatrics* (*St. Louis*) **84**, 527–33.

Wilmore, J. H. & Sigerseth, P. O. (1967). Physical work capacity of young girls 7–13 years of age. *J. Appl. Physiol.* **22**, 923–8.

Wohlfeil, T. (1928). Über den Energieverbrauch bei sportlicher Körperarbeit (Kanufahren). *Arch. Hyg.* (*Berlin*) **100**, 393–411.

Wojtczak-Jaroszowa, J. & Banaszkiewicz, A. (1974). Physical work capacity during the day and at night. *Ergonomics* **17**, 193–8.

Wolanski, N. (1966). The secular trend: micro-evolution, physiological adaptation and migration, and their causative factors. *Seventh International Congress of Nutrition, Hamburg.*

Wolanski, N. (1969). An approach to the problem of inheritance of systolic and diastolic blood pressure. *Genetica Polonica* **10**, 263–8.

Wolanski, N. (1970). Review article: genetic and ecological factors in human growth. *Human Biol.* **42**, 349–68.

Wolanski, N. (1972). About the theory of the limited direction of development. *Acta Med. Auxol.* **3**, 201–15.

Wolanski, N. & Pyzuk, M. (1972). Morpho-physiological characters and physical work capacity in 15–72 years old inhabitants of low mountains, Pieniny range, Poland. *Human Biol.* **44**, 595–611.

Wolanski, N. & Pyzuk, M. (1973). *Studies in Human Ecology.* Vol. I. Warsaw, Poland: PWN Polish Scientific Publishers.

Wolff, H. S. (1958). The integrating pneumotachograph: A new instrument for the measurement of energy expenditures by indirect calorimetry. *Quart. J. Exp. Physiol.* **43**, 270–83.

Wolff, H. (1966). Physiological measurement on human subjects in the field, with special reference to a new approach to data storage. In: *Human Adaptability and its Methodology.* Ed: Yoshimura, H. & Weiner, J.S. Tokyo: Japan Society for the Promotion of Sciences.

Wright, G. R., Bompa, T. & Shephard, R. J. (1976). Physiological evaluation of a winter training programme for oarsmen. *J. Sports Med. Phys. Fitness* **16**, 22–37.

Wyndham, C. H. (1966). Southern African ethnic adaptation to temperature and exercise. In: *The Biology of Human Adaptability*, pp. 201–44. Ed: Baker, P. T. & Weiner, J. S. Oxford, UK: Clarendon Press.

Wyndham, C. H. (1969). Physiological requirements for world-class performances in endurance running. *S. Afr. Med. J.* **43**, 996–1002.

Wyndham, C. H. (1973). The work capacity of rural and urban Bantu in South Africa. *S. Afr. Med. J.* **47**, 1239–44.

Wyndham, C. H. & Morrison, J. F. (1958). Adjustment to cold of Bushmen in the Kalahari desert. *J. Appl. Physiol.* **13**, 219–25.

Wyndham, C. H. & Sluis-Cremer, G. (1968). The capacity for physical work of white miners in South Africa. II. The rates of oxygen consumption during a step test. *S. Afr. Med. J.* **42**, 841–4.

Wyndham, C. H. & Strydom, N. B. (1971). Mechanical efficiency of a champion walker. *S. Afr. Med. J.* **45**, 551–3.

Wyndham, C. H. & Strydom, N. B. (1972). Körperliche Arbeit bei höher Temperatur. In: *Zentrale Themen der Sportmedizin.* Ed: Hollmann, W. Berlin, Germany: Springer Verlag.

Wyndham, C. H., Bouwer, W., V.D.M., Patterson, H. E. & Devine, M. G. (1953). Working efficiency of Africans in heat. *Arch. Industr. Hyg. Occup. Med.* **7**, 234–40.

Wyndham, C. H., Strydom, N. B., Maritz, J. S., Morrison, J. F., Peter, J. & Potgieter, Z. V. (1959). Maximum oxygen intake and maximum heart rate during strenuous work. *J. Appl. Physiol.* **14**, 927–36.

Wyndham, C. H., Strydom, N. B., Morrison, J. F., Peter, J., Williams, C. G., Bredell, G. A. G. & Joffe, A. (1963). Differences between ethnic groups in physical working capacity. *J. Appl. Physiol.* **18**, 361–6.

Wyndham, C. H., Strydom, N. B., Leary, W. P. & Williams, C. G. (1966*a*). A comparison of methods of assessing the maximum oxygen intake. *Int. Z. angew. Physiol.* **22**, 285–95.

Wyndham, C. H., Strydom, N. B., Leary, W. P. & Williams, C. G. (1966*b*). Studies of the maximum capacity of men for physical effort. *Int. Z. angew. Physiol.* **22**, 296–303.

Wyndham, C. H., Strydom, N. B., Morrison, J. F., Williams, J. F., Bredell, C. G. & Heyns, H. (1966*c*). The capacity for endurance effort of Bantu males from different tribes. *S. Afr. J. Sci.* **62**, 259–63.

Wyndham, C. H., Watson, M. & Sluis-Cremer, G. K. (1970). The relationship between weight and height of South African males of European descent, between the ages of 20 and 60 years. *S. Afr. Med. J.* **44**, 406–9.

Yamakawa, J. & Ishiko, T. (1966). Standardization of physical fitness test for oarsmen. In: *Proceedings of International Congress of Sports Sciences, Tokyo.* Ed: Kato, K. Tokyo: Japanese Union of Sports Sciences.

Yamaoka, S. (1965). Studies on energy metabolism in athletic sports. *Res. J. Phys. Educ.* **9**, 28ff.

Yoshimura, H. (1960). Acclimatization to heat and cold. In: *Essential Problems in Climatic Physiology.* Ed: Yoshimura, H., Ogata, K. & Itoh, S. Kyoto, Japan: Nankodo.

Yoshizawa, S. & Ushihisa, J. (1967). Comparative study on physical fitness of the youth in rural and city districts. *Taiiku-No-Kagaku* **17**, 222–9.

Young, C. M. (1965). Body composition and body weight: criteria of overnutrition. *Canad. Med. Ass. J.* **93**, 900–10.

Young, C. M., Blondin, J., Tensuan, R. & Fryer, J. H. (1963). Body composition studies of 'older' women thirty to seventy years of age. *Ann. N.Y. Acad. Sci.* **110**, 589–607.

Young, F. N. (1964). Terrestrial animals in humid heat: Introduction. In: *Handbook of Physiology*, Section 4. Ed: Dill, D. B. Washington, DC: American Physiological Society.

Young, H. B. (1965). Body composition, culture and sex: two comments. In: *Human Body Composition. Approaches and Applications.* Ed: Brozek, J. New York: Pergamon Press.

Zelenka, V., Seliger, V. & Ondrey, O. (1967). Specific function testing of young football players. *J. Sports Med. Phys. Fitness* **7**, 143–7.

Zinner, S. H., Levy, P. S. & Kass, E. H. (1970). Familial aggregation of blood pressure in childhood. *New Engl. J. Med.* **284**, 401–4.

Index

Note: references in *italic* are to figures